Contemporary
Microbial Ecology

Proceedings of the Second International Symposium on Microbial Ecology
held from 7th-12th September, 1980, under the auspices of
The International Commission on Microbial Ecology
at the University of Warwick, Coventry, UK.

AN ACADEMIC PRESS FAST PUBLICATION

Contemporary Microbial Ecology

Edited by

D.C. ELLWOOD
*Centre for Applied Microbiology and
Research, Porton Down, UK*

J.N. HEDGER
*University College of Wales
Aberystwyth, UK*

M.J. LATHAM
*National Institute for Research in
Dairying, Reading, UK*

J.M. LYNCH
*ARC Letcombe Laboratory
Wantage, UK*

J.H. SLATER
*University of Warwick
Coventry, UK*

1980

ACADEMIC PRESS
A Subsidiary of Harcourt Brace Jovanovich, Publishers
London New York Toronto Sydney San Francisco

ACADEMIC PRESS INC. (LONDON) LTD
24/28 Oval Road,
London NW1

United States Edition published by
ACADEMIC PRESS INC.
111 Fifth Avenue
New York, New York 10003

British Library Cataloguing in Publication Data
International Symposium on Microbial Ecology, *2nd,*
University of Warwick, 1980
Contemporary microbial ecology.
1. Microbial ecology – Congresses
I. Title   II. Ellwood, D C
576'.15   QR100   80-40532
ISBN 0-12-236550-X

Printed in Great Britain

CONTRIBUTORS

Baker, K.F., *U.S. Department of Agriculture SEA-AR, Ornamental Plants Research Laboratory, Oregon State University, Corvallis, Oregon, 97330, U.S.A.*

Bauchop, T., *Applied Biochemistry Division, Department of Scientific and Industrial Research, Palmerston North, New Zealand.*

Bowen, G.D., *Division of Soils, C.S.I.R.O., Glen Osmond, Adelaide, Australia.*

Bull, A.T., *Department of Applied Biology, University of Wales Institute of Sciences and Technology, King Edward VII Avenue, Cardiff, CF1 3NU Wales.*

Carlile, M.J., *Department of Biochemistry, Imperial College of Science and Technology, London SW7 2A2, England.*

Doonan, S.A., *Department of Microbiology, University of Aberdeen, AB9 1AS, Scotland.*

Dow, C.S., *Department of Biological Sciences, University of Warwick, Coventry CV4 7AL, England.*

Godwin, D., *Department of Environmental Sciences, University of Warwick, Coventry CV4 7AL, England.*

Gooday, G., *Department of Microbiology, University of Aberdeen, Aberdeen AB9 1AS, Scotland.*

Gregory, P.H., F.R.S., *Rothamstead Experimental Station, Harpenden, Herts AL5 2JQ, England.*

Halldal, P., *Botanical Laboratory, University of Oslo, Box 1045, Blindern, Oslo 3, Norway.*

Jones, C.W., *Department of Biochemistry, School of Biological Sciences, University of Leicester, Leicester LE1 7RH, England.*

Jones, G.W., *Department of Microbiology, University of Michigan, Ann Arbor, Michigan 48109, U.S.A.*

CONTRIBUTORS

Jørgensen, B.B., *Institute of Ecology and Genetics, University of Aarhus, Ny Munkegade, DK-8000 Aarhus C, Denmark.*

Konings, W.N., *Department of Microbiology, University of Groningen, Kerklaan 30, 9751 NN Haren, The Netherlands.*

Kushner, D.J., *Department of Biology, University of Ottawa, Ottawa, Ontario K1N 6N5.*

Lacey, J., *Rothamstead Experimental Station, Harpenden, Herts AL5 2JQ, England.*

Marshall, K.C., *School of Microbiology, The University of New South Wales, Kensington, New South Wales, 2033, Australia.*

Paul, E.A., *Department of Soil Science, University of Saskatchewan, Saskatoon, Canada S7N OWO.*

Reid, D.S., *Unilever Research, Colworth Laboratory, Sharnbrook, Bedford, England.*

Slater, J.H., *Department of Environmental Sciences, University of Warwick, Coventry CV4 7AL, England.*

Veldkamp, H., *Department of Microbiology, University of Groningen, Kerklaan 30, 9751 NN Haren, The Netherlands.*

Voroney, R.P., *Department of Soil Science, University of Saskatchewan, Saskatoon, Canada S7N OWO.*

Whittenbury, R., *Department of Biological Sciences, University of Warwick, Coventry CV4 7AL, England.*

Williams, F.M., *Department of Biology and Ecology Program, Pennsylvannia State University, University Park, Pennsylvannia 16802, U.S.A.*

PREFACE

   The need for an international symposium in the field of
microbial ecology was first met in 1977 with the organisation
of the First International Microbial Ecology Symposium in
Dunedin, New Zealand. This has now been followed by the
Second International Symposium on Microbial Ecology,
sponsored by the International Commission on Microbial
Ecology, and held at the University of Warwick, England,
from the 7th to 12th September, 1980. The organisation
of this Symposium was largely achieved through a National
Committee whose members were as follows: Professor G.E.
Fogg F.R.S. (Chairman), Dr. Madilyn Fletcher (Secretary),
Professor D.P. Kelly (Treasurer), Dr. M. Bazin, Dr. J.F.
Darbyshire, Professor D.C. Ellwood, Dr. G.D. Floodgate,
Dr. J.N. Hedger, Dr. M.J. Latham, Dr. J.M. Lynch, Dr. E.I.
Newman, Professor J.R. Norris, Dr. R.W.A. Park, Professor
G.J.F. Pugh and Dr. J.H. Slater.
   The National Committee believed that the subject of
microbial ecology was "ripe" for new developments and
insights and hoped that this Symposium would provide the
venue for the exchange of information and ideas which
nurtured such advances. The Committee also believed that by
including aspects of microbiology, such as biochemistry or
genetics, which are not traditionally thought of as
"ecological", an even greater understanding of the interactions of microorganisms with their environment might be
achieved. Thus, twenty main topics were decided upon for
Symposium Sessions, and twenty keynote speakers and sixty-
seven supporting speakers were invited to present papers on
a wide range of topics and thereby provide some "ammunition"
and inspiration for searching and productive discussion by
all participants. This volume contains nineteen keynote

PREFACE

papers presented at the Symposium Sessions and, therefore, reflects the scope of the ideas dealt with at the Symposium.

Our thanks go to the National Committee and to all speakers and to the Sessions Chairmen for their important contributions. Particularly important in the preparation of this volume was the excellent typing assistance of Mrs. Ann Branch and Miss Vicki Narbett and the assistance of the Academic Press staff.

We are also most grateful to a number of organisations, without whose financial assistance, this Symposium would not be possible. The Organising Committee wishes to acknowledge generous financial assistance through gifts or loans from the following:

The Royal Society
The International Union of Biological Sciences
The Society for General Microbiology
The British Ecological Society
The British Mycological Society
The Society for Applied Bacteriology
Imperial Chemical Industries Ltd
Shell Research Ltd
The British Petroleum Company Ltd
Unilever Research
Cadbury Schweppes Ltd
Beckman-RIIC Ltd
Packard Instruments Ltd

The Committee acknowledges the following valuable material assistance to the Symposium organisation:

The Lord Mayor and Chief Executives of the City of Coventry for a Civic Reception

The Vice Chancellor of the University of Warwick for a University Reception

Cockburn, Smithes & Co. Ltd., for the provision of Cockburn's Special Reserve Port

Wm. Teacher & Sons Ltd., for the provision of Teacher's Highland Cream Whisky

Grants of St. James's for a gift of wine

Arthur Guinness, Son & Co. Ltd., for a gift of Guinness

G.E. Fogg   M.M. Fletcher
Bangor   Warwick

D.P. Kelly   May 1980
Warwick

CONTENTS

CONTENTS

INTRODUCTION

It has been our privilege and pleasure to act as editors for this book. In reading the chapters we felt that there were signs of a lack of communication between a number of areas of microbial ecology, perhaps reflecting the broad nature of the subject and the diverse, individual disciplines which need to be drawn together. In particular this lack of communication was evident between microbial physiologists and microbial ecologists. Thus, in order to bridge these gaps, or at least to identify them, we felt that some of these problems ought to be commented upon in this preface. Clearly this approach will reflect our prejudices and ignorance, for which we trust both contributors and the reader will forgive us.

It may be axiomatic but is worth stressing that in natural systems it is a rarity to find a pure population of a single microbial species. This being so, it seems to highlight one of the problems of experimental approach and consequently communication between different breeds of microbial ecologist. The microbial physiologist likes to study organisms in clearly defined systems, one organism at a time. This is done for a number of reasons, not the least of which is due to the principles enunciated by Koch that the disease state relates to invasion and growth of a single microbial species within the body. This concept, and the major technique for isolating pure strains, yielded outstanding advances in the treatment and prevention of human disease but conceptually it can act as a blindfold to the microbiologist seeking to interpret his results in an ecological sense. These arguments are given an extended treatment with respect to degradation by A.T. Bull and are also applicable to the problems of mineralization and element cycling (B.B. Jørgensen) interactions between microbes with the human body (G.W. Jones) and in the rhizosphere (G.D. Bowen) and gut (T. Bauchop). The chapters describing microbial antagonism (K.F. Baker), predator-prey interactions (F.M. Williams) and algal-invertebrate symbioses (G.W. Gooday and S.A. Doonan) indicate that certain areas of microbial ecology study clearly recognise the importance of understanding the significance of relationships between populations of different genetic makeup. But here too, we are remarkably ignorant of the relative importance of the multitude of possible inter-actions, particularly with respect to their significance in nature: what may seem to us to be important is of little consequence to the organisms themselves in natural situations. One further concept that Bull puts forward which ought to be of use, not to mention comfort, to the microbial ecologist is that mixed populations can be treated in exactly the same way as single populations. For example, the kinetics of growth may equally well be examined with a mixed system as a monoculture. Thus we can conceptually treat a series of enzyme reactions, for example, which may be effected by a number of different organisms as being representative of the total population: that is, the microbial ecologist can examine functional activities without, necessarily, being concerned about the details of individual populations.

However, detailed knowledge of how microorganisms react with one another and how these relationships are modified under conditions in the natural habitat is presently a matter of conjecture in the main. The importance of synergistic relationships, for example, in the methanogenic communities, is well understood but the importance, or otherwise, of multimembered communities, for example, in the degradation of plant material and leaf litter has yet to be fully explored. What is the role of fungal-bacterial communities or fungi-fungi interactions in such processes? The importance of such relationships, which are just as likely to include antagonistic as well as mutualistic responses, in determining community structure and overall processes is likely to be of considerable interest in the future (see, for example, K.F. Baker's chapter).

The arguments deployed above about microbial physiology apply equally strongly to microbial genetics. Thus in the chapter by J.H. Slater and D. Godwin the concept of genetic fluidity and transfer between unrelated microbial species growing as mixed cultures is discussed. This again suggests that conceptually it may be right to consider the total gene pool of a given mixed micoflora located in a particular environment in determining the response individuals or

interacting groups of organisms may make to change in
environmental conditions, or whatever else influences the
growth of microorganisms and communities. Microbial
ecologists are only just becoming aware of the gap in
ecological studies which exists because such little attention
has been paid to the genetics and genetic systems of "unusual"
organisms (anything other than *Escherichia coli* and a few
Pseudomonas spp.) or their significance in the natural
environments. Sophisticated techniques are to hand and a
profitable area of development ought to be the bridging of
the gap between microbial geneticists and ecologists. So,
for example, how do these principles relate to microbial
evolution in nature as opposed to the laboratory? What
does genetic fluidity mean in terms of taxonomic groups?
What are the subtle differences at the genetic level, which
makes one microbe more successful than a very close relative?

Another area which ought to be of considerable interest
to the microbial ecologist is that of microbial energetics
and several contributors address themselves to aspects of
this topic (W.N. Konings and H. Veldkamp, C. Jones and E.A.
Paul and R.P. Voroney). Since organisms growing in nature
are frequently growing slowly, there is considerable debate
over the concept of maintenance energy; that is, energy
needed to maintain the necessary functions of a cell and
preserve its viability. The concept derives from the
observation that the yield values for a given carbon
substrate decrease with decreasing growth rate for aerobic
chemostat cultures for a few "laboratory" organisms. But
the question now is: What does this concept actually mean
for organisms in their natural habitat? Considerable
discussion has been generated in recent years because the
energy demands on an organism in terms of turnover of macromolecules and other cellular components are too small to
account for the marked decreases in yield normally observed.
Furthermore there is an indication that maintenance energy
requirements are variable with respect to growth parameters.
Thus, even experimentation with single organisms under
highly defined conditions is needed to understand fully
the energetics of organisms growing slowly.

The uptake of glucose is controlled by a phosphoenolpyruvate dependent phosphotransferase system in some organisms
and this activity may be maximal at low growth rates under
glucose limitation. If this is so, then the maintenance
energy could, in part, be related to this energy demand. In
contrast to this is the suggestion made by Konings and
Veldkamp on the basis of the chemiosmotic hypothesis that
excretion of acids from a cell could provide energy to
charge up the membrane and allow transport of nutrients into

the cell on this basis. But here too there is a significant gap
between our basic understanding of the energetics of microorganisms and its influence on the ecology of organisms.
    Microbiologists working with pure and mixed cultures in
the laboratory generally refer to biomass as the total
amount of living and dead biological material present.   This
presents few problems in terms of measurement, largely
because most, if not all, organisms present are living.
In the ecosystem, biomass refers specifically to living
tissue and this is often difficult to assess.   Until recently
the common method used to observe cells and mycelium
involved the microscope and, knowing cell dimensions and
specific gravity, the biomass could be calculated. However,
microscopic methods are difficult to interpret because
different staining methods can yield different results.
These microscopic methods and alternative techniques are
discussed by Paul and Voroney.   One method that has recently
come to the fore is the measurement of ATP.   In order for
the ATP determinations to be meaningful in absolute terms the
measurement has to be done instantaneously in order to avoid
the problems of ATP decay.   Thus, in order to obtain accurate
data for ATP content of bacteria it would perhaps be more
useful to measure both ATP and ADP.   Because of the difficulties
in obtaining this instantaneous ideal, no method of biomass
determination should be regarded as absolute and the various
methods are only useful in a comparative sense.   Even if an
asbolute method were available, attention would still have
to be given to the method of sampling.   Minimum disturbance
of the sample is essential and provided samples are stored
in suitable conditions then reliable results are obtainable.
For example the sieving of soil samples can alter greatly
the biomass determined by fumigation/respiration, but delay
in measurement of well stored samples for 24 hours does not
affect the measured biomass.   The estimation of microbial
biomass is a fundamental parameter for the microbial
ecologist and yet debate over the most suitable estimation
methods is rampant and surely needs to be resolved.
    While microbiologists increase their understanding of
microbial behaviour in well-defined laboratory systems
(autecology) and this is often the only way to produce
reliable and unambiguous data, the application to natural
environments (synecology) must eventually be attempted.   We
must therefore familiarise ourselves with the physical nature
of environments and the terms used to express them.   For
example, water activity, discussed by D.S. Reid, indicates
the water actually available to microorganisms, particularly
when considering food spoilage, but an alternative terminology
for the same concept is that of water potential which is used

principally in the soils and plant pathology literature. In
the latter consideration, there can be components due to the
retention of water in the voids between mineral particles
(matric potential), the water in plant cells (turgour
potential) and the presence of solutes (osmotic potential).
Unlike water activity, water potential is usually expressed
in units of pressure. This point serves to illustrate the
differences of approach and the need to measure a particular
phenomenon in a way which is relevant to microbiology.
Moreover, it perhaps also illustrates the need to be aware
of such important principles which frequently are relegated
from consideration by the microbial ecologist with his own
particular axe to grind. These principles are, of course,
particularly important to those ecologists examining
organisms which inhabit extreme environments (D.J.
Kushner's chapter, for example).

It is an interesting comparison with the ecology of
large organisms in that microorganisms can get their
nutrients by being attached to a surface and taking up
nutrient from aqueous systems that flow past them. Or in
contrast, seeking out nutrients by moving towards localised
concentrations of "their food". It becomes almost a
philosophical question, is it better to spend resources
in seeking out supplies of food (see M.J Carlile's chapter)
or to stay in one place and allow the nutrient to come to
the organism (see K.C. Marshall's chapter).

In many ecosystems some, if not most, of the microorganisms
present are to be found associated in some way with solid
surfaces in that ecosystem. Organisms growing on surfaces
first have to attach to that surface and this process is
considered in physico-chemical terms by Marshall. It is
important to recognise that organisms frequently have a
domain of control near to the cell wall of the organism
which is achieved by importing polymers of various kinds
into the local environment. For example, if they secrete
negatively charged polymers into their immediate environment,
these polymers can act directly as a bridge to the surface
if that surface is positively charged. If the surface is
negatively charged then divalent cations can act as bridges
to form the bases of attachment to that surface. The role
of lipoteichoic acid in allowing attachment of oral streptococci to the tooth surface has been interpreted in such a
way. Clay particles like the bacterial surface carry a net
electronegative charge but nevertheless the two can be
attracted by a variety of physical mechanisms such as London-
van der Waal's interactions, simple hydrogen bonds, protonation,
cooordination bonds and water bridging. The attachment of
the organism to the clay surface is generally beneficial and

can influence its growth, germination, survival, respiration, metabolism and susceptibility to heavy metal and volatile toxins. The surface probably also affects the interactions between organisms, such as predator-prey relationships (see F.M. Williams chapter), but this area has been explored little. The adhesion of viruses to clays exhibits characteristics of the adhesion of both microbial cells and soluble monomeric and polymeric amino acids; as viruses are usually protected by clays this has important implications in the disposal of sewage sludge and animal waste in soils. The attention given to the role of polymers in organism/surface interactions has been relatively less than that given to charge interactions, particularly in relation to soil surfaces. Whereas clay particles have large surface charges, soils also contain sand and silt particles which are uncharged and hence the only scope for microbial attachment is by interaction with polymers on the cell surface. These various interactions are important in stabilizing soil structure and preventing erosion, thus making soil a suitable medium for plant growth. Experimentation is necessary to characterize the nature of the polymer interaction and its role relative to the charge interactions at the clay surfaces. In general terms, therefore, we need to know much more about the modification of microbial physiology, biochemistry and genetics as a result of their close interaction at surfaces, or interfaces generally, than we do at present understand. For example, what is the significance of surface growth in the process of genetic exchange between organisms growing on surfaces. If different species of organisms are growing on a surface, then some of these organisms may die and their DNA is released by lysis. This DNA could then be cut by restriction enzymes produced by another species in this community and could then transform another organism in the community. If this DNA conferred a growth advantage on to the recipient then the derived organism could outgrow the other species and become a predominant organism in this community. It is tempting to speculate that in this close community of microbial films much of the evolution of species adapted to a specific environmental niche goes on.

Microbial activity undoubtedly plays an essential role in every kind of ecosystem but microbial ecology has been slow to develop as a distinct discipline. This is perhaps largely because the simple approach of description and classification open to the pioneers of plant and animal ecology is difficult and laborious with microorganisms and consequently no general framework to which microbial ecologists can refer has emerged. Now, when techniques and ideas are developing rapidly, it is especially important that workers in this field should meet,

INTRODUCTION

exchange ideas, and begin to build a coherent science from the often beautiful but disconnected fragments of microbial ecology which exist. The International Commission on Microbial Ecology has taken a leading role in working towards this end and it is with its support and encouragement that the Second International Symposium on Microbial Ecology has been held at the University of Warwick, England of which this volume incorporates the keynote topics discussed. Some may be foreign, if not quite anathema, to some microbial ecologists but we sincerely hope that all the topics discussed will contribute to the continuing development of microbial ecology as a substantial discipline.

D.C. Ellwood
Porton Down
Salisbury

M.J. Latham
Reading

J.H. Slater
Warwick

J.N. Hedger
Aberystwyth

J.M. Lynch
Wantage

May 1980

LIGHT AND MICROBIAL ACTIVITIES

Per Halldal

Botanical Laboratory, University of Oslo, Box 1045,
Blindern, Oslo 3, Norway

INTRODUCTION

A broad topic is indeed covered by the title "Light and Microbial Activities". It spans vast areas of research worthy of presentation in a multi-volume Encyclopaedia. The purpose of the keynote paper which I have been asked to present emphasises the importance of light on different types of microbial action and to place them in an ecological perspective.

My presentation will give a general consideration of a few selected topics with which I am familiar from my own research. Thus I will deal with phototopotactic responses in flagellates, in the UV and visible region, the effects of far-UV radiation on the motility and growth of motile flagellates, and the repair mechanisms for far-UV damage. In this treatment I will further discuss how different parts of the spectral region of photobiology from 220 to 1000 nm, affect these responses.

At the present time, there is not complete agreement on the terminology for the subdivision of the spectral regions of photobiology and so the following terms will be used in this paper (Fig. 1)

The term light is usually reserved for that part of the electromagnetic spectrum which lies within the spectral region of human vision, that is, from 380 to 780 nm approximately. The terms ultraviolet (UV) and infrared (IR) light are unfortunately frequently used rather than the subdivisions indicated in Fig. 1. The boundaries between these different spectral regions have been fixed on both physical and physiological bases.

Microbial activities influenced by light can be documented for the whole spectral range, from below 220 nm to 1000 nm.

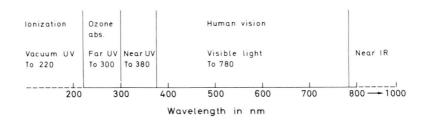

Fig. 1. The terminology of the spectral span of photobiology used in this presentation.

The lowest limit of photobiology is often fixed at 220 nm. At shorter wavelengths the photons cause ionization of oxygen, forming ozone which is an effective absorber of electromagnetic radiation below 220 nm. Thus, photobiological phenomena may be observed below 220 nm in oxygen-free atmospheres but only with a radiation path of a few cm in systems where oxygen is present. The spectral region below 220 nm is therefore frequently called the *vacuum UV*. Between 220 and 300 nm (approximately) the *far UV region* is found. The longer wave-length limit has been fixed by the shortest wavelength of natural sunlight measurable at sea level. Due to the ozone layer of the atmosphere with its maximum concentration at a height of about 25 km, radiation below 300 nm is effectively absorbed (Henderson, 1970). As a consequence, photobiology in the far UV region can only be studied by the use of artificial sources of radiation, such as mercury and xenon lamps. The spectral region between 300 and 380 nm is called the *near UV*. The longer wavelength limit of this region is fixed by the shortest wavelength average human vision. The photobiological effects of near UV radiation are very similar to those of visible light, so that near UV can influence photo-synthesis, photoreactivation and phototaxis. One may also recall that insect vision extends from red through the near UV spectral region. Towards the longer wavelengths the sensitivity of the average human eye ends at 780 nm. However, photobiology in *near IR* is observed in the photosynthetic bacterium *Rhodospirillum rubrum* up to 1000 nm.

Perhaps the most important spectral region is that around 300 nm, both from a photobiological and ecological point of view. From longer wavelengths towards 300 nm important processes such as photosynthesis, photoreactivation and phototaxis end and below 300 nm wavelengths both proteins and nucleic acids show increased absorption. In the far UV region both temporary and permanent lesions to macromolecules using

streptomycin resistant transforming DNA, Rupert (1964) showed
that UV contained in natural sunlight caused DNA damage
(photoinactivation) which was repaired upon the addition of a
photoreactivating enzyme from yeast.

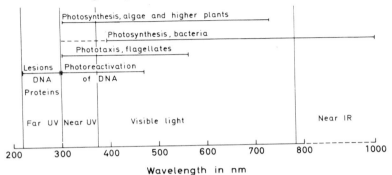

*Fig. 2. Spectral regions of some photobiological activities
over the complete spectrum of photobiology.*

EXAMPLES OF DIFFERENT MICROBIAL PHOTOACTIVITIES

Photosynthesis

   Photosynthesis is studied in the spectral region of
visible light, usually between 400 and 750 nm. A number of
action spectra of photosynthesis, mainly obtained by measuring
oxygen evolution, have been determined in great detail for
this part of the spectrum. However, it is important to point
out that photosynthesis, measured either as oxygen production
or carbon dioxide fixation, has been measured for the complete
spectrum between 220 and 1000 nm. Halldal (1966) recorded that
photosynthetic induction phenomena, determined by oxygen
evolution, could be measured between 220 and 720 nm in *Chlorella*
(Fig. 3) an observation that later was confirmed by Taube and
Halldal (unpublished). The action spectrum of the first
induction phase shows maximum effect at 220 nm, a decrease
towards longer wavelengths with a minimum at 259 nm and a
second, smaller maximum around 275 nm. In the near UV and
visible part of the spectrum the action spectrum of this first
induction phase resembles that of normal photosynthesis.
   Halldal (1966) assumed that quanta directly absorbed by
the chlorophyll molecule were involved in the response in the
far UV region. However, another possibility could be that
quanta absorbed by proteins resulted in energy transfer to
chlorophyll.
   This photosynthetically stimulated oxygen evolution observed
in the far UV region, was of short duration. It was recorded

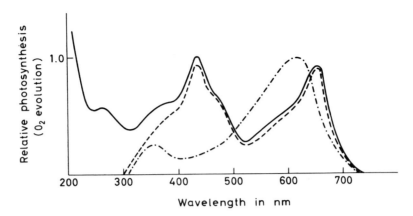

Fig. 3. Action spectra of photosynthesis in algal groups.
The first induction phase in Chlorella ————; Normal
photosynthetic spectral response in Chlorella - - - - -; and
action spectrum of cyanobacteria —·—·—·—·—·—.

for only a few minutes, and was then followed by photosynthetic
inhibition. The action spectrum for this inhibition resembled
that of protein absorption and it was assumed that a lesion
of the proteins in the photosynthetic apparatus resulted in
significant damage, both in the red and green algae investigated
(Halldal, 1964a). Data from delayed growth and growth
inhibition (death) were shown to have action spectra resembling
that of the absorption spectrum of nucleic acid (DNA) (Halldal,
1967).

At wavelengths longer than 300 nm photosynthesis is normally
stimulated in the near UV region at moderate quanta flux levels
and this has been demonstrated for several algal groups (Halldal,
1967; Halldal and Taube, 1972). The photosynthetic action
spectra curves show a sharp increase from 300 nm towards 440
nm in green algae as well as in dinoflagellates, brown algae
and diatoms (Fig. 2 Halldal and Taube, 1972; P. Halldal and
O. Taube, unpublished results). For the cyanobacteria (Fig. 3)
and the red algae, the increase from 300 nm towards longer
wavelengths is somewhat more moderate than for the other algal
groups. A small maximum is usually recorded around 350 to
375 nm and another large maximum at approximately 620 nm for
the cyanobacteria. Thus, in the near UV region of the spectrum,
at moderate radiation levels, photosynthesis is normally
stimulated and the response is very similar to that of visible
light between 380 and 720 nm, a result previously shown by

Meier (1932). *Chlorella* sp. on agar plates were killed by the mercury lines in the far UV, that is, at wavelengths shorter than 302 nm, while no inhibitory effect was observed in the near UV region, where even the mercury line at 313 nm did not affect growth.

The UV component of natural sunlight has in some cases been reported as inhibiting photosynthesis and causing destruction to the photosynthetic apparatus. Gessner (1955) observed that sunlight from which near UV radiation was filtered by means of a glass plate, was less destructive to the photosynthetic apparatus than unfiltered light. Steemann Nielsen (1964) whilst measuring photosynthesis by the fixation of ^{14}C-carbon dioxide by natural phytoplankton populations, reported a similar inhibitory effect of unfiltered light compared with light that had passed through a glass plate. It should, however, be pointed out that a clear glass plate, in addition to causing a reduction of the near UV radiation also reduces the total light level by at least 10% and probably by considerably more. According to the data of Gessner (1955) and Steemann Nielsen (1964) there is no indication that a similar reduction of total light level was used in the control experiments. Thus a possible positive effect due to the significant reduction in the total light level cannot be excluded.

The sharp decrease in photosynthetic rates from 440 towards 300 nm in green algae, dinoflagellates and diatoms ("diatom group") combined with the marked reduction in intensity of solar radiation over the same wavelength span (Henderson, 1970), has the effect of producing a moderately low rate of photosynthesis in the near UV region. For the cyanobacteria the rate must be even lower (Fig. 3).

For all algal groups the highest photosynthetic rates are found in the visible spectrum between 400 and 680 nm, though the complete spectral span of photosynthesis is from 300 to 750 nm. The "diatom group" has a photosynthetic rate peak in the blue region, approximately 440 nm, and in the red region at 675 nm. The cyanobacteria have a distinct peak in the yellow region at 620 nm (Fig. 3). Halldal (1979) discussed the spectral responses of photosynthesis of these algal groups with regard to underwater light conditions in coastal and oceanic waters. Diatoms, dinoflagellates, coccolithophorids and brown algae were treated as one group, the diatom group. The spectral response of photosynthesis of the green algae was so similar to the response of the diatom group that they may be considered together for the purposes of this discussion. The other group was called the "red algae group" and for the present discussion the cyanobacteria may represent this group, as it does in Fig. 3. It is surprising to note how ineffectively

the diatom group utilizes natural underwater light and this
is particularly true for coastal waters (Fig. 4). At about
the compensation point for photosynthesis, where respiration
balances carbon dioxide fixation, neither the blue nor the
red chlorophyll peaks are excited. The maximum light
penetration into the water occurs at a wavelength region where
photosynthesis reaches a minimum. In contrast to this the red
algal group shows a maximum photosynthetic response at this
wavelength region.

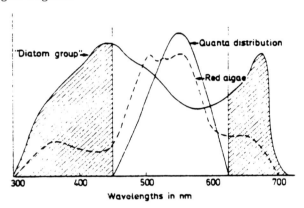

*Fig. 4. Action spectra of photosynthesis for the diatom
group and red algae, compared with the light spectral distribution at 25 m depth in coastal waters from west coast of
Sweden where about 1% of the light remains "compensation
depth" (data from Halldal, 1974).*

Finally under photosynthesis should be mentioned the
unique photosynthetic spectral response of the purple bacteria.
These microorganisms show phototactic response and perform
photosynthesis up to 1000 nm (Milatz and Manton, 1953).
Presumably this bacterium will photosynthesise over the
complete range to 300 nm, though the response in the near UV
region has, to my knowledge, not been analysed. Maybe the
first phases of photosynthesis occur at 220 nm as has been
shown for the green alga *Chlorella*; should this be the case,
the spectral span of their photosynthesis would be from 220
to 1000 nm, or more than 50% greater than shown for the algae.

Phototaxis

Two different types of phototaxis will be considered,
phototopotaxis, which is found in photosynthetic flagellates,
and photophobotaxis demonstrated by the photosynthetic
bacteria. The two distinct modes of response to light and

direction are illustrated in Fig. 5.

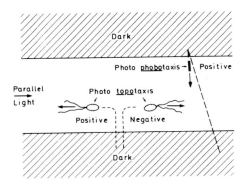

*Fig. 5.   Illustration of two different types of phototaxis,
the photophobic response of photosynthetic bacteria and the
phototopic response of motile flagellates.*

The photophobotactic response is exemplified by the behaviour
of a photosynthetic bacterium, for example *Rhodospirillum
rubrum*. A bacterium responds positively by not showing any
response initially upon entering into a beam of parallel
light. However, when the organism leaves the light region to
enter a dark field, it reverses its direction of movement
if it continues to exhibit a positive photophobotactic response.
In this way the bacteria congregate in the light region but
do not "detect" the light direction. This response is
closely linked to photosynthesis and both action spectra are
identical (Milatz and Manten, 1953).
      A phototopotactically responding flagellate, when entering
the parallel light from the dark area, responds to the light
direction. An organism exhibits a positive response by moving
towards the light source. An organism shows a negative
response by moving away from the light source (Fig. 5).   In
the green flagellates, photosynthesis is probably only
indirectly linked to phototaxis since the action spectrum of
photosynthesis is very different from that of phototaxis
(compare Figs. 3 and 8). Halldal (1959, 1960, 1962, 1964a)
has analysed several of the factors involved in the change
between negative and positive phototopotaxis. The literature
was reviewed by Nultsch (1970) and by Hand and Davenport (1970).
      The change in response from positive to negative phototopo-
taxis and *vice versa* is a rather slow process. In short-time
experiments a positively responding organism will approach
the light source irrespective of light intensity as illustrated

in Fig. 6.

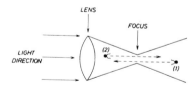

Fig. 6. The behaviour of a positively phototactic organism (1) and a negatively phototactic organism (2) in a focused beam. Light intensity in focus ~ 200 000 lux (after Halldal, 1959).

For several organisms it has been shown that both positively and negatively responding cells may pass a light focus of an intensity up to 200 000 lux, that is an intensity approximately double that of direct sunlight. A positively responding organism placed to the right of the focus in Fig. 6 will first swim towards increasing light levels and approach the light source. After passing the focus, the organism will swim towards the light source, but towards decreasing light intensities. The opposite situation may be observed for a negatively responding cell. This cell, placed to the left of the focus swims away from the light source, but towards increasing light levels. Passing the focus it responds in what we may call a "normal manner", so both cells will behave "normally" to the right of the focus where light direction and intensity usually are similar to normal light conditions in nature. If a population of photosynthetic responding phototopotactic flagellates is placed in a light gradient of moderate intensity for some time, for example, a couple of hours, they will accumulate at a very distinct place (Fig. 7). This fixed position is determined by the intensity of the light and may be retained for several days in a balanced nutrient solution. The main factor for the determination of this position is presumably to optimise the photosynthetic process but other factors such as the $Ca^{++}/Mg^{++}/K^+$ ratios have also been shown to be of importance, as well as the quality of the light to which the population is exposed (Halldal, 1959, 1960).

The action spectra of phototopotaxis have been determined for a great number of species from different algal groups, including motile green algae, green flagellates, dinoflagellates (Halldal, 1958, 1962, 1963, 1964b). Recent analyses of such spectra confirm that most responses are restricted to the blue

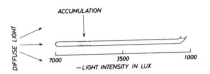

Fig. 7. The accumulation of phototactic organisms in long-time experiments in properly balanced medium e.g. sea water (after Halldal, 1959).

part of the visible spectrum with a maximum at 490 nm. The action spectra for positive and negative response for the green flagellate *Platymonas subcordiformis* have been determined over the whole spectral region of response, from 220 to 570 nm (Halldal, 1961a) (Fig. 8).

Previously it was pointed out that the first phase of photosynthesis in *Chlorella* sp. could be effected by radiation in the far UV region. Such radiation was also shown to direct the phototopotactic swimming in *Platymonas subcordiformis*. In this part of the spectrum it seems as if the response was directed by quanta absorbed by a protein and Halldal (1961a) assumed that a carotenoprotein was the photoreceptive pigment.

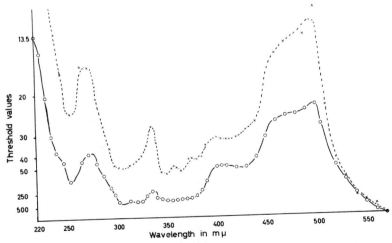

Fig. 8. Action spectra of positive (x) and negative (o) photo-topotaxis in the green flagellate Platymonas subcordiformis (After Halldal, 1961a).

This assumption has not been verified. Under the experimental
conditions used, the response in the positive direction was
shown to be approximately twice as fast as the response in
the negative direction. This indicates a more efficient
steering mechanism towards a light source which has been
discussed by Halldal (1963, 1964a).

The positive effect of radiation in the far UV region
is only temporary since directed swimming was soon followed
by immobilization. The action spectrum of this immobilization
has been measured (Halldal, 1961b), results which suggested
that lesions to proteins were responsible (Fig. 9). After
the first immobilization obtained by gentle far UV radiation,
it was shown that the cells resumed moving after about one
day. However, after three to four days a second stage of
immobilization was observed. The action spectrum of the
second immobilization reflected the absorption curve of DNA.

If the cells, however, immediately after the first far UV
radiation were irradiated with near UV and visible light,
the first, but not the second immobilization, could be observed
showing photoreactivation of the DNA damage (Fig. 9, lower
part). This is similar to the photoreactivation observed
in *Streptomyces griseus* by Rupert (1964). Halldal (1961b)
suggested that a flavoprotein was involved but this assumption
has not been confirmed.

CONCLUSIONS

*The relationship, or lack of it, of microbial activities
to environmental radiant conditions.*

It is surprising how many examples can be given of a poor,
in many cases extremely poor, relationship ("tuning") between
spectral response of photoactivities and the radiant environmental conditions where these activities take place. This is
true for responses both in the near UV, the visible and the
infrared regions of the spectrum, and holds for nearly all the
examples which have been given here, photosynthesis, photoreactivation and phototaxis.

Photosynthesis Photosynthetic microorganisms, such as
photosynthetic bacteria, cyanobacteria and different algal
groups, living on soil or water surfaces are exposed to solar
radiation with a spectrum that shows a sharp increase from
300 nm towards 475 nm. From that wavelength the curve on an
energy basis is rather flat (Henderson, 1970). If the
spectrum is plotted on a quanta basis the rise continues into
the infrared part of the spectrum (Halldal, 1979). The *in
vivo* absorption spectra of the purple bacteria, as well as
those of cyanobacteria and green algae are poorly "tuned" to
this spectrum (see Fig. 3). It is interesting to note, however,

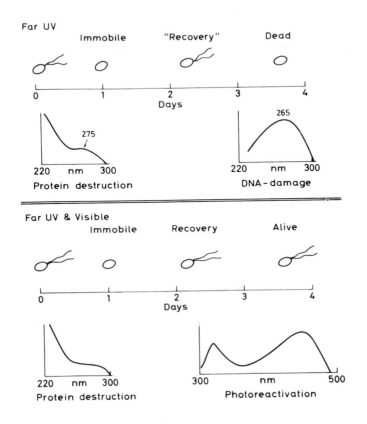

Fig. 9. The effect of far UV irradiation and the effect of near UV and visible light after irradiation, on motility and growth of the green flagellate Platymonas sp. For explanation see text (after Halldal, 1961b).

that the spectral distribution of solar radiation around dusk and dawn, first reported by Johnson, Salisbury and Connor, (1967) ("dusk and dawn effect"), has maxima in the blue and red regions of the spectrum and this fits rather well with the absorption and action spectra of algae in the diatom group (Halldal, 1974). The ecological importance of this fit is, however, rather uncertain. In contrast there is an extremely good fit between the spectral response of photosynthesis of cyanobacteria and red algae and light quanta distribution underwater in coastal areas and inland lakes. However, these light conditions give a poor fit with the spectral properties

of diatoms. The photosynthetic purple bacteria have their
most efficient photosynthetic response at wavelengths longer
than 700 nm. As practically no radiation in this spectral
region penetrates to depths below a few cm, this absorption
region of bacteriochlorophyll is not normally used under
natural conditions.

Photoinactivation - Photoreactivation The wavelength region
300 nm is very important in this respect. This is the
shortest wavelength of natural radiation from the sun and
lesions to proteins and nucleic acids are observed when these
materials are exposed to such radiation. Microorganisms
have different means of protection from protein damage by
autorepair or morphological protection. Far UV damage to
DNA may be photoreactivated with irradiation between 300 and
approximately 500 nm. Furthermore, the main part of the
visible spectrum, between 500 and 780 nm, and the whole part
of the near IR region where photobiology can be observed (to
1000 nm) are not used in this photorepair process.

Phototaxis Phototopotaxis in flagellates and motile green
algae can be observed between 300 and approximately 600 nm
under natural conditions. For this photoresponse only a
small part of the spectral region of photobiology is utilized
and this is particularly true in coastal waters and inland
lakes. At rather moderate depths most of the radiation below
450 nm is normally absorbed due to the yellow substance of
these water types (Jerlov, 1968). Thus, motile flagellated
algae utilize efficiently, for their photoresponse, only the
region between 450 nm and 550 nm. As the action spectrum of
photophobotaxis in the purple bacteria is the same as their
action spectrum of photosynthesis, I refer to Photosynthesis
(above) for comments on this response.

REFERENCES

Gessner, P. (1955). *In* "Hydrobotanik", Vol 1, Gustav
    Fischer Verlag, Stuttgart.
Halldal, P. (1958). Action spectra of phototaxis and related
    problems in Volvocales, Ulva gaments and Dinophyceae.
    Physiologia Plantarum 11, 742-752.
Halldal, P. (1960). Action spectra of induced phototactic
    response changes in *Platymonas*, *Physiologia Plantarum* 13,
    726-735.
Halldal, P. (1961a). Ultraviolet action spectra of positive
    and negative phototaxis in *Platymonas subcordiformis*,
    Physiologia Plantarum 14, 133-139.
Halldal, P. (1961b). Photoinactivation and their reversals
    in growth and motility of the green alga *Platymonas*
    (Volvocales), *Physiologia Plantarum* 14, 558-575.

Halldal, P. (1962). Taxes. *In* "Physiology and Biochemistry of Algae" (Ed. R.A. Lewin), pp. 583-593. Academic Press, New York.

Halldal, P. (1963). Zur Frage des Photoreceptors bei der Topophototaxis der Flagellaten, *Berichte der Deutschen Botanische Gesellshaft* 76, 323-327.

Halldal, P. (1964a). Ultraviolet action spectra of photosynthesis and photosynthetic inhibition in a green and a red alga, *Physiologia Plantarum* 17, 414-421.

Halldal, P. (1964b). Phototaxis in protozoa. *In* "Biochemistry and Physiology of Protozoa" (Ed. R. Hutner), Vol. 3, pp. 277-296. Academic Press, New York.

Halldal, P. (1966). Induction phenomena and action spectra analyses of photosynthesis in ultraviolet and visible light studied in green and blue-green algae, and in isolated chloroplast fragments, *Zeitschrift für Pflanzenphysiologie* 54, 28-44.

Halldal, P. (1967). Ultraviolet action spectra in algology. A review, *Photochemistry and Photobiology* 6, 445-460.

Halldal, P. (1974). Light and photosynthesis of different algal groups. *In* "Optical Aspects of Oceanography" (Ed. N.G. Jerlov and E. Steemann Nielsen). Academic Press, New York.

Halldal, P. (1979). Biological Energy Capture: Special Features of Marine Plants. Keynote speech at "Marine Science and Ocean Policy Symposium". Santa Barbara, California.

Halldal, P. and Taube, O. (1972). Ultraviolet action and photoreactivation in algae. *In* "Photophysiology" (Ed. A.C. Giese), 7, pp. 163-188. Academic Press, New York.

Hand, W.G. and Davenport, D. (1970). The experimental analyses of phototaxis and photosynthesis in flagellates. *In* "Photobiology of Microorganisms" (Ed. P. Halldal) pp. 253-282. Wiley-Interscience, London.

Henderson, S.T. (1970). "Daylight and its Spectrum". Adam Hilger Ltd., London.

Jerlov, N.G. (1968). "Optical Oceanography" Elsevier Publishing Co , Amsterdam.

Johnson, T.B., Salisbury, F.B. and Connor, G.I. (1967). Ratio of blue to red light. A brief increase following sunset, *Science* 155, 1663-1665.

Meier, F.E. (1932). Lethal action of ultraviolet light on a unicellular green alga, *Smithsonian Miscellaneous Collections* 87, 1-11.

Milatz, J.M.W. and Manten, A. (1953). The quantitative determination of the spectral distribution of phototactic sensitivity in the purple bacterium *Rhodospirillum rubrum, Biochemica et Biophysica Acta* 11, 1727.

Nultsch, W. (1970). Photomotion of microorganisms and its interaction with photosynthesis. *In* "Photobiology of Microorganisms" (Ed. P. Halldal), pp. 231-251. Wiley-Interscience, London.

Rupert, C.S. (1964). Photoreactivation of ultraviolet damage. *In* "Photophysiology" (Ed. A.C. Giese) Vol. 2, pp. 283-327. Academic Press, New York.

Steemann Nielsen, E. (1964). On a complication in marine productivity work due to the influence of ultraviolet light, *Journal du Conseil International pour Exploration de la Mer* 22, 130-135.

WATER ACTIVITY AS THE CRITERION OF WATER AVAILABILITY

D.S. Reid

Unilever Research, Colworth Laboratory
Sharnbrook, Bedford, England

INTRODUCTION

Water is one of the most important compounds on this planet, and life and the processes of life have evolved in an environment where water is the principal liquid solvent available. It is possible that water, with its unique properties, is an essential prerequisite for the evolution of life. In order to adequately describe the environment in which a microorganism exists, some measure is required which indicates the amount and form of water in that environment. This measure has to relate to the biochemical reactions which comprise the metabolic processes of the organism. It has to give some indication of the suitability of the solvent environment for the reaction sequences necessary for normal metabolism.

The measure required has to be, in some way, an indication of the water available to the microorganism for its essential life processes. In early work, the "water content" of the environment was the most widely used measure. As Mossell (1975) and Corry (1978) have pointed out, this is not a very satisfactory procedure as the properties of, for example, food systems, at the same water content can be very different. Another measure popular with early workers was osmotic pressure. This has been discussed at some length by Brown (1976) who commented particularly on the difficulty of accurately measuring the osmotic pressure in many systems. Scott (1957) brought some order to the field by suggesting that a suitable measure of water availability was the thermodynamic water activity, a_w, of the equilibrium system. Brown (1976) and Troller and Christian (1978) have discussed in some detail the advantages and disadvantages of the use of water activity as a measure of water availability, and the following section considers how well a_w serves as a partial description of the aqueous environment.

WHAT IS a_w?

The thermodynamic definition of activity, as stated by Lewis and Randell (1961) is:

Activity is, at a given temperature, the ratio of the fugacity, f, of a substance in some given state, and its fugacity, f^o, in some state which, for convenience, has been chosen as a standard state:

$$a = {}^f/f^o \qquad (1)$$

The quantity termed fugacity has the form of a corrected vapour pressure. For all practical purposes, the fugacity of water, f_w, can be replaced by the partial vapour pressure of the water, p_w. So long as the total pressure is not markedly in excess of 1 atmosphere the water activity, a_w, can be expressed as:

$$a_w = {}^f w/f^o_w = {}^p w/p^o_w \qquad (2)$$

Normally the standard state chosen is pure water at 1 atmosphere total pressure, where its vapour pressure is p^o_w. Reid (1976) has considered some of the consequences of this simple definition as applied to microbiological systems. The activity a_w is simply related to the chemical potential, μ_w, by:

$$\mu_w = \mu^o_w + RT\ln a_w + \overline{V}_w P \qquad (3)$$

Where R is the gas constant, T is the absolute temperature, $\overline{V}_w$ is the partial molar volume of the water and P the pressure. Gravitational effects are ignored. The chemical potential, μ, can be identified with the partial molar free energy. At equilibrium the chemical potential of a substance must be identical in all phases which are in contact.

Standard thermodynamic relationships connect activity and chemical potential with osmotic pressure, water potential and the other colligative properties of a system. For example, the water potential ψ, is defined as:

$$\psi = (\mu_w - \mu^o_w) / \overline{V}_w \qquad (4)$$

This measure, popular with plant physiologists, can also be expressed as:

$$\psi = \frac{RT}{\overline{V}} \ln a_w + P \qquad (5)$$

Thus, as Brown (1976) has pointed out, describing "water availability" by any of these measures is really equivalent. Since a_w is usually the easiest to measure, most workers have standardised on this parameter.

Thus a_w is a measure of the relative escaping tendency of water from the system, compared to pure water, and can be

adequately described by the equilibrium relative humidity
(E.R.H.) which the system could maintain. If two systems have
the same a_w then, at the same temperature and at one atmosphere
total pressure, they must establish above themselves the same
partial pressure of water vapour. If a sample with a given a_w
is exposed to an environment of higher a_w, water will tend to
transfer into the sample, thus raising its a_w. Similarly, an
environment with a lower a_w will tend to remove water from the
sample. In a way, a_w is a description of a hypothetical
situation in which a sample is in contact with its environment
only by means of vapour transfer paths (Fig. 1).

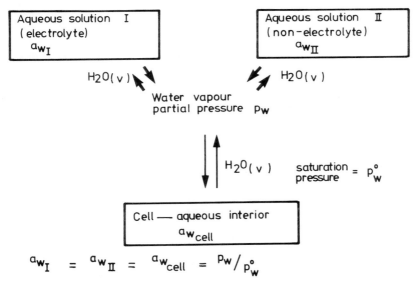

$$a_{w_I} \; = \; a_{w_{II}} \; = \; a_{w_{cell}} \; = \; P_w / p_w^\circ$$

Fig. 1. A system described adequately by a_w.

It does not follow that because two solutions have an identical
a_w that the molecular state of water is identical in each; it
only follows that the energetics of water escape are, at that
point, identical (that is, the chemical potential of water
in each phase is identical). This is readily illustrated by
a simple example. If carbon tetrachloride is shaken up with
water and allowed to stand, the system separates out into two
layers. Though the compositions of the "water rich" and "water
poor" layers will be very different, the chemical potential of
the water will be the same in each layer as indeed will the
chemical potential of carbon tetrachloride. In Fig. 1, therefore, it would not be expected that the structure (or molecular
arrangements) of the water would be the same in both electrolyte

and non-electrolyte solutions, even though μ_w is identical.
Indeed a saccharide solution and an alcohol solution would
be expected to have different solvent structures (Franks and
Reid, 1973; Franks, 1975). Only in the hypothetical arrangement of Fig. 1, in which solvent vapour transfer is the sole
means of contact between different solutions and the microorganism,
would solutions of all these different types, maintaining identical
water activities, be expected to have identical effects on
microorganisms. In the natural environment, the microorganism,
in order to extract its necessary nutrients from the environment,
has to be in direct contact with the liquid phase. The detailed
structural arrangements of the water molecules and the particular
properties of the solutes which are responsible for the lowering
of the water activity, will affect the response of the microorganism to its environment. It would be an impossible task to
fully characterise the aqueous environment, so a compromise has
to be made by describing the response of microorganisms to a_w
in systems where this a_w has been produced in a variety of ways
by different solute types. In so doing general effects may
then become apparent. It must be recognised, however, that
water activity describes only one aspect of the environment
of microorganisms, namely the water component, and that even
this aspect is described imperfectly. The description does
not give any indication of the structural organisation of the
water nor does it give any information relating to the kinetics
of water transfer. Nevertheless, it is the best measure available
at present and it serves to increase our understanding of the
interrelationship between microorganisms and their environment.

MEASUREMENT OF a_w

Having decided on water activity as a measure of water
availability, simple methods are required for its determination,
either directly or through related quantities. A variety of
methods exist which are described by Troller and Christian (1978),
Gal (1975) and Hardman (1976). These authors all emphasise the
importance of close temperature control of the system, whatever
measuring technique is employed. The more important methods of
water activity determination control can be grouped into three
main classifications. The first group involves the equilibration
of the sample with solutions of known a_w (or E.R.H.) and, after
a suitable period, the sample will have established within it
the same a_w as the control solutions. These are known as the
isopestic methods and a detailed description of one method is
given by Robinson and Sinclair (1934). In order to choose
equilibrating solutions of the appropriate a_w, tables are
required which list the composition of solutions with particular
a_w values, and some examples are given in Table 1.

TABLE 1

Composition of solutions maintaining stated a_w at $25^{\circ}C$

Solute	Concentration (molal)	a_w	Solute	Concentration (molal)	a_w
NaCl	0.15	0.99	Sucrose	4.11	0.90
NaCl	0.61	0.98	Sucrose	4.98	0.85
NaCl	1.20	0.96	$LiCl.H_2O$	sat.	0.12
NaCl	2.31	0.92	$MgCl_2.6H_2O$	sat.	0.33
NaCl	2.83	0.90	K_2CO_3	sat.	0.44
NaCl	4.03	0.85	NaCl	sat.	0.75
Sucrose	0.27	0.99	Li_2SO_4	sat.	0.85
Sucrose	1.03	0.98	KNO_3	sat.	0.94
Sucrose	1.92	0.96	NaH_2PO_4	sat.	0.98

In a variant of this type of method, (for example, Corry, 1974), a particular partial pressure of water is maintained by controlling closely a water reservoir at a predetermined temperature. If the sample temperature is also closely maintained, this has the effect of producing the required relative humidity, (R.H.) above the sample and hence the sample equilibrates to the required a_w. The dew point method of determining the relative humidity above a sample, which consists of identifying the temperature at which the vapour will just condense on to a cooled surface - the partial pressure of water above the sample being the saturation pressure at the dew point temperature, is the reverse of this method of sample equilibration.

A second group of methods is dynamic in character and does not involve sample equilibration (for example, Landrock and Procter, 1951). Instead the sample is placed in contact with environments of different a_w, and the tendency for the sample to gain or lose water is determined. It is assumed that the a_w at which water would be neither gained nor lost is the true a_w value which can be maintained by the sample.

The third group comprises a variety of methods which make use of electronic sensors. In general, the sample is allowed to equilibrate in a sealed, temperature controlled chamber with the sensor. The sensor produces an electrical response which varies with the R.H. of the atmosphere to which it is exposed and which will be equal to the a_w of the water within the sensor, assuming rapid equilibration. A variety of electrical responses have been utilised, such as the electrical conductivity of salts (which varies with water content and the usual salt used is lithium chloride) or the dielectric properties of anodised aluminium oxide. Hardman (1976) presents these methods in more detail.

A MICROORGANISM'S EYE VIEW OF a_w

Let us now consider the response of microorganisms to environments of different a_w, particularly with respect to the growth of microorganisms rather than survival. Growth can be observed over a wide range of a_w values (Table 2). Many organisms are capable of growth at an a_w of 0.999, though none can grow at an a_w of 1.00 (that is pure water) as there are no nutrients available to the organism. As Table 2 shows, some organisms are capable of growth at an a_w of 0.60 although the lower a_w limit for growth is different for bacteria, yeasts and fungi. The range of a_w from just below 1.00 to 0.96 covers the normal fresh water, brackish water and sea water environments in which most microorganisms exist and will not be discussed further. These are the conditions of normal metabolism referred to in the introductory paragraphs. Considerable interest centres on those organisms which are capable of growth under conditions of "water stress", that is in an environment of significantly lowered water activity and in particular on the stratagems available to a cell enabling it to maintain a functional metabolism.

From Table 2 it can be seen that a variety of microorganisms exist which have their growth optima at an a_w close to 0.99 but are capable of growth at low a_w. These are classed as osmotolerant or xerotolerant organisms (Pitt, 1975; Brown, 1976; Measures, 1975). Others exhibit growth at some lower a_w value and these are the osmophilic or xerophilic organisms. Furthermore, an organism's response to a low a_w can depend on whether the reduction in a_w in the external environment is due to high salt levels at high non-electrolyte levels. This is not surprising as the ionic strength of the external medium would be expected to affect the charge distribution on the cell surface. Table 3 illustrates the differing response of microorganisms to lowered environmental a_w produced both by electrolytes and non-electrolytes.

Consider a microorganism suddenly transferred from an environment of relatively high a_w to one with a much lower a_w. The initial response is likely to be a loss of water from the organism to the external environment. After this initial loss of water and consequent decrease in cell volume, a variety of longer term responses is possible. Depending on the composition of the external medium, the organism might transfer certain of the external solute species through the cell membrane into the cell, until a balance of internal and external water activities resulted. Alternatively, the loss of internal water could continue until internal and external water activities came into balance. A third possibility, available to an actively metabolising cell, is the internal synthesis of solutes which would produce a

TABLE 2

a_w *limits for growth of selected microorganisms*

Water Activity a_w		Organism	Medium	
Optimum	Minimum			Reference
		BACTERIA		
<0.99	0.94	*Aerobacter aerogenes*	A	1
<0.99	0.99	*Bacillus cereus* var *mycoides*	A	1
<0.99	0.95	*Bacillus megaterium*	B	2
<0.99	0.95	*Bacillus subtilis*	A	1
<0.99	0.93	*Bacillus subtilis*	C	2
<0.99	0.90	*Bacillus subtilis*	B	2
n.d.	0.92	*Micrococcus* sp.	C	2
n.d.	0.83	*Micrococcus* sp.	B	2
0.96	0.85	*Micrococcus halodenitrificans*	B	3
0.99	0.97	*Pseudomonas fluorescens*	B	4
0.99	0.83	*Staphylococcus aureus*	B	5
0.99	0.75	*Halococcus* sp.	B	3
0.90	0.75	*Halobacterium* sp.	B	3
0.93	0.80	*Ectothiorhodospira halophila*	B	3
		ALGAE		
<0.99	0.95	*Dunaliella tertiolecta*	B	3
0.94	0.75	*Dunaliella viridis*	B	3
		YEASTS		
n.d.	0.94	*Torula utilis*	A	1
n.d.	0.94	*Saccharomyces cerevisiae*	B	6
<0.99	0.86	*Saccharomyces ronxii*	B	6
n.d.	0.85	*Saccharomyces ronxii*	D	6
n.d.	0.61	*Saccharomyces ronxii*	E	3
n.d.	0.92	*Saccharomyces cerevisiae*	D	6
		MOULDS		
0.97	0.78	*Aspergillus nidulans*	A	7
0.98	0.84	*Aspergillus ruger*	A	8
0.94	0.70	*Aspergillus amstelodami*	A	8
0.98	0.78	*Aspergillus flavus*	A	7
0.97 – 0.90	0.61	*Xeromyces bisporus*	A	9

n.d. = not determined.
The media are described in the original papers, the principal
solute employed for water activity control being:
A, controlled atmosphere equilibration; B, NaCl; C, glycerol;
D, glucose; and E, sugars.
References: 1, Burcik, (1950); 2, Marshall, Ohye and Christian,

(1971); 3, Brown, (1976); 4, Limsong and Frazier, (1966); 5, Tatini, (1973); 6, Onishi, (1957); 7, Ayerst, (1969); 8, Scott, (1957); and 9, Pitt and Christian, (1968).

TABLE 3

Variation of response of microorganisms to a_w produced by different means

Organism	Minimum Water Activity		Reference
	Salt	Non-electrolyte (compound)	
Bacillus megaterium	0.95	0.92 (glycerol)	1
Bacillus sphaericus	0.93	0.92 (glycerol)	1
Bacillus subtilis	0.90	0.93 (glycerol)	1
Escherichia coli	0.95	0.93 (glycerol)	1
Micrococcus sp.	0.83	0.92 (glycerol)	1
Saccharomyces cerevisiae	0.94	0.92 (glucose)	2
Saccharomyces ronxii	0.86	0.61 (sugars)	2,3
Pseudomonas fluorescens	0.97	0.95 (glycerol)	1
Staphylococcus aureus	0.85	0.89 (glycerol)	1
Salmonella orienburg	0.95	0.94 (glycerol)	1
Candida guillermondii	0.91	0.87 (glucose)	3
Candida parapsilosis	0.87	0.85 (glucose)	3
Citeromyces matritensis	0.94	0.87 (glucose)	3
Torulopsis candida	0.75	0.65 (sucrose/ glycerol)	3
Torulopsis versatilis	0.91	0.70 (sucrose/ glycerol)	3

The media are described in the original papers. The major solute influencing the water activity is either salt or the indicated non-electrolyte.
References: 1, Marshall, Ohye and Christian, (1971); 2, Onishi, (1957); and 3, Tilbury, (1976).

lowering of the internal water activity and thus equalisation of internal and external water activities. Examples of all these types of response can be found. Gould and Measures (1977) demonstrate that *Bacillus subtilis* can respond to osmotic shock with a combination of all three.

At this point it is relevant to mention another important aspect of the survival of organisms in a low a_w environment; namely cell freezing and cryopreservation. At low temperatures in the presence of ice the a_w is low and cells (if internally unfrozen) tend to lose water and shrink. In slowly frozen cell suspensions, cooled to very low temperatures, cell death frequently results and Meryman (1971) has suggested that this may be due to the dehydration of cells below some minimum cell volume which results in irreparable cell damage. Cryoprotection could consist of adding solutes which prevent this extreme reduction in cell volume and which protect the macromolecules of the cell from concentration denaturation. Regardless of whether this minimum cell volume hypothesis is correct, there may be useful parallels between the response of cells to freezing and the response of cells to low a_w, that is dehydrating, environments at ambient temperatures.

Once a microorganism has come to thermodynamic terms with its environment, by somehow adjusting its internal a_w, there remains the problem of ensuring continued growth and metabolism. Two basic strategies appear to have evolved which allow cell metabolism to proceed in environments of reduced a_w. The first strategy appears to be used only in particular environments of high ionic strength and has been termed mechanism I by Brown (1976). The second strategy has two variants, depending on whether the reduction in a_w is of ionic or non-ionic origin and has been termed mechanism II by Brown (1976).

In the first option a range of microorganisms have evolved with modified cell proteins which function optimally in a high ionic strength environment, such as brine pans or the Dead Sea. These organisms, the halophiles, contain a special range of enzymes which are adapted to work in high concentration sodium chloride solutions. Indeed these enzymes work poorly, if at all, in the more normal low ionic strength environments.

The alternative strategy involves compatible solutes and in this mechanism the solute utilised by the cell to equalise the internal a_w with the external environment is one that does not interfere significantly with the internal metabolic function of the cell. In organisms exhibiting this mechanism the cell proteins are not different in any major way to those of cells capable of growth in the normal a_w range. Indeed the solute which is accumulated in large quantities within the cell to lower the internal water activity also in some way protects the cell enzymes from the effects of lowered water activity.

Table 4 summarises different compatible solute responses which have been elucidated. For ionic systems high levels of sodium ion seem to be inhibitory and so the compatible solute of choice appears to be potassium ion. Cells somehow exclude sodium ion and accumulate potassium ion to balance the external osmotic pressure (Brown, 1976).

Measures (1975 a,b) and Measures and Gould (1976) have shown that, in non-electrolyte systems where high external ionic strength was not a problem, a variety of amino acids were utilised in the compatible solute role. In particular proline was synthesised by a number of organisms in response

TABLE 4

Compatible solute responses in microorganisms

Maximum NaCl concentration for growth	Minimum water activity for growth	Organism	Compatible solute accumulated	Reference
0.7M	0.98	*Ps. aeruginosa*	glutamic acid	1
1.3M	0.96	*Clostridium perfringens*	proline/ glutamic acid	1
1.5M	0.95	*Bacillus megaterium*	proline	1
2.0M	0.93	*Sarcina lutea*	proline	1
2.7M	0.90	*Bacillus subtilis*	proline	1
4.0M	0.85	*Staph. aureus*	proline	1
n.d.	0.61	*Sacc. ronxii*	polyols	2
sat.	0.75	*Dunaliella* sp.	glycerol	3
sat.	0.75	*Halobacterium* sp.	K+	4
sat.	0.75	*Halococcus* sp.	K+	4

n.d. = not determined.
References: 1, Measures, (1975); 2, Brown, (1974); 3, Borowitzka and Brown, (1974); and 4, Brown, (1976).

to severe osmotic stress. In addition, many microorganisms have been found to utilise polyols as compatible solutes, including *Saccharomyces ronxii*. The mechanism of action of compatible solutes is rather obscure. Cryobiological studies have suggested that the mode of action of small molecule cryoprotectants, such as dimethylsulphoxide and glycerol, involves, in part, a reduction of the ionic strength of the internal medium at any given temperature where ice is present (that is at any given a_w). This might be one of the functions of compatible non-electrolyte solutes. Also, it is possible that certain solutes can in some way replace the hydration water of a macromolecule and so help stabilise the conformation of proteins against dehydration.

In this presentation I have tried to point out the ramifications of the use of the water activity concept as a description of the aqueous environment and briefly to outline the range of responses available to a microorganism in a low a_w environment.

REFERENCES

Ayerst, G. (1969). The effects of moisture and temperature on growth and spore germination of some fungi, *Journal of Stored Product Research* 5, 127-141.

Borowitzka, L.J. and Brown, A.D. (1974). The salt relations of marine and halophilic species of the unicellular green alga, *Dunaliella.* The role of glycerol as compatible solute, *Archives of Microbiology* 96, 37-52.

Brown, A.D. (1974). Microbial water relations: features of the intracellular composition of sugar tolerant yeasts, *Journal of Bacteriology* 118, 769-777.

• Brown, A.D. (1976). Microbial water stress, *Bacteriological Reviews* 40, 803-846.

Burcik, E. (1950). The relationship between hydration and growth of bacteria and yeasts (in German), *Archiv für Mikrobiologie* 15, 203-235.

Corry, J.E.L. (1975). The effect of water activity on the heat resistance of bacteria. *In* "Water Relations of Foods" (Ed. R.B. Duckworth) pp. 325-337. Academic Press, London.

Corry, J.E.L. (1978). Relationships of water activity to fungal growth. *In* "Food and Beverage Mycology" (Ed. L.R. Beuchat) pp. 45-82. AVI, Westport.

Franks, F. (1975). The hydrophobic interaction. *In* "Water, A Comprehensive Treatise" (Ed. F. Franks) Vol. 4, pp. 1-94. Plenum, London.

Franks, F. and Reid, D.S. (1973). Therodynamic properties of aqueous non-electrolyte solutions. *In* "Water, A Comprehensive Treatise" (Ed. F. Franks) Vol. 2, pp. 323-380. Plenum, London.

Gal, S. (1975). Recent advances in techniques for the determination of sorption isotherms. *In* "Water Relations of Foods" (Ed. R.B. Duckworth) pp. 139-154, Academic Press, London.

Gould, G. and Measures, J.C. (1977). Water relations in single cells, *Philosophical Transactions of the Royal Society of London* B 278, 151-166.

Hardmann, T.M. (1976). Measurement of water activity. Critical appraisal of methods. *In* "Intermediate Moisture Foods" (Ed. R. Davies, G.C. Birch and K.J. Parker) pp. 75-87. Applied Science, London.

Landrock, A.H. and Proctor, B.E. (1951). A new graphical interpolation method for obtaining humidity equilibria data, *Food Technology* 5, 332-337.

Lewis, G.N. and Randall, M. (1961). Thermodynamics (Revised by K.S. Pitzer and L. Brewer) Second edition, McGraw-Hill, New York.

- Limsong, S. and Frazier, W.C. (1966). Adaption of
  Pseudomonas fluorescens to low levels of water activity
  produced by different solutes, *Applied Microbiology* 14,
  889-901.
- Marshall, B.J., Ohye, D.F. and Christian, J.H.B. (1971).
  Tolerance of bacteria to high concentrations of glycerol
  and NaCl in the growth medium, *Applied Microbiology* 21,
  363-364.

Measures, J.C. (1975a). Role of amino acids in osmoregulation
  of non-halophilic bacteria, *Nature (London)* 257, 398-400.

Measures, J.C. (1975b). Reactions of microbial cells to
  osmotic stress. *In* "L'eau et les Systems Biologique"
  pp. 303-309, Editions du CNRS, Paris.

Measures, J.C. and Gould, G. (1976). Interactions of
  microorganisms with the environment of intermediate moisture
  foods. *In* "Intermediate Moisture Foods" (Ed. R. Davies,
  G.C. Birch and R.J. Parker) pp. 281-296. Applied Science,
  London.

Meryman, H.T. (1971). Osmotic stress as a mechanism of
  freezing injury, *Cryobiology* 8, 489-500.

Mossel, D.A.A. (1975). Water and microorganisms in foods –
  a synthesis. *In* "Water Relations of Foods" (Ed. R.B.
  Duckworth) pp. 347-361. Academic Press, London.

Onishi, H. (1957). Studies on osmophilic yeasts III.
  Classification of osmophilic soy and miso yeasts, *Bulletin
  of the Agricultural Chemistry Society of Japan* 21, 151-156.

Pitt, J.I. (1975). Xerophilic fungi and the spoilage of
  foods of plant origin. *In* "Water Relations of Foods"
  (Ed. R.B. Duckworth) pp. 273-307. Academic Press, London.

Pitt, J.I. and Christian, J.H.B. (1968). Water relations of
  xerophilic fungi isolated from prunes, *Applied Microbiology*
  16, 1853-1858.

Reid, D.S. (1976). Water activity concepts in intermediate
  moisture foods. *In* "Intermediate Moisture Foods" (Ed.
  R. Davies, G.C. Birch and R.J. Parker) pp. 54-63.
  Applied Science, London.

Robinson, R.A. and Sinclair, D.A. (1934). The activity
  co-efficients of the alkali chlorides and of lithium iodide
  in aqueous solution from vapour pressure measurements,
  Journal of the American Chemical Society 56, 1830-1835.

- Scott, W.J. (1957). Water relations of food spoilage microorganisms, *Advances in Food Research* 7, 83-127.

Tatini, S.R. (1973). Influence of food environments on growth
  of *Staphylococcus aureus* and production of various enterotoxins, *Journal of Milk Food Technology* 36, 559-563.

Tilbury, R.H. (1976). The microbial stability of intermediate
  moisture foods with respect to yeasts. *In* "Intermediate
  Moisture Foods (Ed. R. Davies, G.C. Birch and R.J. Parker)

pp. 138-165. Applied Science, London.
Troller, J.A. and Christian, J.H.B. (1978). "Water Activity
and Food". Academic Press, London.

EXTREME ENVIRONMENTS

D.J. Kushner

Department of Biology, University of Ottawa
Ottawa, Ontario K1N 6N5,
Canada

INTRODUCTION: ATTEMPTS AT A DEFINITION

What is an "extreme environment"? Probably there is no generally satisfactory definition. We can all agree that, say, alkaline salt lakes and acid hot springs are extreme. Beyond these and a few other clear examples each person will have favourite candidates. Instead of attempting to settle the matter once and for all, I will consider some environments that have been called "extreme" to see what attributes they have in common, dealing mainly with those terrestrial environments in which living creatures are known to actually live and grow. I say "terrestrial" because there has been much recent interest in the surface of Mars as an undoubted extreme environment, by any of our standards, where one might still expect to find microbial life. The fact that the Viking mission found no life on Mars (Horowitz, 1979) has, alas, removed considerable interest and financial support for this kind of comparison and for other studies of extreme environments.

Some of the environments to be considered will seem strange indeed to the microbiologist concerned mainly with man's environment and with the microorganisms that live on or in him. They will seem less strange to those microbial ecologists who study microbial life in forests, lakes and oceans, the last of which are sometimes considered to be extreme environments.

Several books on this subject have appeared in the last few years (Heinrich, 1976; Ponnamperuma, 1976; Caplan and Ginzburg, 1978; Kushner, 1978a; Shilo, 1979) which should be consulted for a more detailed treatment than is possible here. The review of Alexander (1976) presents a good overview

of the influence of widely varying environments on microbial
populations. Because there are so many examples of microbial
life under harsh and interesting conditions, I apologize in
advance to any colleague whose work or whose favourite extreme
environment I may have neglected.

Some examples

Hot places   These include volcanic fumaroles and soils, acid
and alkaline hot springs, and submarine hot springs, which
in some sites, may reach temperatures well in excess of $100^{\circ}C$
(Tansey and Brock, 1978; Castenholz, 1979). To such places,
ultimately warmed by volcanic action, we must add sun-warmed
soil and rocks (up to $70^{\circ}C$), refuse piles and compost heaps
where microbiological action raises the temperature to $70^{\circ}C$
or higher, and hot water tanks and industrial cooling water
(up to $100^{\circ}C$) (Tansey and Brock, 1978). Though many forms
of microbial life have been found in such environments, as
their temperature increases the forms of life become more
limited. Above about $60^{\circ}C$ no eukaryotic cells are found,
and above $70^{\circ}C$, the temperature at which the cyanobacterium
Synechococcus lividus grows, no photosynthetic microorganisms
at all. In the hottest parts of hot springs only bacteria
are found. Although most of these are still uncharacterized,
some have been the objects of intense study including the
filamentous *Thermus* species (maximum temperature $85^{\circ}C$) and
Sulfolobus acidocaldarius which can grow at pH 1 and about 90°
(Langworthy, 1978; Castenholz, 1979). These observations
strongly suggest that the higher the temperature, the more
difficult it is for structurally complex organisms and cells
to exist.

Salty places   Solar salts, prepared by evaporating sea water,
contain substantial amounts of extremely halophilic bacteria
in the genera *Halobacterium* and *Halococcus*. The development
of microbial populations while sea water is being concentrated,
which manifests itself by the increasing appearance of pink
and red colours as the extreme halophiles grow, has been
remarked on since ancient times. Students of halophiles are
fond of quoting the report cited by Baas-Becking (1931) from
a Chinese treatise some 5000 years old giving directions for
isolating salt from drying seawater. "An embankment is made
and ditches to draw the clear sea water. It is left for a
long time, until the colour becomes red. If the south wind
blows with force ... the salt may grain overnight." Baas-
Becking's paper should also be consulted for interesting
notes on other biological aspects of salt manufacture,
especially the use of *Artemia salina*, the brine shrimp, for
removing bitter particulate matter from the concentrating

salt solution. In addition, halophilic populations have been
studied in natural salt lakes. The most studied of these, but
by no means the only ones, are the Dead Sea and the Great Salt
Lake (Brock, 1979; Kushner, 1978b; Nissenbaum, 1975; Post,
1977). The microbial ecology of a solar pond by the Red Sea
South of Elat, in which the temperature of the lower layers
can rise to $56°C$ and the salinity to 15% has been studied by
Israeli and other scientists (Cohen, Padan and Shilo, 1975;
Krumbrin and Cohen, 1974). More recently, the microbiology
of a series of highly alkaline salt lakes, and Wadi Natrun
lakes in Egypt, with a salinity of over 30% and a pH of 11
has been studied (Imhoff, Sahl, Soliman and Trüper, 1979;
Imhoff and Trüper, 1977). In considering such "salt" lakes
we must remember that "salt" is not always synonymous with
NaCl. The Dead Sea is richer in magnesium than in sodium and
is very high in calcium, but other lakes, including those of
the Wadi Natrun, contain only traces of these divalent cations.
The Don Juan pond in Antarctica, which contains saturated $CaCl_2$,
was at one time reported to contain living bacteria; it now
seems that such bacteria are introduced into the pond from
the outside but do not actually grow there (Horowitz, 1976,
1979).

Past studies have emphasised the halobacteria and halococci
as inhabitants of such lakes but it is useful to recognise
that many other microorganisms, including those of nitrogen,
sulphur and other cycles have been found in these lakes and
would surely merit further study (Post, 1977; Imhoff *et al.*,
1979; Nissenbaum, 1975).

The forms of life found in very salty lakes are limited.
In the Great Salt Lake some metazoans are found (Post, 1977).
Brine shrimps, *A. salina* are found throughout the lake but
there are some indications that they do not carry out their
full life cycle in the most saline part, the North Arm where
salinity is 330 gl^{-1} but hatch in the less saline (120 gl^{-1})
South Arm before being carried to the North Arm (Post, 1977).
Species of the brine fly (genus *Ephydra*) are found throughout
the lake and may have a complete life cycle in even the most
saline waters. In the highly alkaline salt lakes of the
Wadi Natrun many bacteria, photosynthetic bacteria, cyanobacteria and eukaryotic algae (*Dunalliella* sp.) are found.
Some zooflagellates that feed on lower forms are found in these
lakes but no forms higher than these (Imhoff *et al.*, 1979).

To these natural salty places we must add man-made salt
environments, including salted foods and hides. The fermentations involved in soy sauce manufacture take place in about
20% NaCl and soy sauce contains a number of yeasts that
function well in (but usually do not need) high salt or sugar
concentrations. Some of the most studied strains of halo-

bacteria were originally isolated from salted hides (Kushner,
1978b).

The use of concentrated sugar solutions to preserve fruits,
that is in jams and jellies, attest to the general experience
that such solutions prevent microbial growth. When microorganisms, usually fungi or yeasts, do grow only one species may be
present, a good example of the limited species diversity often
held as a characteristic of extreme environments (Alexander,
1976; Brock, 1969).

Dry environments Such environments are not clearly distinct
from those just discussed, since dissolved solutes lower
the activity of water. A cell in high solute concentration
will be exposed to lower water activity (a_w) in the external
medium and water may be withdrawn from the cell. (For a
discussion of the concepts of water activity and water potential,
see Brown, 1976; Kushner, 1978b; Smith, 1978; Griffin and Luard,
1979; Reid, this volume pp 17-29). Microorganisms in natural
environments may be subjected to osmotic stress in liquids with
high solute concentrations or matric stress when they grow on a
surface, say the surface of a soil particle, under dry conditions.
For a given level of dryness, the matric stress seems to have a
more inhibitory effect on microbial growth (Smith, 1978). Some
of the most detailed studies of the minimum water requirement of
microorganism suggest that for microbial growth dry conditions may
be the most extreme of all. Only a few fungal species can grow
at an a_w of about 0.6 and no known microorganism can grow at a
lower a_w (Pitt, 1975; Brown, 1976; Kushner, 1978b; Horowitz, 1979).

Microbial life in dry soils, especially in desert soils, may
be very limited. In the recent past special interest was taken
in the Antarctic Dry Valleys, a very harsh environment that
combines extreme dryness (10% mean relative humidity) with
extreme cold (mean temperatures -20 to -25°C though rock surfaces
may reach considerably higher temperatures (Horowitz, 1979;
Vishniac and Hempfling, 1979). The first observations suggested
that substantial areas of the soil contained no living microorganisms at all (Horowitz, Cameron and Hubbard, 1972). The
implication of these results was that even on our own planet
limitations of water could make certain zones abiotic. Some
evidence has since accumulated, however, that in Antarctic soils
previously considered sterile a microbial ecology might exist:
endolithic cyanobacteria as primary producers (Friedmann and
Ocampo, 1976) and heterotrophic consumers. Evidence for the
last was the isolation of a number of psychrophilic yeast and
fungi, following heterotrophic enrichment at low temperatures,
from soil samples collected aseptically by Dr. W. Vishniac
shortly before his tragic death in Antarctica in December 1973
(Vishniac and Hempfling, 1979).

The surface of the planet Mars is far drier than even the
Antarctic dry valleys, the highest relative humidity on its
surface being 0.02 (Horowitz, 1979). As is well known, no
evidence for microbial (or other) life of any kind was found
on the Martian surface by the Viking expedition. However, the
suggestion was made (Vishniac and Hempfling, 1979) that other
microbial techniques than those used might have succeeded
better, but it is unlikely that such techniques will be
tested at an early date.

Acidic and alkaline places Acid hot springs and alkaline
salt lakes have already been mentioned. Many natural environments have pH values of 3 or less, including lakes, pine soils
and bogs. Lower values are associated with coal mine refuse
piles and, especially, with mining effluents in which the
sulphur-oxidizing bacteria, notably the *Thiobacillus* species,
can alter the pH to 1.5 or lower. Such mining effluents are
not only acidic but also may contain high concentrations of
toxic heavy metals, including radioactive ones. Laboratory
reagents should not be forgotten; as most of us have learned,
microbial contamination of such reagents, including those with
no apparent source of nutrient, is common. Fungi have been
reported to grow in 2.5 N H_2SO_4 containing high $CuSO_4$ concentrations (Langworthy, 1978).

In addition to alkaline lakes, many soils are alkaline, as
are the guts of lepidopterous insects (pH 10 or more), decaying
animal matter, and temporary latrines where urea decomposes on
the soil. Many microorganisms which decompose urea can, in
fact, live and metabolize at pH values of 10-11. A number of
microorganisms including fungi and protozoa can also grow over
a wide range of pH. Some algae, though not the cyanobacteria,
can grow at pH values of 3 or lower, while cyanobacteria and
other algae tolerate high pH values (pH 10 or higher). Some
especially fascinating microorganisms are those that withstand
both low pH and high temperatures. These include the eukaryotic
algae *Cyanidium caldarium* with a pH optimum of 2-3, a temperature
optimum of $45°C$ and a maximum of $55°C$ and the curious lobed
bacterium *Sulfolobus acidocaldarius* previously mentioned
(Langworthy, 1978; Tansey and Brock, 1978).

Heavy metals Certain bacteria, as a consequent of their
metabolic activities, mainly acid formation, produce concentrations of heavy metals which are quite toxic to other forms
of life. In fact, this is one of the most interesting aspects
of the studies of microbial interactions with heavy metals,
especially with mercury which may be responsible for production
of toxic methyl mercury. Bacteria, fungi and algae have been
found to withstand concentrations of heavy metals toxic to
other forms of life, though the hot brines $56°C$ of the Atlantis

Deep in the Red Sea, which contains high concentrations of a
number of heavy metals, may contain no living microorganism
(Ehrlich, 1978).

Other strong chemicals   Scattered reports have appeared
concerning microorganisms resistant to phenols, chlorine,
saturated KCN and LiCl (Kushner, 1964; Alexander, 1976).   These
have received less systematic study than many of the other
extreme conditions, though they are certainly interesting.

High radiation   Although radiation, especially ionizing
radiation, can ultimately destroy all forms of life, some
microorganisms are much more resistant than others.   Examples
include the marine flagellate *Bodo marina* which is extremely
resistant to ultraviolet radiation, and *Micrococcus radiodurans*
which is highly resistant to ionizing radiation.   Though some
reports have found that bacteria in natural radioactive springs
have a high level of radiation resistance, there is in general
no clear correlation between natural levels of exposure and
species resistance (Kushner, 1964; Nasim and James, 1978).

Cold places   Here we consider, mainly, microorganisms that grow
in conditions that are cold to us.   Many observations have been
made on psychrophilic (cold requiring, generally unable to grow
above 20°C) and psychotrophic (cold tolerant, able to grow at
0-5°C and at temperatures above 20°C) (Morita, 1975) microorganisms of soils, caves, fresh water and snow and ice, the last
of which harbour a number of interesting algae (Alexander,
1976; Baross and Morita, 1978).   The most important cold regions
of the biosphere, without doubt, are the oceans, the bulk of
whose waters have temperatures less than 5°C.   In contrast
to mesophilic organisms like *Escherichia coli*, we might
consider that marine microorganisms live in extreme conditions,
but they certainly do not do so alone.   They are joined in
their cold water by a host of more advanced creatures from
invertebrates and fish to mammals (whose physiological response
to temperature and pressure pose very interesting problems for
the comparative physiologist (Hochachka and Somero, 1973)).
Some marine microorganisms live under pressures of more than
1000 atmospheres which are certainly extreme to unadapted human
beings.   However, even such deeps harbour much higher forms of
life, a topic discussed in detail by Marquis (1976) and Marquis
and Matsumura (1978).   The really extreme condition as far as
microbial growth is concerned is the combination of low temperature and high pressure, at which few known microorganisms are
able to grow.

Generally, microbial activity is low or absent in frozen
soils (Baross and Morita, 1978).   The main concern with
microbial life at sub-zero temperature has been with the survival

of frozen cells (Kushner, 1976). Trees, which continue to
carry out photosynthesis and other vital processes in
temperatures of -40°C and lower, may well be more suitable
objects than microorganisms for study of physiological adaptation
to very low temperatures.

Life in low nutrient concentrations It is worthwhile reminding
the microbial physiologist used to growing cells in rich media
that most natural waters are very low in nutrients. This is so
for the sea (Jannasch, 1979) and many natural waters, including
hot springs (Castenholz, 1979). Nevertheless, a great many
different microorganisms grow in such conditions. Even *E. coli*
can undergo metabolic adaptations that permit it to grow in
much lower nutrients than it is accustomed to in nature or in
the laboratory (Koch, 1979). Other bacteria, especially
prosthecate ones such as *Caulobacter* spp. and *Hyphomicrobium*
spp. can grow well in water too low in nutrients to support
growth of many other bacteria (Poindexter, 1979); these bacteria
do not grow well at higher nutrient concentrations.

Aerobic conditions It might be said (and in fact has been) that
these can be regarded as an extreme environment. Oxygen or
its products are certainly toxic for many microorganisms. All
but the strict anaerobes have evolved ways of dealing with
oxygen toxicity. Humans who are not microbiologists, however,
would probably regard an anaerobic environment as a more
"extreme" one than an aerobic environment. Certainly, one
finds a much greater diversity of species aerobically than
anaerobically.

*Environments that are extreme because of their range of
variation* Though the physical characteristics of some
environments are quite stable - a certain depth in the ocean,
say, or a certain part of a hot spring - others may be subject
to very wide fluctuations. These may be variations of
temperature or moisture in the soil or on rocks, or a feast
and famine existence. Alexander (1976) points out that algae
located on the seashore between high and low tides are subject
to dessication, flooding, fresh and salt water, and great
variations in temperature. The ability to survive and, more
important, to grow over a wide range of such conditions can
be a great ecological advantage. Baross and Morita (1978)
have pointed out that in the sea obligate psychrophiles are
found almost exclusively in stable cold environments, whereas
in environments which are seasonally cold but can undergo
large temperature variation one finds predominantly psychotrophic
bacteria which can grow over a wider temperature range.

Lichens, very slow growing and very long-lived organisms,
suffer long periods of drought between excesses of moisture.

Although lichens are sometimes considered as living in low
nutrient conditions there is, in fact, usually an adequate
supply of minerals from their substate for their slow growth
rate as well as of carbon from their photosynthesizing
component. Much of the CO_2 fixed is converted to high
concentrations of polyols in the fungal cytoplasm, which seems
to permit the lichen to withstand alternate wetting and drying
(Smith, 1975, 1979). This strategy is especially interesting
in view of the use of polyols and other "compatible solutes"
by microorganisms growing in high solute concentrations (see
below).

All the above examples have been called extreme environments
and all have been studied with pleasure and profit. However,
consideration of their range and diversity shows that no one
definition of an extreme environment can encompass all these
examples. From the anthropocentric point of view, all
environments but the last qualify as extreme. Any microorganism
may encounter conditions that prevent its growth: the wrong
pH, antibiotics, lack of proper nutrients, and, for pathogens,
host defences including antibodies, iron-binding proteins and
phagocytosis. It hardly seems profitable to characterize any
environmental condition that inhibits microbial growth as
extreme. ╱ If I were to define an extreme environment in terms
of severity of stress, I would begin with the conditions that
just inhibit the growth of a "normal" organism such as *E. coli*,
and then greatly increase the severity of these conditions.
Thus, if 0.5 M NaCl inhibits the growth of *E. coli* (which it
does, though not completely) 5.0 M NaCl would be considered
extreme. This numbers game becomes hard to play when limits
of solubility or of the freezing the boiling points of water
are imposed, but may convey the idea that the stress in an
environment should not be too subtle, if the environment
deserves to be called "extreme". I would also stipululate
that the stress be such that no microorganism can adapt to
it by a single mutation.

Environments are generally held to limit species diversity
(Alexander, 1978; Brock, 1969). By this criterion an extreme
environment would be one in which only microorganisms are
found and, at the limit, only a few species of these. Hot
springs belong to this category as do most very salty lakes.
The sea and aerobic conditions certainly do not.

If a given environment contains only a few species, one
implication is that mutations permitting individuals, native
or foreign, to cope with it, are very rare events indeed.
This is probably so for very hot and very salty environments
which are the ones that have been most studied. Despite many
attempts (Horowitz, (1979) and others reported at meetings but
not published) no one seems to have seriously lowered the salt

concentrations that extreme halophiles require for growth or
greatly raised the salt tolerance or requirement of non-
halophilic bacteria, including marine bacteria. Some success
has been reached in slightly extending the upward limit of
salt tolerance of certain organisms; the most recent being a
Beneckea species (M. Shilo, personal communication).

Upper temperature limits of growth by training at progressively
higher temperature; for example, that of *Bacillus subtilis*
was extended from 55°C to 72°C (Dowben and Weidenmuller, 1968)
though after growth at 37°C the cells could no longer grow at
the highest temperature. Recently attempts were made to convert
strains of *Bacillus subtilis* of greater heat tolerance by the
transfer of genetic material from more thermophilic bacilli
and some success has been reported, though not very convincing
or repeatable success (Johnson, 1979). Some years ago Kushner
and Lisson (1959) were able to extend the upper pH limit of
growth of a *Bacillus cereus* strain slightly. Very recently
Marquis and Bender (1980) increased the pressure resistance
of a strain of *Streptococcus faecalis*.

It is not known if any genetic variation or selection is
involved in the examples cited above. Indeed, very few genetic
studies using markers of any kind have been carried out with
organisms that live in extreme conditions. These reports
recently, however, do show some intriguing but disturbing
results. Some thermophiles have an unusual rate of spontaneous
variability for auxotrophic and other markers (Johnson, 1979);
the same seems true for the genes involved in pigment and
gas vacuole formation and for certain nutritional abilities
in the halobacteria (J. Weber, personal communication). It
would clearly be of interest to study further the effects of
extreme conditions on genetic variability and, more important,
to learn something more of the basic genetics of microorganisms
that live in such conditions.

STRATEGIES OF DEALING WITH EXTREME ENVIRONMENTS

When faced with an adverse external environment, a living
creature has the options of avoiding it or coming to terms
with it. Mammals have elaborate mechanisms to control the
temperature and pH of their blood. Some desert animals, such
as the kangaroo rat, have evolved metabolic pathways that
convert their food into a source of water without the need
for external water (Horowitz, 1979). Single cells, however,
cannot insulate their interiors from temperature or pressure
changes or control the solute concentration to which they are
subjected. They are less able to avoid the extreme condition
than the cells of the whole animals and must deal with it
in other ways. The physiological mechanisms such cells use
have been explored in depth and detail, and have revealed new

and fascinating aspects of cell physiology, which can only be
dealt with briefly here.

High temperature

The most striking attribute of thermophilic microorganisms
is that their proteins can function at temperatures at which
the classical biochemist would normally predict that no protein
could function at all: up to $80^{\circ}C$ and higher for those microorganisms that actually live, as some do, in boiling water. A
considerable body of literature exists on the enzymes of
thermophilic bacteria, especially *Bacillus stearothermophilus,
Clostridium thermoaceticum, Thermus aquaticus*, and other
thermophilic members of these genera. More than 20 enzymes,
purified to homogeneity, have been studied, and several others
that have not been purified (Amelunxen and Murdock, 1978;
Ljungdahl and Sherod, 1976; Middaugh and MacElroy, 1976;
Oshima, 1979; Singleton, 1976; Zuber, 1979). Almost all these
enzymes function at considerably higher temperatures than the
enzymes from mesophilic bacteria that perform the same
function.

We are not dealing here only with thermal stability. Often
the whole temperature-activity curve of the enzyme is shifted
upwards, so that an enzyme which can function at $70^{\circ}C$ cannot
do so below $45^{\circ}C$. The thermodynamic analysis of Zuber (1979)
suggests that some thermophilic enzymes are inefficient
catalysts at lower temperatures and require high temperatures
to make them act efficiently. Detailed comparisons of
thermophilic and mesophilic enzymes have been carried out with
purified glyceraldehyde 3-phosphate dehydrogenase, formyl-
tetrahydrofolate synthetase, ferredoxin, and others. Despite
the large amount of work done, however, it is by no means
clear what specific amino acid residues, sequences of
residues or other groups of these proteins are responsible
for their action at high temperatures. Presumably, non-covalent
interactions between amino acid residues are responsible for
the ability of proteins to function at high temperatures, but
studies thus far cannot ascribe this property to ionic,
hydrogen, hydrophobic bonds or other definite interactions.
Amelunxen and Murdock (1978) and Zuber (1979) point out that
changing even one amino acid in a protein can make a great
difference in its thermal stability. Such results and the
sum of experiments on enzymes of thermophiles, suggest that
the changes that lead to thermal stability may well be subtle
ones. In some cases, however, thermal stability may be
attributed to known chemical changes, even if the mechanism
of stabilization is not clear. Binding of calcium, zinc or
cobalt ions may be responsible for stabilizing certain
proteolytic enzymes of thermophilic *Bacillus* species (Zuber,

1979).

A number of these enzymes are subject to allosteric control at high temperatures, indicating that both activity and regulation can go on under such conditions (Ameluxen and Murdock, 1978).

An especially striking phenomenon in thermophiles is thermoadaptation, whereby a bacterium that can grow over a wide temperature range synthesizes enzymes with greater heat stability at higher temperatures (Amelunxen and Murdock, 1978). These examples are of special interest since such adaptation, involving the formation of different proteins under different growth stresses, seems rare in microorganisms that grow in extreme conditions. Very little is known of the intimate details of such control, which should be a fascinating subject for further study.

Other properties of thermophiles are interesting, though less clearly related to their ability to grow at high temperatures. It was suggested during early studies that these organisms were protected from the consequences of high temperatures by rapid repair mechanisms and by the presence of protective factor within the cell. Little evidence exists for the first mechanism, except that a high rate of protein turnover has been noted in growing thermophilic bacteria (Bubela and Holdsworth, 1966). More evidence exists for the second mechanism. Examples of enzyme stabilization by cation binding have already been cited (Zuber, 1979). Some interesting polyamines (thermine, caldine, and others) never before found in bacteria, have been observed in *T. thermophilus* and other *Thermus* species. Addition of these compounds permits the cell-free extracts to carry out *in vitro* protein synthesis at high temperatures (Oshima, 1979).

Enzymes are not the only cell components that change in thermophiles. Flagella, ribosomes, membranes, nucleic acids and other parts of the thermophiles, including certain phages show increased heat resistance (Amelunxen and Murdock, 1978; Oshima, 1979). It is not always clear to what extent these changes can be ascribed to the protein fraction of such structures and to what extent to other parts. In any case, a number of changes occur in lipids during growth at high temperatures, but their contribution to membrane stability at high temperatures is by no means clear, although much studied (Esser, 1979). The situation with thermoacidophiles (see over) may be clearer.

High intensity of radiation

This is another stress that microorganisms cannot avoid, except for the well known ability of pigments to protect cells from the harmful effects of visible light. Although the suggestion has been made that bacteria such as *M. radiodurans* managed in some way to shield their DNA from radiation damage (reviewed in Kushner, 1964) this is not considered a serious possibility. As far as is known, radiation resistant microorganisms have an unusual ability to repair damaged DNA (Nasim and James, 1978) and this is their chief defence mechanism.

Extremes of pH

Organisms that can grow at extremes of pH can, as far as is known, maintain a pH near neutrality in their cytoplasm. That is, they exclude the harsh external environment as far as their internal workings are concerned (Langworthy, 1978, 1979). However, the cells may also have special reactions with hydrogen or hydroxyl ions outside the cell. *Thermoplasma* species require acidic conditions for stability; cells and membranes are stable at pH 5 but disintergrate at pH 7 (Smith, Langworthy, Mayberry and Houghland, 1973; Langworthy, 1978 1979). Altering the surface charge by blocking free -COOH groups, renders cells insensitive to higher pH values, while attaching positively charged radicals to -COOH groups makes them stable only at high pH. Such results suggest that in the normal cell repulsion of ionized -COOH groups at high pH can lead to cell disruption (Langworthy, 1979).

Bacillus species which grow at alkaline pH values (at moderate temperatures) have special problems in carrying out active transport, since they cannot generate a pH gradient across the membrane with the interior alkaline to the exterior as required by the chemiosmotic hypothesis. Transport of different substances in these organisms seems to depend on symport with sodium ions or on direct energization by ATP (Guffanti, Susman, Blanco and Krulwich, 1978; Guffanti, Monti, Blanco, Ozick and Krulwich, 1979).

Another type of defence against pH extremes has been known for a long time, namely the ability of growing bacteria to produce metabolites that change the pH to a more favourable one. Gale (1943) and Kushner and Lisson (1959) observed that alkali-resistant bacilli acidified the growth medium, but concluded that this was a secondary mechanism of resistance since acidification only began after growth had started.

High solute concentrations

Freshwater microorganisms maintain a lower a_w inside their cells than outside using their walls to control the resulting osmotic pressure. Microorganisms that live in high concentrations of salt or other solutes cannot do the reverse; that is, they cannot exclude substantial amounts of all solutes maintaining an a_w much higher inside than out, since this would require imper-

meability to water. Rather, such microorganisms keep high
internal concentrations of solutes inside their cells, whose
composition is quite different from the outside medium. Among
the most fascinating of such creatures are the salt-tolerant
yeast and algae, especially the green alga *Dunaliella* which,
while growing in high concentrations of NaCl or other solutes,
have high internal concentrations of glycerol or other polyols.
Intracellular enzymes of these organisms are strongly inhibited
by the salt or sugar concentrations found in the external medium
but not by their intracellular solutes. Such solutes, which
have been called "compatible solutes" (Brown, 1976) serve to
maintain the cell's osmotic balance and, at the same time, permit
its internal mechanisms to function.

In the extremely halophilic bacteria which may represent one
of the possible limits of physiological adaptation (Kushner,
1978b), salts are found both inside and outside, but not the
same salts. Cells growing in saturated NaCl contain more than
enough KCl to saturate their internal water. The internal
enzymes function in, and most indeed require, high salt
concentrations for activity, stability or both. For some enzymes,
a specific requirement for KCl for well-regulated activity can
be demonstrated; others function as well with NaCl as with KCl.
The importance of potassium ions is most strikingly demonstrated
for the enzymes and ribosomes involved in protein synthesis, a
process that can take place *in vitro* in high KCl (3 M)
concentrations, but not in high NaCl concentrations. In low
salt concentrations, for example 0.01 M $MgCl_2$, which is able to
maintain *E. coli* ribosomes, most of the proteins of the halo-
bacterial ribosomes dissociate from the RNA, possibly through
mutual repulsion with the nucleic acid since these proteins have
the very unusual property of being largely acidic, as indeed are
many of the proteins of extreme halophiles.

A number of enzymes of extreme halophiles have been studied,
although usually not in a purified form. The salt required for
stability has seriously hampered research into extreme halophile
enzymology; most enzymes are irreversibly inactivated unless
high salt concentrations are present, but the salt can make
enzyme purification by ion exchange chromatography very
difficult indeed. Kinetic studies on unpurified enzymes suggested
that high salt concentrations were needed partly for shielding of
excess negative charges and also partly for maintaining hydrophobic
bonds which, in the absence of large amounts of "salting out"
salts, are not strong enough to maintain the protein in an active
configuration (Lanyi, 1974).

Recently, using affinity chromatography techniques two
dehydrogenases and a ferredoxin from a *Halobacterium* species
have been purified to homogeneity (Werber, Mevarech, Leicht and
Eisenberg, 1978) and found to have the properties predicted by

kinetic studies of unpurified enzymes (Lanyi, 1979), that is
an excess of acidic residues and a need for salts to maintain
hydrophobic interactions.

The purple membrane of the halobacteria which has been the
subject of many reviews and symposia (Caplan and Ginzburg,
1978) is too well known to require detailed comment here.
This system uses the pigment bacteriorhodopsin to export
protons and generate ATP under the influence of light. It has
been a fascinating object of biochemical study, perhaps of
special value in deepening our understanding of membrane
function (Lanyi, 1978). The early hopes that it might somehow
provide a utilizable source of energy have not been realised
(and perhaps not pursued too far). There is no clear indication
of the relation of the purple membranes to the salt requirement
of the halobacteria, or indeed to any aspect of their life.

Extremely halophilic bacteria have certain other unusual
properties not necessarily associated with their life in high
salt concentration which suggest something of their evolutionary
position. These include: unusual di-ether linked phospholipids
as almost the sole lipids of the membrane; large amounts of
satellite DNA; glycoproteins on their cell walls which lack
the peptidoglycan structure; properties of protein-synthesizing
systems, ribosomal proteins and ribosomal RNAs which resemble
in some ways more closely those of eukaryotes than of prokaryotes
(Bayley and Morton, 1978, 1979).

In their lipid composition and several other properties,
including amino acid sequences in ribosomal proteins and RNA
base sequences, these organisms resemble the thermoacidophiles
and the methanogens, organisms that grow in very different
environments indeed (Matheson, Yaguchi, Nazar, Visentin and
Willick, 1978; Nazar, Matheson and Bellemare, 1978). Some
thermoacidophiles possess a fascinating variation of this
structure, a diglycerol tetraether which forms a lipid monolayer
(instead of a bilayer) in the cell membrane (Langworthy, 1979).

The thermoacidophiles, methanogens and extreme halophiles
have been linked together in a group, the "Archaebacteria"
(Magrum, Leuhrsen, and Woese, 1978; Woese, Magrum and Fox,
1978) a name that implies, although without strong evidence,
a very early origin. Indeed, the apparent evolutionary
similarities of these bacteria with eukaryotes may suggest that
they are ancestors of the eukaryotic cells, although it is
hard to understand how, with so many other prokaryotes to arise
from, the eukaryotes should have chosen these.

In most of these unusual chemical properties there is no
obvious adaptive function. It is true that the ether linkage
of lipids of thermoacidophiles are more acid-resistant than
ester-linked lipids, but such lipids have no similar function
to serve in the extreme halophiles (which are, in any case,

quite sensitive to acidic conditions) nor in the methanogens.

Moderately halophilic bacteria, that is, those that can grow in the salt range about 0.5 - 3.0 M, in contrast to the extreme halophiles that grow in the range 3.0 - 5.0 M, have been much less studied. As I pointed out earlier (Kushner, 1978b), these organisms are of interest partly because of their salt requirement but largely because they are able to carry out the kind of adaptation that permits them to grow over a wide range of salt concentrations. We showed earlier (Forsyth and Kushner, 1970) that, for two species, *Vibrio costicola* and *Micrococcus* (now *Paracoccus*) *halodenitrificans* this ability did not involve the selection of more or less salt-resistant populations. Knowledge on the internal solutes of moderate halophiles is much less complete than is that of salt-tolerant yeast and algae. Some seem able to exclude most salts (as measured by sodium and potassium contents) when grown in high NaCl concentrations and others to maintain quite high internal salt concentrations (Kushner, 1978b). Of these few enzymes studied, the majority seem to have greatest activity but least stability, in the absence of salt and also to function reasonably well in substantial salt concentrations. As pointed out earlier, in contrast to the situation in the extreme halophiles, it is not possible to obtain a coherent picture of the intracellular ionic conditions from the behaviour of the enzymes of moderate halophiles (Kushner, 1978b).

Some very recent studies suggest that the cell membrane may be the best site for determining some of the mysteries of moderate halophiles. Despite any uncertainties over the state of intracellular ions, we *know* what the external salt concentration is. We have found (D.J. Kushner and R.A. MacLeod, unpublished observations) that *V. costicola* can transport α-amino-isobutyric acid (AIB) and other amino acids over a wide range of salt concentrations, the upper limit of this range depending on the salt concentration at which cells are grown. Cells grown in 1 M with NaCl can transport AIB at concentrations as high as 4 M NaCl, whereas cells grown in 0.5 M NaCl are practically unable to do so, although they can still carry out transport in 1 M NaCl. Salt-resistant transport can be developed in a few hours by incubating 0.5 M grown cells in growth media containing 1 M NaCl, but does not seem to involve protein synthesis since its acquisition is not blocked by chloramphenicol. This is one of the very few examples of physiological adaptation to changing salt concentrations in a halophilic microorganism of which I am aware. The very few studies that have been carried out on enzymes of moderate halophiles grown at different salt concentrations show no difference in salt response of these enzymes. It is interesting that this particular adaptation apparently does not involve

synthesis of a new protein. Recent results have shown that
in *V. costicola* the salt concentration can affect the sensitivity
to streptomycin (Kogut and Madira, 1978) and the protein turnover
which takes place at quite a high rate in all conditions
(Hipkiss, Armstrong and Kushner, 1980) but the adaptive
significance of these changes are not clear.

OTHER MECHANISMS

These have been studied less intensively and will be dealt
with in even less detail.

Toxic chemicals

In the case of toxic heavy metals, the mechanism probably
involves sequestration from the sensitive site but it does
not always involve exclusion from the cell. Many microorganisms
that are resistant to heavy metals concentrate them, sometimes
on the cell surface or on external layers (Ehrlich, 1978). The
site is usually uncertain and the mechanism of concentration
has been little studied. Some of our work (Ramamoorthy,
Springthorpe and Kushner, 1977; Laube, Ramamoorthy and Kushner,
1979) suggests that binding of heavy metals to algae and bacteria
may greatly influence their removal from sediment and entry into
the food chain. To my knowledge, enzyme systems especially
resistant to heavy metals have not been reported in such
organisms.

Oxygen certainly constitutes an extreme environment for
certain microorganisms, that is a highly toxic one. Mechanisms
of oxygen toxicity, through direct action and through formation
of highly toxic compounds such as H_2O_2 and the superoxide
radical have been discussed in detail elsewhere (Fridovich,
1975; Kuenen, Hassan, Krinsky, Morris, Pfennig, Schlegel,
Shilo, Vogels, Weser and Wolfe, 1979), as have the mechanisms
of resisting this toxic action which depend partly on enzymes
(catalase and superoxide dismutase (Fridovich, 1975)) that
destroy the toxic forms and partly on the cells' reducing
power and mechanisms for dealing with excess oxygen (Morris
1979).

Cold and starvation

Here we deal with more subtle stresses, possibly extreme,
certainly universal. In many ways resistance to cold, which
in practice means the ability to grow between $0^{\circ}C$ and $10^{\circ}C$,
an ability shared by very many microorganisms though not by
E. coli, is more subtle than resistance to heat. It is worth
remembering that low temperatures, above freezing, decrease but
usually do not abolish enzyme activity. When cells such as
E. coli stop growing below about $10^{\circ}C$ it is not because their
enzymes do not function. The reasons may involve failures of

membrane function or, as the work of Ingraham with cold-
sensitive mutants has suggested, improper assembly of ribosomes
or too strict regulation of biosynthetic enzymes. To an
organism that can grow at $0^{\circ}C$, however, a temperature of $20^{\circ}C$
or even $10^{\circ}C$ may provide an environmental stress that inactivates
a key enzyme or causes a lethal change in cell structure (Inniss
and Ingraham, 1978). Cold then becomes a requirement for
avoiding such stresses.

Cells deal with low nutrient conditions by metabolic and
structural alterations. It is worth noting that even *E. coli*
and *Pseudomonas aeruginosa*, bacteria not normally thought to
grow under low nutrient conditions, can undergo metabolic
alterations at low nutrient conditions. Levels of catabolic
enzymes greatly increase, due to release from catabolic
repression. Mixed substrates can be used simultaneously, a
very different situation to that which we observe in the
normal rich batch culture (Koch, 1971, 1979; Matin, 1979).
In addition, cells become smaller so that the surface-to-
volume ratio increases. Cells that must grow at low nutrient
conditions can show very high substrate affinity. They can
also adopt strategies, such as attachment to particles, which
permit them to take advantage of the richer nutrient conditions
that prevail on surfaces (Marshall, 1976, 1979). This is
strikingly illustrated by the prosthecate bacteria (Poindexter,
1979), such as *Caulobacter* species. The special case of
Bdellovibrio species which get their nutrient by entering other
cells, has also been considered a response to an extreme
environment (Rittenberg, 1979). Finally, dormancy may be
thought of an as excellent response to low nutrient conditions:
the ability to form spores, cysts, or other resting forms and
to wait for better times (Hanson, 1976).

EVOLUTIONARY RELATIONSHIPS

Although evolutionary relationships have been much discussed,
I think it not unfair to say we know practically nothing about
them. In discussing the extreme halophiles, Bayley and Morton
(1979) wrote "As almost nothing is known of their genetics, and
their DNA sequences show no homology to those of other bacteria,
the problem of understanding the evolution of the *Halobacterium*
is difficult." It is indeed. These organisms seem related to
thermoacidophiles and methanogens in certain highly unusual
biochemical markers and in protein and in RNA sequences, but not
in physiological functions. They do not seem related to other
salt tolerant bacteria in their protein sequences (Matheson,
Yaguchi, Nazar, Visentin and Willick, 1978) nor in their RNA
sequences (M. Yaguchi, personal communication). I would suggest
that it is among the more and less salt tolerant bacteria other
than the extreme halophiles that one should look for genetic

relationships. There are certainly enough of these, in
laboratory cultures and in nature, to make such a comparison
feasible.

Thermophiles may provide more hope. The thermophilic and
non-thermophilic bacilli and clostridia are at least in the
same genera (though admittedly ones with wide genetic variation:
G + C of 32-53% for bacilli and 25-45% for clostridia). More
information on genetic relations between thermophilic and
mesophilic members of each genus would be very welcome.

Reviewing the subject shows that practically nothing is
known of the genetic markers, and of ways of effecting transfer
of these markers, in microorganisms that live in extreme
environments. It is disturbing that the markers thus far
studied in the extreme halophiles and thermophiles seem so
subject to spontaneous variation. Such organisms appear
determined to guard their secrets well. Hopefully, however,
the need to understand better their evolutionary relationships
will inspire more extensive genetic and comparative studies
among such microorganisms.

CONCLUSION: HUMAN ECOLOGY OF STUDIES ON EXTREME ENVIRONMENTS

The progress of science depends on a number of factors:
flashes of insight, humble persistent curiosity, ease or
difficulty of experimental methods, availability of funds,
the need for peer approval, the demand for scientific
"productivity". I have personally known a few competent
young enzymologists who once worked with halophilic enzymes
but who, because of the difficulties involved in their purification, felt compelled to switch over to enzymes in yeast
or *E. coli*, organisms about which a tremendous amount is
known, in which sophisticated genetic manipulation are
possible and whose enzymes do not need high salt concentrations
for stability. The newer techniques developed recently
(Werber *et al.*, 1978) may overcome some of this reluctance.
Technical difficulties haunt research on some of the most
interesting organisms. The flagella of halobacteria require
salt for stability and this has virtually precluded study of
their chemistry. It is not necessary to have tried, for one
to imagine the difficulty of searching in the electron microscope for organelles among a mass of salt crystals. In
contrast, the gas vacuoles of *Halobacterium halobium* are
stable in the absence of salt, a fact that has permitted
studies of their structure and chemistry. Isolation of the
purple membrane depends on the fact that it remains stable
in low salt concentrations at which most other cell structures
disintergrate. Would it have been studied so much otherwise?

The recent surge of interest in the halobacteria can be
ascribed to two things: Mars and the purple membrane. Interest

in the latter is not apt to decrease while some of its
mysteries remain unsolved, whether or not it has any real role
in the life of the cell on which it resides. Mars, however,
seems to have failed us, or we have failed it. We are in a
post-Mars stage of looking at extreme environments. Perhaps
it is good to return to earth and to be concerned with more
earthly problems. Work on halophiles started because of
their involvement in food spoilage. More and more interest
is now being expressed in salt tolerance and resistance to
dryness with the hope, among others, of finding ways to
make plants grow in dry environments.

Previously much more detailed work was done on the salt
requirement of the extreme halophiles than with the salt
tolerance of other organisms. There are signs that the
directions are now changing: a good thing, in my opinion.

Studies of life under cold conditions is of great relevance
to life in the sea and in cold countries such as Canada.
Work with heavy metals is intimately tied to industrial
pollution and to strategies of dealing with it.

Thermophiles can be, and have been, studied because of
their importance in industrial processes. Many other applied
aspects can doubtless be suggested by readers of this volume.

With all these wholesome thoughts, however, let us not
forget the intrinsic lure of studying extreme environments
and the creatures that live in them. We must always hope
and expect that scientists who are interested in the limits
at which life can exist will find the support, financial
and intellectual, to delve into the secrets of microorganisms
growing in conditions where no one suspects anything can.
Such creatures - this is an article of faith - cannot be
fundamentally different from other living cells, but studying
them may reveal some truly surprising and exciting facets of
the definition of life.

ACKNOWLEDGEMENTS

This chapter was written while the author was Visiting
Professor at the Department of Microbiology, Macdonald
Campus of McGill University, Ste Anne de Bellevue, Que.

REFERENCES

Alexander, M. (1976). Natural selection and the ecology of
    microbial adaptation in a biosphere. *In* "Extreme
    Environments: Mechanisms of Microbial Adaptation" (Ed.
    M.R. Heinrich), pp. 3-25. Academic Press, New York.
Amelunxen, R.E. and Murdock, A.L. (1968). Microbial life
    at high temperatures: Mechanisms and molecular aspects. *In*
    "Microbial Life in Extreme Environments" (Ed. D.J. Kushner),
    pp. 217-278. Academic Press, London.

Baas-Becking, L.G.M. (1931). Historical notes on salt and
    salt manufacture, *Scientific Monthly* 32, 434-446.
Baross, J.A. and Morita, R.Y. (1978). Microbial life at
    low temperatures: ecological aspects. *In* "Microbial
    Life in Extreme Environments" (Ed. D.J. Kushner), pp. 9-71.
    Academic Press, London.
Bayley, S.T. and Morton, R.A. (1978). Recent developments in
    the molecular biology of extremely halophilic bacteria,
    CRC Critical Reviews of Microbiology 6, 151-205.
Bayley, S.T. and Morton, R.A. (1979). Biochemical evolution
    of halobacteria. *In* "Strategies of Microbial Life in
    Extreme Environments" (Ed. M. Shilo), pp. 109-124. Berlin:
    Dahlem Konferenzen. Verlag Chemie, Weinheim, New York.
Brock, T.D. (1969). Microbial growth under extreme conditions.
    In "Microbial Growth" 19 Symposium of the Society for
    General Microbiology (Ed. P.M. Meadow and S.J. Pirt),
    pp. 15-41. Cambridge University Press, Cambridge.
Brock, T. (1979). Ecology of saline lakes. *In* "Strategies
    of Microbial Life in Extreme Environments" (Ed. M. Shilo),
    pp. 29-47. Berlin: Dahlem Konferenzen. Verlag Chemie,
    Weinheim, New York.
Brown, A.D. (1976). Microbial water stress, *Bacteriological
    Reviews* 40, 803-846.
Bubela, B. and Holdsworth, E.S. (1966). Amino acid uptake,
    protein and nucleic acid synthesis and turnover in
    *Bacillus stearothermophilus, Biochemica et Biophysica
    Acta* 123, 364-375.
Caplan, S.R. and Ginzburg, M. (1978). "Energetics and
    Structure of Halophilic Microorganisms" Elsevier North-
    Holland, Amsterdam.
Castenholz, R.W. (1979). Evolution and ecology of thermo-
    philic microorganisms. *In* "Strategies of Microbial Life
    in Extreme Environments" (Ed. M. Shilo), pp. 373-392.
    Berlin: Dahlem Konferenzen. Verlag Chemie, Weinheim, New
    York.
Cohen, Y., Padan, E. and Shilo M. (1975). Facultative
    anoxygenic photosynthesis in the cyanobacteria *Oscillatoria
    limnetica, Journal of Bacteriology* 123, 855-861.
Dowben, R.M. and Weidenmuller, R. (1968). Adaptation of
    mesophilic bacteria to grow at elevated temperatures,
    Biochemica et Biophysica Acta 158, 255-261.
Ehrlich, H.L. (1978). How microbes cope with heavy metals,
    arsenic and antimony in their environments, *In* "Microbial
    Life in Extreme Environments" (Ed. D.J. Kushner), pp. 381-408.
    Academic Press, London.
Esser, A.F. (1979). Physical chemistry of thermostable
    membranes. *In* "Strategies of Microbial Life in Extreme
    Environments" (Ed. M. Shilo), pp. 433-453. Berlin: Dahlem

Konferenzen. Verlag Chemie, Weinheim, New York.

Forsyth, M.P. and Kushner, D.J. (1970). Nutrition and distribution of satl response in populations of moderately halophilic bacteria, *Canadian Journal of Microbiology* 16, 253-261.

Fridovich, I. (1975). Superoxide dismutase, *Annual Reviews of Biochemistry* 44, 147-159.

Friedmann, E.I. and Ocampo, R. (1976). Endolithic blue-green algae in the Dry Valleys: primary producers in the Antarctic desert ecosystem, *Science* 193, 1247-1249.

Gale, E.F. (1943). Factors influencing the enzymatic activities of bacteria, *Bacteriological Reviews* 7, 139-173.

Griffin, D.M. and Luard, E.J. (1979). Water stress and microbial ecology. *In* "Strategies of Microbial Life in Extreme Environments" (Ed. M.Shilo), pp. 49-63. Berlin: Dahlem Konferenzen. Verlag Chemie, Weinheim, New York.

Guffanti, A.A., Monti, L.G., Blanco, R., Ozick, D. and Krulwich, T.A. (1979). β-Galactoside transport in an alkaline-tolerant strain of *Bacillus circulans*, *Journal of General Microbiology* 112, 161-169.

Guffanti, A.A., Susman, P., Blanco, R. and Krulwich, T.A. (1978). The protonmotive force and α-aminobutyric acid transport in an obligately alkalophilic bacterium, *Journal of Biological Chemistry* 253, 708-715.

Hanson, R.S. (1976). Dormant and resistant stages of procaryotic cells. *In* "Chemical Evolution of the Giant Plants" (Ed. C. Ponnamperuma), pp. 107-120. Academic Press, New York.

Heinrich, M.R. (1976). "Extreme Environments: Mechanisms of Microbial Adaptation" Academic Press, New York.

Hipkiss, A.R., Armstrong, D., and Kushner, D.J. (1980). Protein turnover in a moderately halophilic bacterium, *Canadian Journal of Microbiology*, in press.

Hochachka, P.W. and Somero, G.N. (1973). "Strategies of Biochemical Adaptation" W.B. Saunders, London and Philadelphia.

Horowitz, N.H. (1976). Life in extreme environments: Biological water requirements. *In* "Chemical Evolution of the Giant Plants" (Ed. C. Ponnamperuma), pp. 121-128. Academic Press, New York.

Horowitz, N.H. (1979). Biological water requirements. *In* "Strategies of Microbial Life in Extreme Environments" (Ed. M. Shilo), pp. 15-27. Berlin: Dahlem Konferenzen. Verlag Chemie, Weinheim, New York.

Horowitz, N.H., Cameron, R.E. and Hubbard, J.S. (1972). Microbiology of the dry valleys of Antarctica, *Science* 176. 242-245.

Imhoff, J.F., Sahl, H.G., Soliman, G.S.H. and Truper, H.G.
(1979). The Wadi Natrun: Chemical composition and microbial mass developments in alkaline brines of eutrophic
desert lakes, *Geomicrobiology Journal* 1, 219-234.

Inniss, W.E. and Ingraham, J.L. (1978). Microbial life at
low temperatures: Mechanisms and molecular aspects. *In*
"Microbial Life in Extreme Environments" (Ed. D.J. Kushner),
pp. 73-103. Academic Press, London.

Jannasch, H.W. (1979). Microbial Ecology of aquatic nutrient
habitats. *In* "Strategies of Microbial Life in Extreme
Environments" (Ed. M. Shilo), pp. 243-260. Berlin: Dahlem
Konferenzen. Verlag Chemie, Weinheim, New York.

Johnson, E.J. (1979). Thermophile genetics and the genetic
determinants of thermophily. *In* "Strategies of Microbial
Life in Extreme Environments" (Ed. M. Shilo), pp. 471-487.
Berlin: Dahlem Konferenzen. Verlag Chemie, Weinheim, New
York.

Koch, A.L. (1971). The adaptive response of *Escherichia coli*
to a feast and famine existence, *Advances in Microbial
Physiology* 6, 147-217.

Koch, A.L. (1979). Microbial growth in low concentrations of
nutrients. *In* "Strategies of Microbial Life in Extreme
Environments" (Ed. M. Shilo), pp. 261-279. Berlin: Dahlem
Konferenzen. Verlag Chemie, Weinheim, New York.

Kogut, M. and Madira, W.M. (1978). Dihydrostrptomycin as
a probe to study the effects of salt concentration during
growth on cell constituents in a moderate halophile. *In*
"Energetics and Structure of Halophilic Microorganisms"
(Ed. S.R. Caplan and M. Ginsburg), pp. 521-527. Elsevier-
North-Holland, Amsterdam.

Krumbein, W.E. and Cohen, Y. (1974). Biogene, klastische
und evaporitsche sedimentation in einem mesothermen
monomiktishchen ufernahen See (Golf von Aquaba), *Geologische
Rundschau* 63, 1036-1065.

Kuenen, J.G., Hassan, H.M., Krinsky, N.I., Morris, J.G.,
Pfennig, N., Schlegel, H., Shilo, M., Vogels, G.D., Weser,
U. and Wolfe, R. (1979). Oxygen toxicity: Group report.
In " Strategies of Microbial Life in Extreme Environments"
(Ed. M. Shilo), pp. 223-241. Berlin: Dahlem Konferenzen.
Verlag Chemie, Weinheim, New York.

Kushner, D.J. (1964). Microbial resistance to harsh and
destructive environmental conditions, *Experimental
Chemotherapy* 2, 113-168.

Kushner, D.J. (1976). Microbial life at low temperatures.
In "Chemical Evolution of the Giant Plants" (Ed. C.
Ponnamperuma), pp. 85-93. Academic Press, New York.

Kushner, D.J. (1978a). "Microbial Life in Extreme
Environments" Academic Press, London.

⊙Kushner, D.J. (1978b). Life in high salt and solute
    concentrations: halophilic bacteria. *In* "Microbial Life
    in Extreme Environments" (Ed. D.J. Kushner), pp. 318-368.
    Academic Press, London.
Kushner, D.J. and Lisson, T.A. (1959). Alkali resistance in
    Bacillus cereus, *Journal of General Microbiology* 21, 96-108.
Langworthy, T.A. (1978). Microbial life in extreme pH values.
    In "Microbial Life in Extreme Environments" (Ed. D.J.
    Kushner), pp. 279-315. Academic Press, London.
Langworthy, T.A. (1979). Membrane structure of thermoacido-
    philic bacteria. *In* "Strategies of Microbial Life in
    Extreme Environments" (Ed. M. Shilo), pp. 417-432. Berlin:
    Dahlem Konferenzen. Verlag Chemie, Weinheim, New York.
Lanyi, J.K. (1974). Salt-dependent properties of proteins
    from extremely halophilic bacteria, *Bacteriological
    Reviews* 38, 272-290.
Lanyi, J.K. (1978). Light energy conversion in *Halobacterium
    halobium*, *Microbiological Reviews* 42, 682-706.
Lanyi, J.K. (1979). Physiocochemical aspects of salt-
    dependence in halobacteria. *In* " Strategies of Microbial
    Life in Extreme Environments" (Ed. M. Shilo), pp. 93-107.
    Berlin: Dahlem Konferenzen. Verlag Chemie, Weinheim, New
    York.
Laube, V., Ramamoorthy, S., and Kushner, D.J. (1979).
    Mobilization and accumulation of sediment bound heavy
    metals by algae, *Bulletin of Environmental Contamination
    and Toxicology* 21, 763-770.
Ljungdahl, L.G., and Sherod, D. (1976). Proteins from
    thermophilic microorganisms. *In* "Extreme Environments:
    Mechanisms of Microbial Adaptation" (Ed. M.R. Heinrich),
    pp. 147-188. Academic Press, New York.
Magrum, L.J., Leuhrsen, K.R., and Woese, C.R. (1978). Are
    extreme halophiles actually bacteria?, *Journal of Molecular
    Evolution* 11, 1-8.
Marquis, R.E. (1976). High-pressure microbial physiology,
    Advances in Microbial Physiology 14, 159-241.
Marquis, R.E. and Bender, G.R. (1980). Isolation of a
    variant of *Streptococcus faecalis* with enhanced baro-
    tolerance, *Canadian Journal of Microbiology*, in press.
Marquis, R.E. and Matsumura, P. (1978). Microbial life under
    pressure. *In* "Microbial Life in Extreme Enviornments"
    (Ed. D.J. Kushner), pp. 105-158. Academic Press, London.
Marshall, K.C. (1976). "Interfaces in Microbial Ecology"
    Harvard University Press, Cambridge.
Marshall, K.C. (1979). Growth at interfaces. *In* "Strategies
    of Microbial Life in Extreme Environments" (Ed. M. Shilo),
    pp. 281-290. Berlin: Dahlem Konferenzen. Verlag Chemie,
    Weinheim, New York.

Matheson, A.T., Yaguchi, M., Nazar, R.N., Visentin, L.P. and Willick, G.E. (1978). The structure of ribosomes from moderate and extreme halophilic bacteria. *In* "Energetics and Structure of Halophilic Microorganisms" (Ed. S.R. Caplan and M. Ginsburg), pp. 481-501. Elsevier-North-Holland, Amsterdam.

Matin, A. (1979). Microbial regulatory mechanisms at low nutrient concentrations as studied in chemostat. *In* "Strategies of Microbial Life in Extreme Environments" (Ed. M. Shilo), pp. 323-339. Berlin: Dahlem Konferenzen. Verlag Chemie, Weinheim, New York.

Middaugh, C.R. and MacElroy, R.D. (1976). Kinetic behaviour of a thermophilic enzyme in response to temperature perturbations. *In* "Extreme Environments: Mechanisms of Microbial Adaptation" (Ed. M.R. Heinrich), pp. 201-212. Academic Press, New York.

Morita, R.Y. (1975). Psychrophilic bacteria, *Bacteriological Reviews* 39, 144-167.

Morris, J.G. (1979). Nature of oxygen toxicity in anaerobic microorganisms. *In* "Strategies of Microbial Life in Extreme Environments" (Ed. M. Shilo), pp.149-161. Berlin: Dahlem Konferenzen. Verlag Chemie, Weinheim, New York.

Nasim, A. and James, A.P. (1978). Life under conditions of high irradiation. *In* "Microbial Life in Extreme Environments" (Ed. D.J. Kushner), pp. 409-439. Academic Press, London.

Nazar, R.N., Matheson, A.T. and Bellemare, G. (1978). Nucleotide sequence of *Halobacterium cutirubrum* ribosomal 5S ribonucleic acid, *Journal of Biological Chemistry* 253, 5464-5469.

Nissenbaum, A. (1975). The microbiology and biogeochemistry of the Dead Sea, *Microbial Ecology* 2, 139-161.

Oshima, T. (1979). Molecular basis for unusual thermostabilities of cell constituents from an extreme thermophile, *Thermus thermophilus*. *In* "Strategies of Microbial Life in Extreme Environments" (Ed. M. Shilo), pp. 455-470. Berlin: Dahlem Konferenzen. Verlag Chemie, Weinheim, New York.

Pitt, J.I. (1975). Xerophilic fungi and the spoilage of foods of plant origin. *In* "Water Relations of Foods" (Ed. R.B. Duckworth), pp. 273-307. Academic Press, London.

Poindexter, J.S. (1979). Morphological adaptation to low nutrient concentrations. *In* "Strategies of Microbial Life in Extreme Environments" (Ed. M. Shilo), pp. 341-356. Berlin: Dahlam Konferenzen. Verlag Chemie, Weinheim, New York.

Ponnamperuma, C. (1976). "Chemical Evolution of the Giant Plants" Academic Press, New York.

Post, F.J. (1977). The microbial ecology of the Great Salt Lake, *Microbial Ecology* 3, 143-165.

Ramamoorthy, S., Springthorpe, S., and Kushner, D.J. (1977). Competition for mercury between river sediment and bacteria, *Bulletin of Environmental Contamination and Toxicology* 17, 505-511.

Rittnberg, S.C. (1979). Bdellovibrio: A model of biological interactions in nutrient-impoverished environments? *In* "Strategies of Microbial Life in Extreme Environments" (Ed. M. Shilo), pp. 305-322. Berlin: Dahlam Konferenzen. Verlag Chemie, Weinheim, New York.

Shilo, M. (1979). "Strategies of Microbial Life in Extreme Environments" Berlin: Dahlam Konferenzen. Verlag Chemie, Weinheim, New York.

Singleton, Jr., R. (1976). A comparison of the amino acid compositions of proteins from thermophilic and nonthermophilic origins. *In* "Extreme Environments: Mechanisms of Microbial Adaptation" (Ed. M.R. Heinrich), pp. 189-200. Academic Press, New York.

Smith, D.C. (1975). Symbiosis and the biology of lichenised fungi. Symposium of the Society of Experimental Biology 29, 373-405.

Smith, D.C. (1979). Is a lichen a good model of biological interactions in nutrient-limited environments? *In* "Strategies of Microbial Life in Extreme Environments" (Ed. M. Shilo), pp. 291-303. Berlin: Dahlem Konferenzen. Verlag Chemie, Weinheim, New York.

Smith, D.W. (1978). Water relations of microorganisms in nature. *In* "Microbial Life in Extreme Environments" (Ed. D.J. Kushner), pp. 369-380. Academic Press, London.

Smith, P.F., Langworthy, T.A., Mayberry, W.R. and Houghland, A.E. (1973). Characterization of the membranes of *Thermoplasma acidophilum, Journal of Bacteriology* 116, 1019-1028.

Tansey, M.R., and Brock, T.D. (1978). Microbial life at high temperatures: ecological aspects. *In* "Microbial Life in Extreme Environments" (Ed. D.J. Kushner), pp. 159-216. Academic Press, London.

Vishniac, H.S. and Hempfling, W.P. (1979). Evidence of an indigenous microbiota (Yeast) in the Dry Valleys of Antarctica, *Journal of General Microbiology* 112, 301-314.

Werber, M.M., Mevarech, M., Leicht, W. and Eisenberg, H. (1978). Structure-function relationships in proteins and enzymes of *Halobacterium* of the Dead Sea. *In* "Energetics and Structure of Halphilic Microorganisms" (Ed. S.E. Caplan and M. Ginsburg), pp. 427-445. Elsevier-North-Holland, Amsterdam.

Woese, C.R., Magrum, L.J. and Fox, G.E. (1978). Archaebacteria,

Journal of Molecular Evolution 11, 245-253.
Zuber, H. (1979). Structure and function of enzymes from
    thermophilic microorganisms. *In* "Strategies of Microbial
    Life in Extreme Environments" (Ed. M. Shilo), pp. 393-415.
    Berlin: Dahlem Konferenzen. Verlag Chemie, Weinheim, New
    York.

POSITIONING MECHANISMS - THE ROLE OF MOTILITY, TAXIS AND TROPISM IN THE LIFE OF MICROORGANISMS

Michael J. Carlile

*Department of Biochemistry,
Imperial College of Science and Technology,
London SW7 2AZ, England*

INTRODUCTION

Microorganisms may be moved passively for long distances. They can be dispersed through the air by eddy diffusion and by wind and in aquatic environments and the soil by water movements. They may also be transported through air, water and soil on or in the bodies of insects and other animals. Many microorganisms are also motile, moving actively through the expenditure of energy. The rates of transport and distances travelled in this way are small compared with those resulting from passive movement, so motility tends to have been neglected by microbial ecologists. Active movement, however, can be guided by means of a sensory system, so that an organism can select a suitable environment. In many organisms it is likely that motility and passive movement are complementary and that active movement influenced by a sensory system is important at the beginning and end of long passive journeys. The present review will consider the types of motility shown by microorganisms, the environmental factors (including the activities of other organisms) which may influence movement, and how such movements may benefit a motile organism, bringing it to the most favourable position for growth, genetic recombination or dispersal. Guided movements of microorganisms have been considered previously by the present author in the contexts of general biology (Carlile, 1975), cell fusion (Carlile and Gooday, 1978), the mechanisms of sensory transduction (Carlile, 1980a) and evolution (Carlile, 1980b). Here the emphasis will be ecological.

MAJOR TYPES OF ACTIVE MOVEMENT

Microorganisms can swim by a variety of mechanisms and can

glide or crawl on a surface. Moreover, the highly polarised
growth of fungal hyphae has some of the attributes of active
movement. The most widespread types of motility are indicated
below. Some microorganisms show more than one form of motility,
the mode adopted being dependent on the stage in the life cycle
and on environmental conditions.

Swimming of flagellate bacteria

The bacterial flagellum and how it moves the bacterial cell
has been intensively studied in recent years, especially with
Escherichia coli and the closely related *Salmonella typhimurium*
(Macnab, 1978, 1979a). These organisms swim smoothly for about
1-2 seconds, after which a transient change in the direction of
flagellum rotation causes them to tumble. Smooth swimming is
then resumed in a randomly changed direction. The path of the
bacterium is hence a random walk. A population of bacteria
moving in this way will spread slowly in a manner analogous
to the diffusion of molecules (Adler and Dahl, 1967). Non-
motile bacteria will also spread as a result of Brownian
movement, though more slowly than will motile bacteria. In
some bacteria, such as *Spirillum* sp., reversal of flagellum
rotation produces reversal of swimming direction instead of
a tumble. However, Brownian movement will cause gradual
rotation of the cell so here too the path taken will be a
random walk. The swimming speeds of flagellate bacteria
range from 20-60 μmol s^{-1}, but migration speeds are much less
because of frequent changes of direction.

Swimming of spirochaetes

The flagella of spirochaetes are located inside the sheath
that surrounds the cell, and the way that they act to produce
"corkscrew" and other movements of the cell is discussed by
Canale-Parola (1978) and by Berg, Bromley and Charon (1978).
Whereas swimming speeds of typically flagellate bacteria are
greatly reduced in viscous media, spirochaetes swim as well
or better in such media than in those of low viscosity
(Canale-Parola, 1978). Indeed, "viscotaxis", a tendency
to move into regions of high viscosity, has been demonstrated
(Petrino and Deutsch, 1978). The spirochaete mode of
locomotion would seem to be an adaptation to life in relatively
viscous media such as mud and tissues and the tendency to move
into more viscous regions may facilitate infection.

Swimming of flagellate and ciliate eukaryotic microorganisms

Flagellates are propelled by one or a few flagella and
ciliates by many cilia. Eukaryotic flagella and cilia are
closely similar organelles (Satir and Ojakian, 1979; Holwill,

1980) but are wholly different from bacterial flagella.
Swimming speeds in flagellate microorganisms range from
values similar to those of bacteria to over 100 μm s^{-1}
(Holwill, 1974) but ciliates swim rather faster, some moving
at over 1 mm s^{-1} (Sleigh and Blake, 1977). Since most
eukaryotic microorganisms are much larger than bacteria they
are little affected by Brownian movement, hence steering and
precise orientation are possible. They usually swim in a gentle
helix which corrects any tendency to yaw; they can hence
persist in a given direction for long distances. There is a
lack of detailed studies of the path of eukaryotic microorganisms
comparable to those carried out with flagellate bacteria.
Apparently spontaneous direction changes, either random turns
or reversals, do however occur so it is probable that here too
random walks are common.

Gliding on surfaces

Gliding movements (Halfen, 1979) are defined as those which
do not involve any changes in cell shape or any locomotor
organelles visible by light microscopy. Contact with a solid
substratum is necessary. Gliding movements are the main and
perhaps the only form of motility in the cyanobacteria (blue-
green algae), the myxobacteria, and among eukaryotes, the
diatoms (Harper, 1977) and desmids. They also occur in
organisms capable of other forms of locomotion, such as
spirochaetes and various eukaryotic flagellates, when in
contact with solid substrates. Motility rates are generally
low but speeds in excess of 10 μm s^{-1} have been observed in
the cyanobacterium *Oscillatoria princeps* and in some diatoms.

Amoeboid crawling on surfaces

Amoeboid movement, which is widespread in eukaryotic cells,
involves protoplasmic streaming and changes of cell shape.
In addition adhesion to the substratum must occur, with the
formation of new contacts at the anterior end and the breaking
of contact at the posterior. The size and morphology of
amoeboid organisms varies greatly however, and there is a
consequent diversity in the details of amoeboid movement.
In general large amoebae move faster than small. The
versatility of amoeboid movement is shown by the cellular
slime moulds in which, when solitary amoebae have located and
consumed all the bacteria present, populations of up to 10^5
amoebae can cooperate to form a grex ("pseudoplasmodium")
about 1 mm in length and capable of migrating for a day or
more to a site suitable for fruiting and spore dissemination
(Poff and Whitaker, 1979).

Polarised cell growth

Many eukaryotic organisms show highly polarised cell growth. In fungal hyphae protoplasmic streaming can bring to the hyphal apex the materials needed for growth from distances of several millimetres, permitting rapid extension of the hyphal tip (Trinci, 1978). At the same time protoplasm may be withdrawn from hyphae in areas depleted of nutrients. Thus, a movement of the organism has in effect occurred, the agents of the movement being localised growth and protoplasmic streaming (Carlile, 1979). Fungal hyphae spread over and often into a substratum, and can rise above the substratum to facilitate the dispersal of spores. Extension rates of $1 \mu m s^{-1}$ can occur with the hyphae of *Neurospora crassa* and the aerial sporangiophores of *Phycomyces blakesleeanus*, but in most fungi rates range from about one-fortieth to one-quarter of this value. Polarised growth can also occur in prokaryotes, especially hyphal forms such as the Streptomycetes, but the rates of advance are slow, and since protoplasmic streaming does not occur in prokaryotes, such growth is in no way analogous to motility.

MECHANISMS OF ACCUMULATION AND DISPERSAL

Motile microorganisms may accumulate or aggregate at some locations and avoid others. Accumulation may be due to the trapping of organisms that have arrived at a site by random movement, or to actual attraction, the path taken being modified by the response of an organism's sensory system to environmental stimuli. Other stimuli may result in an organism being repelled from a site. Many biologists refer to attraction as positive *taxis* (pl. taxes) and repulsion as negative taxis, and this usage will be used in the present article. Others, especially photobiologists, have adopted a terminology (Diehn, Feinleib, Hanpt, Hildebrand, Lenci and Nultsch, 1977) which restricts use of the term to what is designated below as *topotaxis*. The various types of motile response to environmental factors are discussed below, and in more detail elsewhere (Carlile, 1980a). Terminology and semantics have been considered by Carlile (1975, 1980a) and by Diehn *et al.*, (1977).

Environmental effects on rate of movement − orthokinesis

Environmental factors may increase or decrease the rate at which organisms move, an effect termed *orthokinesis* or sometimes simply "kinesis". The effect of light on the rate of movement of photosynthetic organisms has received most attention (Häder, 1979; Häder and Nultsch, 1979). In general low or moderate light intensities increase the rate of movement

(positive photokinesis) as compared with darkness, and in
some organisms light is essential for movement. Other ortho-
kinetic effects are less studied. Some chemicals, however,
cause an extreme form of orthokinesis, namely cessation of
movement, either directly or by initiating the onset of a
non-motile phase in the life cycle. Negative orthokinesis
can hence cause accumulation of microorganisms through
trapping those that have reached a site by random movement or
taxis and may well be important in terminating the travels
of microorganisms that have reached an appropriate destination.
The zoospores of some plant pathogenic fungi, for example,
cease moving and encyst on reaching the root surface,
encystment being promoted by amino acids and other substances
common in root exudates.

Temporal sensing - klinokinetic and phobic responses

Klinokinetic responses are those in which the rate of
turning is influenced by stimulus intensity, and can result
in an organism moving towards or away from the source of the
stimulus. The most intensive studies of a klinokinetic
response are those that have been carried out on the chemotaxis
of *Escherichia coli* and *Salmonella typhimurium* (Adler, 1975;
Berg, 1975; Goy and Springer, 1978; Hazelbauer and Parkinson,
1977; Koshland, 1979; Macnab, 1978, 1979b). These bacteria
possess a sensory system that monitors changes in stimulus
intensity with time and influences the frequency with which
flagellum reversals, and hence tumbling and random directional
changes, occur. Thus if a cell is experiencing an increasing
concentration of a favourable stimulant, tumbling is suppressed,
whereas an increasing concentration of a noxious stimulant
enhances tumbling. Hence the random walk is biassed and
cells tend to move towards favourable situations and away from
harmful ones. The resulting migration rate is about one-tenth
of the swimming speed of the organism (Macnab, 1978). A
"memory" is essential for klinokinetics, but it has to be a
short one - cells need to adapt to a changed stimulus
intensity and to treat departures from it as a new stimulus.

Klinokinetics is an appropriate tactic for an organism as
small as *E. coli*, which owing to Brownian movement cannot
persist for long in a given direction. Organisms that are
able to do so can utilise temporal sensing in other ways.
They can, for example, reverse on encountering unfavourable
conditions. Such *phobic responses* are most obvious when the
stimulus gradient is very steep, as for example at a light/
dark boundary. When a phobic response is a turn of small
magnitude and consistent in direction in relation to the
previous path of an organism, a succession of phobic responses
can correct the path so that orientation with respect to

stimulus direction can result. This effect, known as
klinotaxis, causes *Euglena* sp. to move towards the light.
Hence temporal sensing can under some circumstances produce
precise orientation.

Spatial sensing: topotaxis and tropism

Topotaxis is orientated movement resulting from spatial
sensing - the simultaneous comparison of stimulus intensity
at different points on the surface of an organism. Spatial
sensing is unlikely to be important in very small organisms
since in these only very steep gradients would produce
discernable intensity differences between the front and rear
of an organism, or in fast-moving organisms where second-to-
second comparisons will be more useful. Topotaxis is hence
likely to be common in relatively large organisms that move
slowly in relation to body length, such as crawling or
gliding eukaryotes, and less frequent in gliding prokaryotes.
Whether it occurs in swimming microorganisms, which commonly
move several body lengths per second, is unclear. Berg and
Purcell (1977) consider the physics of temporal and spatial
chemosensing in detail and calculate that spatial sensing is
a feasible mechanism for the response of the amoebae of
Dictyostelium discoideum (diameter 10 µm; velocity 0.2 µm
s^{-1}) to the attractant acrasin.

Tropism, growth orientated with respect to the source of
a stimulus, involves the bending or bulging of a hypha, and
must be based on side-to-side comparisons in stimulus
intensity at the hyphal apex or at an intercalary growth zone.
Tropism is common in eukaryotes but it is unclear as to
whether it occurs in hyphal prokaryotes, the cell diameters
of which are usually probably too small for spatial stimulus
intensity comparisons. Both spatial and temporal sensing
will normally involve adaptation, which will permit a
sensitive response over a wide range of stimulus intensities.

Interaction between orthokinetics, klinokinesis and topotaxis

Orthokinetics, klinokinesis and topotaxis are not mutually
exclusive and in nature will be combined to yield optimal
results. Thus an organism may exhibit klinokinesis when a
chemical gradient is too gentle to permit spatial comparison,
followed by topotaxis as the source is approached and a
steeper gradient encountered. Positive orthokinesis by
itself has a mildly repellent effect, but computer simulation
studies (Rohlf and Davenport, 1969) show that when combined
with klinokinesis or topotaxis increases their effectiveness.

Pattern swimming

A Petri dish containing a very dense suspension of

microorganisms may develop a tesselated appearance in which
areas densely populated with cells are separated from each
other by narrow zones in which few cells occur. Microscopic
examination shows that at the centre of each area cells are
moving vertically, then spreading outwards, descending and
moving to the centre once more. The phenomenon is often seen
with algae and protozoa, and recently with zoospores of the
fungus *Phytophthora palmivora* at densities of about 10^6 ml^{-1}
(J.N. Cameron and M.J. Carlile, unpublished observations).
The most likely explanation for pattern swimming (Plesset,
Whipple and Winet, 1975) is that cells, slightly denser than
water, move vertically by a tactic response. The upper layer
of water plus cells then becomes denser than the lower, and
as a result of Rayleigh-Taylor instability, a downward
movement results. The effect is almost certainly a laboratory
artifact of no significance in nature, but can alert the
investigator to the occurrence of a taxis that produces upward
movement, such as negative geotaxis, positive phototaxis, or
positive aerotaxis.

TYPES OF ENVIRONMENTAL STIMULI

A wide variety of environmental factors, as listed below,
can cause taxis or tropism in microorganisms. Major features
of the responses to some of these factors, and the possible
role of the responses in the life of microorganisms, will
be indicated.

Light

Phototaxis is very common in photosynthetic microorganisms
(Häder, 1979; Häder and Nultsch, 1979; Hildebrand, 1978; Lenci
and Colombetti, 1978). Taxis is positive at low light
intensities and often negative at high intensities; these
reactions presumably bring organisms to intensities optimal
for photosynthesis. Phototaxis also occurs in a few non-
photosynthetic microorganisms, such as myxomycetes and cellular
slime moulds (Poff and Whitaker, 1979). In myxomycetes, such
as *Physarum nudum*, young plasmodia may show negative phototaxis
(Bialczyk, 1979) which is succeeded by positive phototaxis
when the plasmodium is ready to sporulate (Bialczyk and
Rakoczy, 1975). The migrating stage (grex, "slugs" or
pseudoplasmodia) which precedes sporulation in the cellular,
slime mould *Dictyostelium discoideum* also shows positive
phototaxis (Poff and Whitaker, 1979). Positive phototropism
is very common in the sporulating phase of fungi (Carlile,
1970) where it plays an important role in spore dispersal
(Ingold, 1971). In fungi with explosive spore-discharge
mechanisms it points the "fungus-guns" in the right direction
for launching spores into the open air, and in many without

such mechanisms it guides the growth of spore-bearing structures
past obstacles, as with the sporangiophore of *Phycomyces
blakesleeanus*, which is receiving intensive study (Cerdá-
Olmedo, 1977; Foster, 1977; Russo, 1980; Russo and Galland,
1980). In some plant pathogenic fungi the hyphae (germ-tubes)
that emerge from germinating spores show negative phototropism;
this may guide them into their hosts (Carlile, 1970).

Chemicals

Chemotaxis and chemotropism are very widespread in
microorganisms. Chet and Mitchell (1976) have produced a
general view of microbial chemotactic behaviour from an
ecological point of view and Levandowsky and Hauser (1978) and
Bean (1979) review the chemosensory responses of swimming algae
and protozoa. The chemotactic and chemotropic responses involved
in mating in eukaryotic microorganisms have been discussed by
Gooday (1978) and Kochert (1978).

Recent studies on bacterial chemotaxis have centred on
Escherichia coli and *Salmonella typhimurium*; with these
organisms a determined effort is being made to elucidate the
entire sensory transduction pathway from stimulus to response
in molecular terms (Adler, 1975; Berg, 1975; Goy and Springer,
1978; Hazelbauer and Parkinson, 1977; Koshland, 1979; Macnab,
1978, 1979b). These bacteria show positive chemotaxis to
oxygen and to many amino acids and sugars. They are repelled
by extremes of pH, certain ions and a range of organic substances
including hydrophobic amino acids at high concentrations and
various fatty acids. The latter, some of which are products of
anaerobic metabolism, may serve as a signal for cells to escape
from overcrowded conditions. The chemotactic patterns indicated
above suggest that *E. coli* can move towards nutrients and away
from harmful conditions. Both cooperative and competitive
interactions occur between attractants. A high-background level
of ribose, for example, will prevent chemotaxis to galactose,
but not to serine. The effects of sub-optimal concentrations
of attractants are additive, and the direction of the response
to a combined gradient of an attractant and a repellent will
depend on the magnitude of the positive and negative stimuli
involved. Chet and Mitchell (1976) discuss the more limited
studies on the chemotactic responses of other bacteria, and
Canale-Parola (1978) the chemotaxis of spirochaetes. Increasing
attention is now being devoted to *Bacillus subtilis* (for
example, Ordal, Villani and Rosendahl, 1979) and the study of
the chemotactic responses of photosynthetic bacteria has been
resumed (Armitage, Keighley and Evans, 1979). Photosynthetic
bacteria show chemotaxis under both heterotrophic and phototrophic
conditions; the response patterns of the non-sulphur bacterium
Rhodopseudomonas sphaeroides and the sulphur bacterium

Chromatium vinosus differ markedly, with the latter being attracted to sulphide. The plant pathogenic *Pseudomonas lachrymans* shows chemotaxis to plant exudates (Chet, Zilberstein and Henis, 1973) and chemotaxis may be involved in the infection of legume roots by symbiotic nitrogen-fixing bacteria of the genus Rhizobium (Currier and Stroebel, 1977; E. Orlowska and M.J. Carlile, unpublished observations).

Many eukaryotic microorganisms show chemotaxis towards a range of nutrients. The zoospores of the rumen phycomycete *Neocallimastix frontalis*, for example, show chemotaxis to several carbohydrates (Orpin and Bountiff, 1978) and the plasmodia of the myxomycete *Physarum polycephalum* to various sugars (Knowles and Carlile, 1978) and amino acids (Chet, Naveh and Henis, 1977). Clues other than major nutrients may however suffice to guide predators to their prey and parasites to appropriate hosts. Thus amoebae of the cellular slime mould *Dictyostelium discoideum* respond to folic acid and related compounds secreted by bacteria (Bonner, 1977) and the zoospores of the plant pathogenic fungus *Phytophthora palmivora* are attracted by a range of fatty acids, alcohols and aldehydes, some at least of which are commonly produced by plants (Cameron and Carlile, 1978). In addition to being attracted to bacteria, amoebae of *D. discoideum* show "social taxes", away from each other during the feeding stage (Keating and Bonner, 1977) and towards each other to form a grex after feeding has ended (Bonner, 1977; Newell, 1977; Darmon and Brachet, 1978). The attractant involved in *D. discoideum* and some members of the same genus is 3',5'-cyclic AMP, but different "acrasins" occur in other cellular slime moulds (Bonner, 1977; Poff and Whitaker, 1979). The attractants involved in mating are active at very low concentrations and are highly specific. Female gametes of the chytrid *Allomyces* for example, produce the sex attractant sirenin; male gametes of *Allomyces* show positive chemotaxis to this compound, but to no other substance, in the range 10^{-10} to 10^{-5} M (Carlile and Machilis, 1965).

In non-motile eukaryotic microorganisms chemotropism takes the place of chemotaxis and fulfils similar roles. Hyphae emerging from the germinating spores of some lower fungi grow towards nutrients or into plant hosts (Gooday, 1975). In higher fungi hyphae may grow towards or away from each other (that is, show positive or negative autotropism) depending on circumstances (Gooday, 1975); both forms of behaviour are necessary in establishing the typical colony of a higher fungus, with efficiently spaced but often anastomosing hyphae. Negative autotropism involving a gaseous repellent prevents *Phycomyces blakesleeanus* sporangiophores from colliding (Russo, Halloran and Gallori, 1977), and attractants (gaseous

in *Mucor mucedo*) are involved in mating in a range of fungi
(Gooday, 1975, 1978).

Gravity

Negative geotaxis (a tendency to swim upwards) is widespread
in eukaryotic microorganisms, occurring in many algae and
protozoa (Carlile, 1980a).   Here its role is probably to
counteract gradual sedimentation of organisms more dense than
water.   It also occurs in the zoospores of the plant pathogenic
fungus *Phytophthora* (Cameron and Carlile, 1977), where a
possible function is to keep the zoospores in the upper aerobic
layers of soil that are occupied by plant roots. Geotropism has
an important role in fruiting body development in some basidio-
mycetes, since the spore-bearing gills or pores of these fungi
must have a precise vertical orientation if efficient spore
liberation is to occur (Ingold, 1971).   The sporangiophores
of *Phycomyces blakesleeanus* also show negative geotropism
(Russo, 1980; Russo and Galland, 1980).

Mechanical forces

Positive rheotaxis (upstream swimming) has been observed
in zoospores of the plant pathogenic fungus *Phytophthora
capsici* (Katsura and Miyata, 1971).   Its role is unknown, but
it may prevent zoospores from being washed deep into the
soil away from plant roots.   On the other hand, the role of
induced upwind flying in insect behaviour (Kennedy, 1978)
suggests that upstream swimming, if induced by plant exudates
or other signals, could act as a substitute for or adjuct to
chemotaxis.   Eukaryotic microorganisms also show avoidance
reactions or collision reactions.   Zoospores of *Phytophthora*
sp. change direction when a collision seems imminent (J.N.
Cameron and M.J. Carlile, unpublished observations) and
direction reversal following collision has been intensively
studied in *Paramecium* sp. (Nelson and Kung, 1978).

The movement of gliding and crawling microorganisms is
influenced by properties of the substratum (Carlile, 1975),
an effect known as contact guidance.   The most intensive
studies of contact guidance have been carried out on fibroblasts,
where it has been established that both the shape of the
substratum and its adhesiveness are of importance (Dunn and
Ebendahl, 1978).   Sensitivity of amoebae to vibrations may be
of importance to amoebae in locating prey.   Kolle-Kralik and
Ruff (1967) have shown that *Amoebae proteus* moves towards
vibrations of 50 Hz s^{-1}, approximately the beating frequency
of the cilia of *Tetrahymena pyriformis*, its normal prey.

Hyphae also respond to various physical forces.   The
growth of the germ-tubes of the plant pathogenic fungus
Puccinia coronata is orientated with respect to the surface

pattern of the underlying leaf cuticle, an effect termed
thigmotropism (Dickinson, 1977) which may be of significance
in infection. The sporangiophore of *Phycomyces blakesleeanus*
shows amemotropism, bending into a gentle breeze (Cohen, Jan,
Matricon and Delbrück, 1975). Some fungi have been shown to
exhibit upstream or downstream growth (Carlile, 1975). The
species concerned, however, are terrestrial rather than aquatic,
so the effect is probably an indirect consequence of chemotropic
sensitivity rather than one of significance in nature.

Heat

Thermotactic responses are known in a wide variety of
microorganisms (Carlile, 1980a). In some instances they
presumably lead organisms to optimal temperatures for growth;
the very sensitive positive thermotaxis of the grex of
Dictyostelium discoideum (Poff and Whitaker, 1979), however,
probably facilitates migration to sites suitable for spore
dispersal.

Electricity

Responses to ionic electric currents (*galvanotaxis*) and
electric fields (*electrotaxis*) are widespread in microorganisms.
They may be of significance in nature, since weak currents
and charges are common. They may, on the other hand, be a
consequence of the probably universal involvement of ion
fluxes and transmembrane potential changes in sensory transduction
(Carlile, 1980a) and be essentially a laboratory artifact.

Magnetism

Magnetotaxis has recently been demonstrated in bacteria
from freshwater and marine muds (Blakemore, 1975; Frankel,
Blakemore and Wolfe, 1979). The organisms concerned are
anaerobic and microaerophilic, and the vertical component of
the earth's magnetic field is greater than the horizontal in
the locations where they have been found. It was therefore
suggested that the effect served to direct the organisms
downwards into areas of low oxygen tension favourable for
their growth.

Water

Positive hydrotropism may guide the germ-tubes of plant
pathogenic fungi towards stomata (Dickinson, 1960). Humidity
gradients may also affect the migration of the myxomycete
Physarum nudum, movement being towards damp conditions in
young plasmodia and away from them in plasmodia about to
spore (Rakoczy, 1963).

Interaction between stimuli

As indicated above, useful information about the environment can be conveyed to organisms by a variety of stimuli. Organisms are hence often simultaneously sensitive to many types of environmental factors. *Escherichia coli* has receptors for a wide variety of attractant and repellent chemicals (Adler, 1975), and can integrate the information received from the various stimuli to give an appropriate motor response. The sporangiophore of *Phycomyces blakesleeanus* responds to light, gravity, the presence of other sporangiophores and mechanical barriers and air currents (Russo, 1980; Russo and Galland, 1980). Negative geotropism is effective in complete darkness but is suppressed by low light intensities; phototropism, if feasible, is a better method of getting into the open than negative geotropism, since obstacles can be avoided. Behavioral responses to a variety of stimuli have also been shown in *Paramecium* sp. (Nelson and Kung, 1978).

The sensory responses of microorganisms may change during the life-cycle. The male gamates of *Allomyces* show chemotaxis only to sirenin; within a few minutes of fertilisation having occurred this sensitivity is replaced by chemotaxis to amino acids, which presumably leads to zygote to sites suitable for germination (Machlis, 1969). The amoebae of *Dictyostelium discoideum* are at first sensitive to folic acid secreted by bacteria, then to 3',5'-cyclic AMP secreted by other amoebae, after which the resulting grex responds to light and warmth (Poff and Whitaker, 1979).

The factors which cause taxis and tropism may also be involved in differentiation. Thus in *P. blakesleeanus* light accelerates sporangiophore initiation and subsequent sporangium production, and in doing so interacts with gases produced by the fungus (Russo, Galland, Toselli and Volpi, 1980). Tactic sensitivity may also be influenced by circadian or by tidal rhythms (Harper, 1977).

THE ROLE AND SIGNIFICANCE OF MOTILITY, TAXIS AND TROPISM

As indicated in the previous section, microorganisms show taxis or tropism in response to a wide variety of physical and chemical stimuli. Speculation on the role of the various responses in the life of the organism is inevitable, and in some instances the conclusions reached are almost certainly correct. Thus the chemotaxis of the male gamates of *Allomyces*, which occurs in response to sirenin emitted by the female gametangia and gametes but to no other compound, can only have a role of fertilisation. The function of other sexual and social taxes and tropisms also seem self-evident. When, however, an organism shows taxis to or from a wide range of

simple substances, the significance of each response is less
obvious. Thus the zoospores of the plant pathogenic fungus
Phytophthora show positive chemotaxis to various amino acids,
ethanol and other alcohols, fatty acids and aldehydes (Cameron
and Carlile, 1978), many of which are produced by plants. What
is the relative importance of these substances in bringing
about infection, and what is the role of the negative geotaxis
of *Phytophthora* zoospores (Cameron and Carlile, 1977), the
negative chemotaxis to hydrogen ions and other monovalent
cations (Allen and Harvey, 1974) and electrotaxis (Katsura
and Miyata, 1971; Khew and Zentmyer, 1974)? Elucidation of
these problems requires that studies should be carried out not
only under circumstances optimal for the analysis of behaviour
but also with experimental designs in which natural conditions
are simulated. This has to some extent been done with zoospores
of plant pathogenic fungi. Chang-Ho and Hickman (1970)
fractionated pea-root exudates and examined the effectiveness
of the fractions in attracting zoospores of the pea-infecting
fungus *Pythium aphanidermatum*. Allen and Newhook (1973) used
not only classical capillary tests but also a rhizosphere
model in their study on the chemotaxis of *Phytophthora
cinnamomi* zoospores. They also studied motility and taxis
of zoospores in "ideal soils" composed of glass microbeads
of various sizes (Young, Newhook and Allen, 1979). They
simulated natural soil water movements and found that flow
rates much higher than the swimming speed of zoospores did
not prevent the zoospores from subsequently swimming and that
low flow rates permitted chemotaxis. Mehrotra (1970) was able
to demonstrate attraction of zoospores of two species of
Phytophthora to plant roots in non-sterile soils. More such
studies are needed, both with fungal zoospores and with other
microorganisms showing tactic and tropic responses.

   Experiments of the type indicated above should help to
establish the role of a sensory response in nature. This
leaves unanswered the question of the importance of the
response. For example, is a specific chemotactic response
essential for survival under certain conditions, or does it
merely convey a marginal advantage in competition with strains
which lack the response but which perhaps have other competitive
features absent in the chemotactic strain? Smith and Doetsch
(1969) inoculated liquid cultures with a mixture of equal
number of a motile and a non-motile strain. They found that
if the cultures were shaken a 1:1 ratio between the two
populations was maintained but that in unshaken cultures the
motile strain was favoured, giving at 24 h counts 10-30 times
as high as those of the non-motile strain. The relative
success of the motile strain in unshaken cultures was
probably due to positive aerotaxis, but since only a motility

mutant was used, the possible contribution of other sensory responses and random motility was not assessed. Pilgram and Williams (1976) found that a mutant of *Proteus mirabilis* that had lost the ability to respond chemotactically to amino acids grew as well as the wild type in a liquid medium, in which the establishment of concentration gradients was unlikely (due to occasional shaking and presumably also convection), but more slowly than the wild type when the medium was rendered semi-solid with 0.2% (w/v) agar, which would permit gradients to be formed. When mixed cultures were employed, the chemotactic strain outgrew the mutant in the semi-solid but not in the liquid cultures. Chemotaxis is hence advantageous to *P. mirabilis* under circumstances in which gradients can form. The demonstration that in *Escherichia coli* specific chemotactic mutants, general chemotactic mutants and motility mutants are all obtainable suggests that the above approach should be extended. Plant-infecting microorganisms would be particularly useful in such studies, since competition experiments could be carried out under natural conditions.

CONCLUSIONS

Many microorganisms are motile and many show taxis or tropism in response to a range of stimuli. Rapid progress is now being made on the genetical, biochemical and biophysical basis of microbial motility and sensory responses and possible roles can be suggested. More studies however, are needed in which sensory responses are investigated under simulated natural conditions and in which the importance of motility and sensory responses are evaluated by employing mutants in competition experiments. Meanwhile, it is reasonable to conclude that motility, taxis and tropism are of value in the natural environment; such complex systems could not otherwise have been evolved and retained.

REFERENCES

Adler, J. (1975). Chemotaxis in bacteria, *Annual Review of Biochemistry* 44, 341-356.

Adler, J. and Dahl, M.M. (1967). A method for measuring the motility of bacteria and for comparing random and non-random activity, *Journal of General Microbiology* 46, 161-173.

Allen, R.N. and Harvey, J.D. (1974). Negative chemotaxis of zoospores of *Phytophthora cinnamomi*, *Journal of General Microbiology* 84, 28-38.

Allen, R.N. and Newhook, F.J. (1973). Chemotaxis of zoospores of *Phytophthora cinnamomi* to ethanol in capillaries of soil pore dimensions, *Transactions of the British Mycological Society* 61, 287-302.

Armitage, J.P., Keighley, P. and Evans, M.C.W. (1979).
    Chemotaxis in photosynthetic bacteria, *FEMS Microbiology
    Letters* 6, 99-102.
Bean, B. (1979). Chemotaxis in unicellular eukaryotes. *In*
    "Physiology of Movements" (Ed. W. Haupt and M.E. Feinleib)
    pp. 335-354. Springer Verlag, Heidelberg.
Berg, H.C. (1975). Chemotaxis in bacteria, *Annua. Review of
    Biophysics and Bioengineering* 4, 119-136.
Berg, H.C., Bromley, D.B. and Charon, N.W. (1978).
    Leptospiral motility. *In* "Relations between Structure
    and Function in the Prokaryotic Cell". Society for
    General Microbiology Symposium, Vol. 28 (Ed. R.Y. Stanier,
    H.J. Rogers and B.J. Ward) pp. 267-294. Cambridge
    University Press, Cambridge.
Berg, H.C. and Purcell, E.M. (1977). Physics of chemoreception,
    Biophysical Journal 20, 193-219.
Bialczyk, J. (1979). An action spectrum for light avoidance
    by *Physarum nudum* plasmodia, *Photochemistry and Photo-
    biology* 30, 301-303.
Bialczyk, J. and Rakoczy, L. (1975). Phototaxis of myxomycetes.
    Behaviour of old plasmodia in white light. *Bulletin de
    l'Academie Polonaise des Sciences. Serie des Sciences
    Biologiques* 23, 571-575.
Blakemore, R. (1975). Magnetotactic bacteria, *Science* 190,
    377-379.
Bonner, J.T. (1977). Some aspects of chemotaxis using the
    cellular slime moulds as an example, *Mycologia* 69, 443-459.
Cameron, J.N. and Carlile, M.J. (1977). Negative geotaxis
    of fungal zoospores, *Journal of General Microbiology*
    98, 599-602.
Cameron, J.N. and Carlile, M.J. (1978). Fatty acids,
    aldehydes and alcohols as attractants for zoospores of
    Phytophthora palmivora, Nature 271, 448-449.
Canale-Parola, E. (1978). Motility and chemotaxis of
    spirochaetes, *Annual Review of Microbiology* 32, 69-99.
Carlile, M.J. (1970). The photoresponses of fungi. *In*
    "Photobiology of Microorganisms" (Ed. P. Halldal) pp. 309-
    344. Wiley, London.
Carlile, M.J. (1975). Taxes and tropisms: diversity,
    biological significance and evolution, *In* "Primitive
    Sensory and Communication Systems: the Taxes and Tropisms
    of Microorganisms and Cells" (Ed. M.J. Carlile) pp. 1-28.
    Academic Press, London.
Carlile, M.J. (1979). Bacterial, fungal and slime mould
    colonies, *In* "Biology and Systematics of Colonial Organisms"
    (Ed. G. Larwood and B.R. Rosen) pp. 3-27. Academic Press,
    London.
Carlile, M.J. (1980a). Sensory transduction in aneural

organisms. *In* "Photoreception and Sensory Transduction
in Aneural Organisms" (Ed. G. Colombetti and F. Lenci).
Plenum Press, New York.

Carlile, M.J. (1980b). From prokaryote to eukaryote: gains
and losses. *In* "The Eukaryotic Microbial Cell". Society
for General Microbiology Symposium, Vol. 30. (Ed. G.W.
Gooday, D. Lloyd and A.P.J. Trinci) pp. 1-40. Cambridge
University Press, Cambridge.

Carlile, M.J. and Gooday, G.W. (1978). Cell fusion in
myxomycetes and fungi. *In* "Membrane Fusion" (Ed. G. Poste
and G.L. Nicholson) pp. 219-265. Elsevier North-Holland
Biomedical Press, New York.

Carlile, M.J. and Machlis, L. (1965). The response of male
gametes of Allomyces to the sexual hormone sirenin,
American Journal of Botany 52, 478-483.

Cerdá-Olmedo, E. (1977). Behavioral genetics of Phycomyces,
Annual Review of Microbiology 31, 535-547.

Chang-Ho, Y. and Hickman, C.J. (1970). Some factors involved
in the accumulation of phycomycete zoospores on plant
roots. *In* "Root Diseases and Soil Borne Pathogens"
(Ed. T.A. Tousson, R.V. Bega and P.E. Nelson) pp. 103-108.
University of California Press, Berkeley.

Chet, I. and Mitchell, R. (1976). Ecological aspects of
microbial chemotactic behaviour, *Annual Review of
Microbiology* 30, 221-239.

Chet, I., Naveh, A. and Henis, Y. (1977). Chemotaxis of
Physarum polycephalum towards carbohydrates, amino acids
and nucleotides, *Journal of General Microbiology* 102,
145-148.

Chet, I., Zilberstein, Y. and Henis, Y. (1973). Chemotaxis
of *Pseudomonas lachrymans* to plant extracts and to water
droplets collected from the leaf surfaces of resistant
and susceptible plants, *Physiological Plant Pathology*
3, 473-479.

Cohen, R.J., Jan, Y.N., Matricon, J. and Delbrück, M. (1975).
Avoidance response, house response and wind responses of
the sporangiophore of Phycomyces, *Journal of General
Physiology* 66, 67-95.

Currier, W.W. and Stroebel, G.A. (1977). Chemotaxis of
Rhizobium spp. to a glycoprotein produced by Birdsfoot
Trefoil roots, *Science* 196, 434-436.

Darmon, M. and Brachet, P. (1978). Chemotaxis and
differentiation during aggregation of *Dictyostelium
discoideum* amoebae. *In* "Taxis and Behaviour" (Ed. G.L.
Hazelbauer) pp. 101-139. Chapman and Hall, London.

Dickinson, S. (1960). The mechanical ability to breach the
host barriers. *In* "Plant Pathology - an Advanced
Treatise" (Ed. J.G. Horsfall and A.E. Dimond) Vol. 2,

pp. 203-232. Academic Press, New York.

Dickinson, S. (1977). Studies on the physiology of obligate parasitism. X. Induction of responses to a thigmotropic stimulus, *Phytopathologisches Zeitschrift* <u>89</u>, 97-115.

Diehn, B., Feinleib, M., Haupt, W., Hildebrand, E., Lenci, F. and Nultsch, W. (1977). Terminology of behavioral responses of motile microorganisms, *Photochemistry and Photobiology* <u>26</u>, 559-560.

Dunn, G.A. and Ebendahl, T. (1978). Some aspects of contact guidance, *Zoon* <u>6</u>, 65-68.

Foster, K.W. (1977). Phototropism of coprophilous zygomcyetes, *Annual Review of Biophysics and Bioengineering* <u>6</u>, 419-443.

Frankel, R.B., Blakemore, R.P. and Wolfe, R.S. (1979). Magnetite in magnetotatic bacteria, *Science* <u>203</u>, 1355-1356.

Gooday, G.W. (1975). Chemotaxis and chemotropism in fungi and algae. *In* "Primitive Sensory and Communication Systems: the Taxes and Tropisms of Microorganisms and Cells" (Ed. M.J. Carlile) pp. 155-204. Academic Press, London.

Gooday, G.W. (1978). Microbial hormones. *In* "Companion to Microbiology" (Ed. A.T. Bull and P.M. Meadow) pp. 207-219. Longmans, London.

Goy, M.F. and Springer, M.S. (1978). In search of the linkage between receptor and response: the role of a protein methylation reaction in bacterial chemotaxis. *In* "Taxis and Behaviour" (Ed. G.L. Hazelbauer) pp. 1-34. Chapman and Hall, London.

Häder, D.P. (1979). Photomovement. *In* "Physiology of Movements" (Ed. W. Haupt and M.E. Feinleib) pp. 268-309. Springer Verlag, Heidelberg.

Häder, D.P. and Nultsch, W. (1979). Photomovement of motile microorganisms, *Photochemistry and Photobiology* <u>29</u>, 423-437.

Halfen, L.N. (1979). Gliding movements. *In* "Physiology of Movements" (Ed. W. Haupt and M.E. Feinleib) pp. 250-267. Springer Verlag, Heidelberg.

Harper, M.A. (1977). Movements. *In* "The Biology of Diatoms" (Ed. D. Warner) pp. 224-249. Blackwell Scientific Publications, Oxford.

Hazelbauer, G.L. and Parkinson, J.S. (1977). Bacterial chemotaxis. *In* "Microbial Interactions" (Ed. J.L. Ressig) pp. 59-98. Chapman and Hall, London.

Hildebrand, E. (1978). Bacterial phototaxis. *In* "Taxis and Behaviour" (Ed. G.L. Hazelbauer) pp. 35-73. Chapman and Hall, London.

Holwill, M.E.J. (1974). Hydrodynamic aspects of ciliary and flagellar movement. *In* "Cilia and Flagella" (Ed. M.A. Sleigh), pp. 143-175. Academic Press, London.

Holwill, M.E.J. (1980). Movement of cilia. *In* "The Eukaryotic Microbial Cell", Society for General Microbiology Symposium, Vol. 30, (Ed. G.W. Gooday, D. Lloyd and A.P.J. Trinci), pp. 273-300. Cambridge University Press, Cambridge.

Ingold, C.T. (1971). "Fungal Spores: their Liberation and Dispersal". Clarendon Press, Oxford.

Katsura, K. and Miyata, Y. (1971). Swimming behaviour of *Phytophthora capsici* zoospores. *In* "Morphological and Biochemical Events in Plant-Parasite Interaction" (Ed. S. Akai and S. Ouchi) pp. 107-128. The Phytopathological Society of Japan, Tokyo.

Keating, M.T. and Bonner, J.T. (1977). Negative chemotaxis in cellular slime moulds, *Journal of Bacteriology* 130, 144-147.

Kennedy, J.S. (1978). The concepts of olfactory "arrestment" and "attraction", *Physiological Entomology* 3, 91-98.

Khew, K.L. and Zentmyer, G.A. (1974). Electrotactic response of zoospores of seven species of *Phytophthora, Phytopathology* 64, 500-507.

Knowles, D.J.C. and Carlile, M.J. (1978). The chemotactic response of plasmodia of the myxomycete *Physarum polycephalum* to sugars and related compounds, *Journal of General Microbiology* 108, 17-25.

Kochert, G. (1978). Sexual pheromones in algae and fungi, *Annual Review of Plant Physiology* 29, 461-468.

Kolle-Kralik, U. and Ruff, P.W. (1967). Vibrotaxis von *Amoeba proteus* (Pallas) im vergleich mit der Zilienschlagfrequenz der Beutethiere, *Protistologica* 3, 319-323.

Koshland, D.E. (1979). Bacterial chemotaxis. *In* "The Bacteria" (Ed. J.R. Sokatch and L.N. Ornston), Vol. 7, pp. 111-168. Academic Press, New York.

Lenci, F. and Colombetti, G. (1978). Photobehaviour of microorganisms: a biophysical approach, *Annual Review of Biophysics and Bioengineering* 7, 341-361.

Levandowsky, M. and Hauser, D.C.R. (1978). Chemosensory responses of swimming algae and protozoa, *International Review of Cytology* 53, 145-210.

Machlis, L. (1969). Fertilization-induced chemotaxis in the zygotes of the watermold *Allomyces, Physiologia Plantarum* 22, 392-400.

Macnab, R.M. (1978). Bacterial motility and chemotaxis: the molecular biology of a behavioural system, *C.R.C. Critical Reviews in Biochemistry* 5, 291-341.

Macnab, R.M. (1979a). Bacterial flagella. *In* "Physiology of Movements" (Ed. W. Haupt and M.E. Feinleib) pp. 207-223. Springer Verlag, Heidelberg.

Macnab, R.M. (1979b). Chemotaxis in bacteria. *In* "Physiology of Movements" (Ed. W. Haupt and M.E. Feinleib) pp. 310–334. Springer Verlag, Heidelberg.

Mehrotra, R.S. (1970). Techniques for demonstrating accumulation of zoospores of *Phytophthora* species on roots in soil, *Canadian Journal of Botany* 48, 879–882.

Nelson, D.L. and Kung, C. (1978). Behaviour of Paramecium: chemical, physiological and genetical studies. *In* "Taxis and Behaviour" (Ed. G.L. Hazelbaur) pp. 75–100. Chapman and Hall, London.

Newell, P.C. (1977). Aggregation and cell surface receptors in cellular slime moulds. *In* "Microbial Interactions" (Ed. J.L. Ressig) pp. 1–57. Chapman and Hall, London.

Ordal, G.W., Villani, D.P. and Rosendahl, M.S. (1979). Chemotaxis towards sugars by *Bacillus subtilis, Journal of General Microbiology* 115, 167–172.

Orpin, C.G. and Bountiff, L. (1978). Zoospore chemotaxis in the rumen phycomycete *Neocallimastix frontalis, Journal of General Microbiology* 104, 113–122.

Petrino, M.G. and Deutsch, R.N. (1978). Viscotaxis, a new behavioural response of *heptospira interogans (biflexa)* strain B16, *Journal of General Microbiology* 109, 113–117.

Pilgram, W.K. and Williams, F.D. (1976). Survival value of chemotaxis in mixed cultures, *Canadian Journal of Microbiology* 22, 1771–1773.

Plesset, M.S., Whipple, G.G. and Winet, H. (1975). Analysis of the steady state of bioconvection in swarms of swimming microorganisms. *In* "Swimming and Flying in Nature" (Ed. T.Y.T. Wu, C.J. Brokaw and C. Brennan) Vol, 1, pp. 339–360. Plenum Press, New York.

Poff, K.L. and Whitaker, B.D. (1979). Movement of slime moulds. *In* "Physiology of Movements" (Ed. W. Haupt and M.E. Feinleib) pp. 355–382. Springer Verlag, Heidelberg.

Rakoczy, L. (1963). Application of crossed light and humidity gradients for the investigation of slime-moulds, *Acta Societatis Botanicorum Polaniae* 32, 393–403.

Rohlf, J.F. and Davenport, D. (1969). Simulation of simple models of animal behaviour with a digital computer, *Journal of Theoretical Biology* 23, 400–424.

Russo, V.E.A. (1980). Sensory transduction in phototropism: genetic and physiological analysis in Phycomyces. *In* "Photoreception and Sensory Transduction in Aneural Organisms" (Ed. G. Colombetti and F. Lenci). Plenum Press, New York.

Russo, V.E.A. and Galland, P. (1980). Sensory physiology of *Phycomyces blakesleeanus. In* "Sensory Physiology and Structure of Molecules". (Ed. P. Hemmerich). Springer Verlag, Berlin.

Russo, V.E.A., Galland, P., Toselli, M. and Volpi, L. (1980).

Blue light induced differentiation in *Phycomyces blakes-leeanus*. *In* "The Blue Light Syndrome". (Ed. H. Senger). Springer Verlag, Berlin.

Russo, V.E.A., Halloran, B. and Gallori, E. (1977). Ethylene is involved in the autochemotropism of *Phycomyces Planta* 134, 61-67.

Satir, P. and Ojakian, G.K. (1979). Plant cilia. *In* "Physiology of Movements" (Ed. W. Haupt and M.E. Feinleib) pp. 224-249. Springer Verlag, Heidelberg.

Sleigh, M.A. and Blake, J.R. (1977). Methods of ciliary propulsion and their size limitations. *In* "Scale Effects in Animal haeomation" (Ed. T.J. Pedley) pp. 243-256. Academic Press, London.

Smith, J.L. and Deutsch, R.N. (1969). Studies in negative chemotaxis and the survival value of motility in *Pseudomonas fluorescens*, *Journal of General Microbiology* 55, 379-391.

Trinci, A.P.J. (1978). Wall and hyphal growth. *Science Progress, Oxford* 65, 75-99.

Young, B.R., Newhook, F.J. and Allen, R.N. (1979). Motility and chemotactic response of *Phytophthora cinnamomi* zoospores in "ideal soils", *Transactions of the British Mycological Society* 72, 395-401.

THE ECOLOGICAL SIGNIFICANCE OF MICROBIAL DISPERSAL SYSTEMS

J. Lacey and P.H. Gregory

Rothamstead Experimental Station
Harpenden, Herts AL5 2JQ, England

INTRODUCTION

Microbes travel by various routes. Some are self-propelled, some carried by fomities, some by water, and yet others by insect or other vectors. We are concerned solely with airborne microbes whose dispersal comprises three phases:

i)   launching into the air from the parent substratum;

ii)   transport in air currents;

iii) deposition on a new substratum which may or may not be suitable for colonization.

Attempts to control dispersal usually concentrate on the deposition phase using either protective chemicals or barriers such as filters. Control may also be attempted in the transport phase as with laminar-flow ventilation systems but control of the launching phase is often hampered by our ignorance of the launching mechanism. In general, it seems that the smaller the organism the less the mechanism is understood.

OBSTACLES TO LAUNCHING OF MICROBES INTO THE ATMOSPHERE

The naive idea that the smaller the organism the more easily it becomes airborne is erroneous. Surfaces exposed to the atmosphere are surrounded by a boundary layer of air moving in streamlines, speed increasing with height but at the surface held stationary by molecular forces. Unattached microbial cells must traverse this layer to reach the turbulent air above before dispersal can occur. The thickness of the boundary varies with wind speed and turbulence. Eddies may sometimes penetrate into this layer to remove

particles immersed within it but the smaller the particle,
the less likely is its removal. The mechanism by which
microbes located in or below the boundary layer become airborne need careful explanation.

ACTIVE LAUNCHING

The advantages of dispersal by wind have evidently been a
major factor in the evolution of fungi, many of which are
well adapted to airborne dispersal. Their launching mechanisms have been well studied (Ingold, 1971) and different
types of active mechanisms have been described which are not
found in simpler organisms. Kinetic energy supplied by the
fungus itself, often in response to change in environmental
conditions, propels the spores across the surface boundary
layer into the mobile atmosphere above. Some mechanisms
require water including the "squirt-gun" found in many
Ascomycetes; the squirting mechanisms of some Phycomycetes
and the imperfect fungus *Nigrospora*, the rounding of turgid
cells; and basidiospore liberation. Several imperfect fungi,
for example *Deightoniella torulosa* (Meredith, 1961), depend
on the sudden expansion caused by a gas bubble forming in a
drying and previously contracting cell. However in the
majority of microorganisms launching is passive, dependent on
energy provided by the environment.

WIND-ACTIVATED LAUNCHING

Some microbes have their propagules on stalks projecting
into the moving air from where they can blow away, a process
which has been named "deflation". This is seen for example
in *Penicillium* spp. and many moulds, in Myxomycetes and
Acrasiomycetes. The projecting dry spore chains of some
actinomycetes serve the same function but this mechanism for
launching dry spores into air does not seem to occur elsewhere in the bacteria. Spores of mildews, rusts, smuts and
other fungi infecting aerial parts of plants are already in
a position favourable for "deflation". So too are the "poly-
hedral bodies" of some viruses shed from insects suspended
from the twigs of forest trees (Gregory, 1973).

RAFT TRANSPORT

Much bacterial transport depends on carriage on relatively
large rafts, usually comprising fragments of the substratum
such as skin scales, plant epidermal fragments or soil
particles, each carrying a mixed load of microbial passengers.
Their launching and dispersal is thus dependent on the
launching, transport and deposition of the rafts rather than
on the properties of individual passenger organisms. Man
himself contributes greatly to this process.

The human body surface is covered by about 10^8 scales, each about 30 x 30 x 3-5 μm, which may be colonized by a range of microorganisms (Noble and Pitcher, 1978). A complete layer of scales is lost about every four days. Their removal is aided by friction with clothing or with towels after bathing while large numbers may be shed in bed and become airborne during bed making. They may also be resuspended from dust deposits on floors but the role of body convection in their launching is often neglected.

The naked human body emits about 10^5 J h^{-1} as heat which warms the adjacent air making it rise, first in a laminar layer but within 1 m as a turbulent stream. Flow reaches a maximum speed of 0.5 m sec^{-1} at head level and continues for about 0.5 m above the head (Lewis, Poster, Mullen, Cox and Clark, 1969; Clark, Cox and Lewis, 1970). This airflow contains significantly more microorganisms than the ambient air. Some may come from spore rich air near the ground but many from bacterial colonies on skin scales released from the body. This rising airflow forms about 10% of each inspiration so that some of these microbes may be inhaled, while others escape into the ambient air perhaps to play a role in disease transmission. Most important of these is probably *Staphylococcus aureus*.

LAUNCHING BY MECHANICAL DISTURBANCE

Mechanical disturbance leading to the launching of microbes into the air may occur frequently, for instance where leaves and stems in dense vegetation knock together. Superficial microorganisms may then be knocked off or detached from their sporophores to become airborne although within dense vegetation they may fail to rise above the canopy (Legg and Bainbridge, 1978). Rain falling on vegetation has a similar effect in the "tap and puff" mechanism described below. However, disturbance of the natural environment is a characteristic human activity, most of all when he uses machines to harvest, handle and process agricultural crops. In these situations, the most numerous airborne organisms are fungi and actinomycetes which produce large numbers of spores that are easily detached from their sporophores. These give rise to some of the largest recorded concentrations of airborne spores. Some of the circumstances under which large concentrations of these organisms arise and the types involved can be seen by reference to handling of the barley crop. By contrast, much smaller numbers of airborne microbes (for example, *Coccidioides immitis* and *Histoplasma capsulatum*) can still, in other situations, have a dramatic effect on local communities.

Harvesting and processing the barley crop

The barley plant ripening in the field is colonized by a
rapidly increasing microflora comprising bacteria yeasts
and field fungi, mostly species of *Cladosporium*, *Alternaria*,
Epicoccum and *Verticillum*. The combine harvester shakes their
propagules into the air through vibration of the cutter bar
and the threshing mechanism to form a cloud of dust containing
up to 2×10^8 spores m^{-3} air close to front and rear of the
harvester and up to 3×10^7 m^{-3} in the breathing zone of
drivers, of whom a quarter complain of respiratory symptoms
(Darke, Knowelden, Lacey and Ward, 1976).

Besides releasing many microbes into the air, combine
harvesters serve to redistribute propagules on the grain
surface. This is especially important for storage fungi,
found only in very small numbers before harvest (4.3% of
grains), but on up to 60% of grains after passage through
the combine harvester (Flannigan, 1978).

More spores are launched from the grain every time it is
subsequently handled – from combine to trailer, into driers,
into and out of store and in granneries and docks. Before
and during storage, so long as the grain is kept dry, the
qualitative composition of the microflora changes little
with field fungi predominant. However, if the grain contains
more than about 14% water, storage microbes develop, their
numbers and types differing according to the conditions of
storage, especially water content, degree of spontaneous
heating and aeration. These differences are reflected in
changes in the composition of the air spora, the magnitude
of which depends on amount of moulding, degree of disturbance
and ventilation of the building. Some of these differences
are illustrated in Table 1.

With vigorous disturbance and poor ventilation, concentrations of 10^9 – 10^{10} spores m^{-3} air are not uncommon with 50%
of more "actinomycetes + bacteria", including species of
Micropolyspora, *Thermoactinomyces*, *Saccharomonospora* and
Streptomyces. Species of *Aspergillus* and *Penicillium* are
usually the predominant fungi but exceptionally yeasts or the
plant pathogenic *Ustilago* may be numerous. Sometimes the
large quantities handled may offset the presence of only few
spores in the grain, provided spores are liberated into the
air more rapidly than they are deposited.

After storage, if the grain is used for malting, the
storage microflora may be replaced by *Aspergillus clavatus*
producing dense clouds of spores when grain is turned (Lacey,
1980).

Heavy dust deposits in grain stores contain abundant microorganisms and other debris. These deposits can become

TABLE 1

Airborne Spore Concentration Associated with Storage of Barley Grain

(-, none detected; data form Lacey, 1980)

Spore type	Spore concentration ($\times 10^{-6}$ m^{-3})	Unsealed moist barley silo	
	Barn-moving conventionally stored grain by auger	Still conditions	Unloading
Actinomycetes + bacteria	33.2	11.0	1792
Total fungi	43.8	9.2	1071
Aspergillus/Penicillium	29.4	8.7	826
Mucor	-	0.4	68.9
Cladosporium	5.3	0.1	1.9
Yeasts	6.0	0.04	154
Alternaria	1.2	-	-
Ustilago	1.5	-	-

resuspended when more grain is handled, adding to the airborne spore load, contaminating the incoming grain and perhaps
contributing to dust explosions.

Mechanical disturbance can also spread plant and human
pathogenic fungi,(for example, *Aspergillus fumigatus*) and
cause respiratory complaints in workers resulting variously
from the heavy dust load inhaled from immediate Type I or
delayed Type III allergy from activation of complement by
the alternative pathway from bacterial endotoxins or possibly
from mycotoxins. Many spores are also released into the
general environment by ventilation air currents, ensuring
the wide distribution of the species concerned and perhaps
accounting for the unusual occurrence of more thermophilic
actinomycetes in outdoor air in winter than in summer
(D.R. Henden, personal communication). Launching of microbes
by mechanical disturbance is similarly important during the
handling of hay, sugar-cane, bagasse, cork and even timber,
sometimes with similar consequences (Lacey, 1974).

Coccidioides and *Histoplasma*

Coccidioides immitis and *Histoplasma capsulatum* are fungi
which lack known forcible discharge mechanisms and are found
in soil; both infect many people annually causing respiratory
disease which is sometimes fatal. *C. immitis* occurs in
desert soil in southwestern U.S.A. while *H. capsulatum* is
more widespread, favoured by faecel enrichment of soil by
birds and bats. *C. immitus* may become airborne in dust
storms, by vehicles passing along dirt roads and during
agricultural operations such as discing and harrowing or
harvesting dust laden plants. Only 10 to 100 spores are
needed to infect dogs and in the endemic areas as many as 90%
of the human population may be infected (Levine, 1969).
Similarly, most outbreaks of histoplasmosis can be attributed
to the disturbance of contaminated sites. Thus clearance of
a blackbird roost in an urban area resulted in some 6,000
people becoming infected out of a population of 30,000,
most of them in an area extending 0.8 km downwind of the
site (Edmonds, 1979). However, 6 spores m^{-3} air is the
maximum concentration found during sampling in endemic areas.
(Rooks, 1954).

LAUNCHING IN AEROSOLS RESULTING FROM HUMAN ACTIVITY

Any activity resulting in breaking a liquid film can lead
to formation of airborne droplets in a range of sizes. These
remain airborne for varying periods, either as liquid drops
or, after drying, as droplet nuclei in which microorganisms
may be protected by dissolved or suspended solids. The
smaller the particle, the longer it remains airborne, the

more likely it is to penetrate deeply into lungs or inhalation and often the more infective it is. Even talking, coughing and sneezing result in the projection of many droplets from the mouth into the air which can carry bacteria or viruses. Sneezing, the most vigorous action of the three, projects perhaps 10^6 droplets <100 μm diameter into the air as well as many thousands of larger droplets which settle rapidly. These mostly come from saliva at the front of the mouth which may be contaminated with mucous secretion from the pharynx during coughing (Lidwell, 1967). Coughing and talking tend to produce fewer, larger droplets than sneezing but from all three actions the median particle diameter of droplet nuclei carrying bacteria is aerodynamically equivalent to a unit density spere of 13 μm diameter which would be deposited in the upper airways on inhalation.

Aerosols may also be formed during many of the processes used by man to improve his comfort, during his work and in the disposal of waste. Many of these processes have important consequences for microbial dispersal and human and animal health.

Launching of faecal microbes

Intensive animal production inevitably results in the accumulation of large quantities of slurry with its attendant load of enteric bacteria, viruses and protozoa which may include human pathogens. Cattle and pigs respectively produce 8 and 4 times as much excrement per head as humans and up to 35,000 cattle may sometimes be kept within a square mile. The problems of storing and disposing of such quantities of slurry without creating health hazards or polluting waterways are immense.

Within buildings housing animals, microbes may be launched by splashing (see below), perhaps accounting for the greater incidence of respiratory infection in housed pigs than in those in the open (Donham, Rubino, Thedell and Kammermeyer, 1977). However opportunities for launching in aerosols are probably greatest when slurry is sprayed onto fields. Droplets may be formed directly by atomisation and, after the ground is wetted, by splashing in a thin film of liquid. The smaller droplets rapidly dry down to droplet nuclei whose size depends on the size of the initial droplet and its content of dissolved and suspended solids. These may carry bacteria such as *Escherichia*, *Leptospire* and *Salmonella* and perhaps also viruses. *Salmonella* sp. can survive continuous feed digestion but not batch digestion of sewage (Coker and Davis, 1978). Its survival in slurry is determined by the species of *Salmonella*, size of inoculum, temperature of storage and solids content of the slurry

(Jones, 1976).

Aerosols are similarly generated in sewage treatment plants and coliform bacteria have been detected in air downwind. Viruses may also be launched from these sources (Adams and Spendlove, 1970). However, the water closet is a potent source of aerosols in most buildings. Flushing a lavatory pan containing 10^{11} - 10^{12} bacteria with clean water produced over 70,000 particles containing viable propagules per m^3 level with the front of the seat and 700 m^3 1.2 m higher. Decreasing the inoculum did not decrease the airborne load proportionately and surprisingly large loads were sometimes found with small inocula. Most particles (87%) were smaller than 4 μm (mean about 2 μm) allowing long persistence in the air and deep penetration into lungs and each particle contained an average of 9.5 bacteria (Darlow and Bale, 1959).

Air conditioning

Humidifiers incorporated in air conditioning systems can be important sources of aerosols causing extrinsic allergic alveolitis or humidifier fever. Much confusion surrounds the role of different microorganisms in this disease, emphasising the need for precise descriptions of the environment, characterization of the aerosol and identification of the origins of the contaminating microflora. Humidification may be achieved by atomising cold water into the airstream or by injection of steam. Sometimes the humidifier may be warmed by being sited close to a boiler and often its water becomes colonised by bacteria, actinomycetes, fungi and protozoa. However, this water may not be the only source of microorganisms. Often there are filters or baffles downstream of the humidifiers that become wet and colonised. The air conditioning system then serves to launch and distribute cells, spores, cysts, fragments and metabolic products into the atmosphere throughout the whole building. (Banaszak, Thiede and Fink, 1970; Sweet, Anderson, Callies and Coates, 1971; Solomon, 1974; Pickering, Moore, Lacey, Holford-Strevens and Pepys, 1976; Arnow, Fink, Schlueter, Barboriak, Mallison, Said, Martin, Unger, Scanlon and Kurup, 1978).

Several organisms have been implicated in humidifier disease but their relative importance cannot be assessed from published evidence. Records are few of the number and species of organisms in humidifier water and none indicate whether the water was warm enough for the growth of thermophilic actinomycetes implicated in humidifier fever (greater than 30-35°C). Data on the numbers of organisms originating from humidifier systems relative to numbers usually present in the ambient air are also inadequate. Solomon (1974)

recorded an increase from 500 to 14,000 fungi m^{-3} following
cold mist humidification but this fell far short of the
numbers usually associated with allergic alveolitis (10^6 –
10^{10} m^{-3}) although probably sufficient to cause immediate
allergy. There is now convincing evidence that amoebae are
sometimes implicated in humidifier fever (MRC Symposium, 1977)
but numbers of *Naegleria gruberi* or *Acanthamoeba* cysts in
air have not been recorded. Being 5–12 μm diameter, they are
not well suited for deep penetration in the lung and the
form of antigen, whether cell or cyst fragment or metabolite,
has still to be identified although circumstantial evidence
favours the last.

Large outbreaks of Legionnaires' disease, a form on pneu-
monia caused by the bacterium *Legionella pneumophila*, suggest
exposure to a common airborne source of infection without
secondary person-to-person spread. There is a strong
probability that air-conditioning systems are implicated in
the spread of disease as well as, perhaps, dusty conditions
and soil disturbance (Anon, 1977). The bacterium has been
isolated from cooling tower water of air conditioning systems
(Morris, Patton, Feeley, Johnson, Gorman, Martin, Skaliy,
Politi and Mallison, 1979) where there is ample opportunity
for aersolization and it is possible that humidifiers could
also be colonized.

Laboratory hazards

Many events in the microbiology laboratory can generate
respirable aerosols laden with microbes, for example,
collapsing bubbles, opening culture vessels, dropping liquid
and culture vessels, vibrating inoculation needles, with-
drawing pipettes from liquid centrifuging, homogenising
and sizzling of inoculating loops (Darlow, 1972). Where
pathogenic microbes are being handled this can be very
hazardous. Wedum and Kruse (1969) record nearly 3,000
cases of laboratory-acquired infection and many deaths.
Infections by most pathogens studied were recorded and up
to 80% could be attributed to a specific action or source.
Most frequent laboratory infections are brucellosis,
infectious hepatitis, tuberculosis, Q fever, tularemia,
typhoid, Venezeulan equine encephalitis and coccidioidomycosis.
Pasteurella tularensis and *Coxiella burnetti* (the cause of
Q fever) require only 10 units to infect when inhaled and
the latter can become extensively airborne throughout labor-
atories from infected chicken embryos, (Edmonds, 1979).
The risk is not restricted to those working directly with
the pathogens but extends also to ancillary staff especially
in animal houses where many pathogens may be excreted in
urine which may splash and form aerosols or, after drying,

become airborne as dusts.

LAUNCHING BY BURSTING BUBBLES

Normally, air currents moving slowly over smooth water
containing many microorganisms will fail to pick up any
microbial load. However, with increasing wind speed, waves
are formed and high winds may even blow the tops of waves
off as a spray and so launch microbe-laden droplets. Wind
and rain are powerful agents for creating microbial aerosols.
White-capped waves at sea and breakers on shore are particularly rich aerosol generators, as has been described by
Blanchard (1978). Air bubbles may be forced to depths of
many centimetres, collect a microbial concentrate at the air/
water interface as they ascend to the surface then burst,
launching microbially enriched droplets into the wind. On
land, waterfalls must act similarly, though on a more limited
scale.

LAUNCHING BY RAIN SPLASH

Falling rain drops also generate aerosols, the effect
depending on the surface ("target") at the point of impact.
Quantitatively the effect depends on the size of the drop,
the terminal velocity at which it falls, and on the thickness of the target film (Gregory, Guthrie and Bunce, 1959).
Drops larger than approximately 5 mm diameter break up while
falling through the air, so there is an upper limit to both
the diameter and kinetic energy of the incident rain drop.
However, in drip from trees wetted by rain or fog, larger
drops can reach ground after a short fall at less than
terminal velocity. Depending on the target surface at point
of impact, several possibilities arise.

Rain drop hitting a dry solid surface

The drop flattens and flows out radially from the point
of impact at high speed, pushing the surface boundary air
layer outwards and propelling any dry detached particles on
the surface as a puff into the air, thus clearing the surface
over the radius of a centimetre or two. This phenomenon
probably accounts for most of the transient increase in
concentration of fungus spores such as *Cladosporium* and rusts
in the air, that several observers have recorded at the onset
of heavy rain (Gregory, 1973). This is aided by mechanical
shaking when a rain drop taps a leaf, the two constituting
the "puff and tap" mechanism of Hirst and Stedman, (1963).

But whereas "puff" may aerosolize loose particles from a
dry surface, it will fail to detach firmly adherent microbes
which are often embedded in mucilage. However, continuous
rain wets the surface and frees these also for liberation

by rain splash.

Rain drop hitting a thin liquid film

On hitting a thin layer of liquid of the order of 1 mm
thick, such as a wet surface the incident drop spreads
radially (as on a dry one). As shown in remarkable photo-
graphs by Worthington and Cole (1897), the incident drop
meets the target film, sweeping it outwards and upwards to
form a crater-like structure, perhaps a centimetre in
diameter. Under surface tension the rim of the crater breaks
up, first into about 15 rods, which in turn break into chains
of droplets which are then propelled into the air by the
residual depleted motion derived from the falling drop.

All this happens in about one hundredth of a second after
impact. The largest splash droplets move in a trajectory
under gravity and can hardly be described as airborne.
Within a second or so the smaller splash droplets evaporate
to "droplet-nuclei" (Wells, 1955) and thenceforth behave as
true windborne aerosol particles. Experiments in still air
(Gregory *et al.*, 1959) show that an incident drop 5 mm in
diameter hitting a target film 0.1 mm thick after falling
7.4 metres will produce over 5,000 droplets capable of
carrying inoculum. The droplet size covers a wide range.

The droplets comprise liquid from both the incident drop
and from the surface film, consequently microbes in the surface film will be carried off in the splash droplets; likewise those in the incident drop, whether scrubbed from the
atmosphere through which the rain has fallen, or washed off
leaves by drip from trees will also be scattered as splash
droplets.

Faulwetter (1917), studying black arm disease of cotton,
found rainsplash dominant in the dispersal of *Xanthomonas
malvacaerum*. It is also important in the dispersal of other
crop pathogens, for example, septoria blight of celery,
Phytophthora pod rot of cacao and many other diseases.

Drops falling into deeper liquid

As shown by Worthington and Cole (1897), a drop falling
into a deeper layer of liquid at first sinks and spreads
out over the bottom of a shallow pit, it then rebounds at
the head of a column of liquid before falling back and
forming a crater which shoots off droplets much as described
above. Sometimes the rim of the crater contracts and closes
as a dome, forming a bubble which may be shattered by a
further rebound of the column.

In the field these events happen with enormous frequency.
Calculation shows that the fall of 1 mm of rain could launch
about 10^9 potentially microble-bearing droplets from each

1 m^2 of land surface.

THE WAY FORWARD

Launching mechanisms in the fungi are mostly well understood, basidiospore discharge being a notable exception, but they are less well known in the smaller microorganisms. At the same time, much that is known is often neglected, for example, Williams (1979). Speculation to produce verifiable hypotheses is needed to correct some of our current vagueness. As an example the following speculation on the launching of foot-and-mouth virus is seriously offered for test.

Speculations on foot-and-mouth virus

Foot-and-mouth disease (FMD) of farm animals is known to spread rapidly within a herd and measures for preventing spread to other herds are based on controlling movements of animals and men to prevent transport of inoculum by fomites and on immediate slaughter of diseased animals.

Spread of FMD virus within a herd has never been difficult to account for as numerous opportunities exist (though spread may well be aided by inhalation or ingestion of splashed urine droplets). However, accounting for infection between widely separated herds is more difficult. During the 1967 epidemic in Britain it became clear that the virus could be windborne for distances up to 100 km. Spread from an infected herd was mostly downwind and over 90% of the spread between herds was apparently airborne (Smith and Hugh-Jones, 1969; Wright, 1969). Rainfall increased deposition of the virus and may have been essential for infection beyond 10 km.

Sellers and Parker (1969) sampled air in loose boxes. The total output of airborne virus from one infected pig in 5 days was estimated at 10^6 ID50 units and this was 30 times greater than the output from cattle or sheep. Airborne infectious particles in the loose box were mostly greater than 6 μm aerodynamic equivalent diameter and concentrations were greatest before lesions appeared.

Hurst (1968) gave circumstantial evidence that drift could have carried virus across the sea to infect areas on the English coast and he suggested that the virus particles might be attached to spores of moulds such as *Cladosporium* which would be compatible with the 6 μm particles recorded by Sellers and Parker (1969).

Suggested launching mechanisms include breathing and coughing, or rain splash on urine or lesions. Breathing and coughing seem quantitatively inadequate while rain would not occur in a loose box and virus is airborne before lesions appear. A launching hypothesis compatible with these varied facts is thus needed which must also be compatible with the

probability that the most likely airborne infection track
often involved pigs (housed or semi-housed animals) as the
infection source (L.P. Smith, personal communication).
Chance observation suggested the following speculative
scenario.

The standing urinating animal as a potent aerosol generator

(i)   When urination starts, a target film or pool forms
on the ground.

(ii)   As urination continues, incident liquid hitting the
target liquid throws up vast numbers of droplets with a wide
size-spectrum. Research on sizes and numbers is needed but
some clues can be gathered from studies on single falling
drops which, however, are likely to produce more droplets
(volume for volume) than falling urine. Droplets in the
range 50 μm to 200 μm are likely to be projected above the
surface boundary layer of air (Gregory *et al.*, 1959). Their
ascent will usually be aided by convection above the pool
which is initially at body temperature.

(iii) Splash droplets will evaporate in seconds to droplet
nuclei (Wells, 1955). The equivalent diameter of a droplet
nucleus derived from urine with 4% solids will be respectively
40 μm and 15 μm from droplets 200 m and 50 m diameter.

(iv)   Droplets and droplet nuclei derived from splash in
the open will be diffused at once by wind and air currents.
Droplets from housed animals will be carried away by ventilation, the house acting as a convective emitter. Virus is
known to be excreted abundantly in urine, and the droplet
nuclei will have, as it were, a double dose of virus from
the incident drop and from the target film.

(v)   Local spread may well be aided by inhalation or
ingestion of splashed urine droplets. Meanwhile the windborne cloud of virus-laden droplet nuclei will travel downwind, rapidly diluted by eddies and chance dry deposition,
and likely to become relatively innocuous until concentrated
again by rain falling through the diffusing cloud and
"scrubbing out" suspended particles.

(vi)   The efficiency with which rain falling through a
particulate cloud scrubs out suspended particles depends on
rate of rainfall and on both raindrop and particle size,
the greatest efficiency being with raindrops approximately
2 mm diameter (Langmuir, 1948).

(vii) Infection of a distant herd has been attributed to
virus deposited from rain on herbage and fodder, subsequently
ingested (Smith and Hugh-Jones, 1969). However, because
prolonged rain must wash much deposited virus to ground, an
alternative route is proposed here, also based on splash.
Raindrops laden with virus falling over land occupied by the

recipient herd will splash on an already wet surface, each
impact propelling into the air between 1,000 and 5,000 droplets of a size capable of being inhaled (Gregory *et al.*,
1959).

(viii) Such virus-contaminated droplets could be inhaled
in the field directly, or drawn into animal houses by the
ordinary ventilation route based on convection. This suppo-
sition is consonant with the observation that recipient
animals are equally at risk, whether housed or in the field
(Wright, 1969).

This hypothesis offers a substantial if intermittent
source, a transport mechanism, and reconstitution of a virus-
laden aerosol. If substantiated, the need for the usual
precautions during a foot-and-mouth disease outbreak would
remain, but a new and disturbing dimension would be added
to problems of containing the epidemic.

Control of microbial dispersal

The dispersal of microbes has seldom been considered in
the design of machinery and processes by engineers. Many
produce airborne dusts or aerosols which contain microorganisms. On the farm, new methods either have introduced new
hazards or have replaced one hazard by another. For instance,
damp hay can be baled by pick-up balers and stored with
water contents which, if stacked as loose hay, would permit
spontaneous ignition. However because of greater heat losses,
bales heat only sufficiently to allow growth of thermophilic
fungi and actinomycetes and so give a risk of farmers' lung.
Also the microbial dust hazard from the stationary thresher
has been replaced by that from the combine harvester. In
industry, air classification of domestic refuse in sorting
plants creates much airborne dust carrying propagules of
fungi and bacteria.

In the future, microbial launching must be considered
in the design of new machines and processes if microbial
health hazards are to be decreased. Some possibilities
should be immediately apparent, for instance, disposal of
slurry by trickle irrigation rather than by atomisation or
humidification by steam rather than by cold-mist injection
from a static reservoir. Other processes may be more
difficult to control but every process resulting in the
release of microbe-containing dusts or aerosols needs examination by microbiologists and engineers together to
determine ways in which they can be eliminated or minimised.

REFERENCES

Adams, P.A. and Spendlove, J.C. (1970). Coliform aerosols emitted by sewage treatment plants, *Science* 169, 1218-1220.

Anon. (1977). Legionnaires' Disease, *Lancet* 1, 1265-1266.

Arnow, P.M., Fink, J.N., Schlueter, D.P., Barboriak, J.J., Mallison, G., Said, S.I., Martin, S., Unger, G.F., Scanlon, C.T. and Kurup, V.P. (1978). Early detection of hypersensitivity pneumonitis in office workers, *American Journal of Medicine* 64, 236-242.

Banaszak, E.F., Thiede, W.H. and Fink, J.N. (1970). Hypersensitivity pneumonitis due to contamination of an air conditioner, *New England Journal of Medicine* 283, 271-276.

Blanchard, D.C. (1978). Jet drop enrichment of bacteria, virus and dissolved organic material. *Pure and Applied Geophysics* 116, 302-308.

Clark, R.P., Cox, R.N. and Lewis, H.E. (1970). Particle transport within the human micro-environment, *Journal of Physiology* 208, 43-45P.

Coker, E. and Davis, R. (1978). Sewage sludge to land - a need for more research, *Water* 18, 8-11.

Darke, C.S., Knowelden, J., Lacey, J. and Ward, A.M. (1976). Respiratory diseases of workers harvesting grain, *Thorax* 31, 294-302.

Darlow, H.M. (1972). Safety in the microbiological laboratory: an introduction. In "Safety in Microbiology" (Ed. D.A. Shapton and R.G. Board) *Society for Applied Bacteriology Technical Series* Vol.6, pp. 1-20. Academic Press, London.

Darlow, H.M. and Bale, W.R. (1959). Infective hazards of water closets, *Lancet* 1, 1196-1200.

Donaldson, A.J. (1978). Factors influencing the dispersal, survival and deposition of airborne pathogens of farm animals, *Veterinary Bulletin* 48, 83-94.

Donham, K.J., Rubino, M., Thedall, T.D. and Kammermeyer, J. (1977). Potential health hazards to agricultural workers in swine confinement buildings, *Journal of Occupational Medicine* 19, 383-387.

Edmonds, R.L. (Ed.) (1979). "Aerobiology, the ecological systems approach." *US/IBP Synthesis Series* No.10, Dowden, Hutchinson and Ross Inc., Stroudsberg, Pennsylvania.

Faulwetter, R.F. (1917). Wind blown rain- a factor in disease dissemination, *Journal of Agricultural Research* 10, 639-648.

Flannigan, B. (1978). Primary contamination of barley and wheat grain by storage fungi, *Transactions of the British Mycological Society* 71, 37-42.

Gregory, P.H. (1973). "The microbiology of the atmosphere."
2nd Edition. Leonard Hill - Intertext, Aylesbury.

Gregory, P.H., Guthrie, E.J. and Bunce, M.E. (1959).
Experiments on splash dispersal of fungus spores,
Journal of General Microbiology 20, 328-354.

Hirst, J.M, and Stedman, O.J. (1963). Dry liberation of
fungus spores by raindrops, *Journal of General Microbiology* 33, 335-344.

Hurst, G.W. (1968). Foot-and-mouth disease, *Veterinary
Record* 82, 601-604.

Ingold, C.T. (1971). "Fungal spores: their liberation and
dispersal". Clarendon Press, Oxford.

Jones, P.W. (1976). The effect of temperature, solids
content and pH on the survival of *Salmonella* in cattle
slurry, *British Veterinary Journal* 132, 284-293.

Lacey, J. (1974). Occupational factors in allergy. *In*
"Allergy '74" (Ed. M.A. Gauderton and A.W. Frankland),
pp.303-319. Pitman Medical Publishers, London.

Lacey, J. (1980). The microbiology of grain dust. *In*
"Occupational lung disease: focus on grain dust and health".
(Ed. J.A. Dosman and D.A. Cotton), Academic Press,
New York. (in press).

Langmuir, I. (1948). The production of rain by a chain
reaction in cumulus clouds at temperatures above freezing,
Journal of Meteorology 5, 175-192.

Legg, B.J. and Bainbridge, A. (1978). Air movement within a
crop: spore dispersal and deposition. *In* "Plant disease
epidemiology" (Ed. P.H. Scott and A. Bainbridge) pp.104-110.
Blackwells Scientific Publications, Oxford.

Levine, H.B. (1969). Biological properties of fungal
aerosols. *In* "An introduction to experimental aerobiology"
(Ed. R.L. Dimmick and A.B. Akers), pp.340-346 Wiley
Interscience, New York.

Lewis, H.E., Foster, A.R., Mullen, B.J., Cox, R.N. and
Clark, R.P. (1969). Aerodynamics of the human micro-
environment, *Lancet* 1, 1273-1277.

Lidwell, O.M. (1967). Take-off of bacteria and viruses.
In "Airborne Microbes" (Ed. P.H. Gregory and J.L. Monteith)
Symposium of the Society of General Microbiology Vol.17,
pp. 116-137. Cambridge University Press, Cambridge.

Meredith, D.S. (1961). Spore discharge in *Deightoniella
torulosa* (Syd.) Ellis, *Annals of Botany* N.S. 25,
271-278.

Morris, G.K., Patton, C.M., Feeley, J.C., Johnson, S.,
Gorman, G., Martin, W.T., Skaliy, P., Politi, B.D. and
Mallison, G.F. (1979). Isolation of Legionnaires' disease
bacterium from environmental samples, *Annals of Internal
Medicine* 90, 664-666.

MRC Symposium (1977). Humidifier fever, *Thorax* 32, 653-663.

Noble, W.C. and Pitcher, D.G. (1978). Microbial ecology of the human skin, *Advances in Microbial Ecology* 2, 245-289.

Pickering, C.A.C., Moore, W.K.S., Lacey, J., Holford-Strevens, V.C. and Pepys, J. (1976). Investigations of a respiratory disease associated with an air-conditioning system, *Clinical Allergy* 6, 109-118.

Rooks, R. (1954). Airborne *Histoplasma capsulatum* spores, *Science* 119, 385-386.

Sellers, R.F. and Parker, J. (1969). Airborne excretion of foot-and-mouth disease virus, *Journal of Hygiene, Cambridge* 67, 671-677.

Smith, C.V. (1964). Some evidence of the windborne spread of fowl pest, *Meteorological Magazine* 93, 257-263.

Smith, L.P. and Hugh-Jones, M.E. (1969). The weather factor in foot-and-mouth disease epidemics, *Nature, London* 223, 712-715.

Solomon, W.R. (1974). Fungus aerosols arising from cold-mist vaporizers, *Journal of Allergy and Clinical Immunology* 54, 222-228.

Sweet, L.C., Anderson, J.A., Callies, Q.C. and Coates, E.O. (1971). Hypersensitivity pneumonitis related to a home furnace humidifier, *Journal of Allergy and Clinical Immunology* 48, 171-178.

Wedum, A.G. and Kruse, R.H. (1969). Assessment of risk of human infection in the microbiological laboratory, *Miscellaneous Publication* 30, Department of the Army, Fort Detrich, Maryland, U.S.A.

Wells, W.F. (1955). "Airborne contagion and air hygiene: an ecological study of droplet infections". Harvard University Press, Cambridge, Massachusetts.

Williams, B.M. (1979). The survival of pathogens in slurry and the animal health risks from dispersal to land, *A.D.A.S. Quarterley Review* 32, 59-68.

Worthington, A.M. and Cole, R.S. (1897). Impact with liquid surface, studied by the aid of instantaneous photography, *Philosophical Transactions of the Royal Society, A* 189, 137-148.

Wright, P.B. (1969). Effects of wind and precipitation on the spread of foot-and-mouth disease, *Weather* 24, 204-213.

REACTIONS OF MICROORGANISMS, IONS AND MACROMOLECULES
AT INTERFACES

K.C. Marshall

School of Microbiology
The University of New South Wales
Kensington, N.S.W., 2033, Australia

INTRODUCTION

Interfaces, the boundaries between any two phases in heterogeneous systems, exist in a wide variety of forms in all natural microbial habitats. Solid-liquid, gas-liquid, liquid-liquid and, to a limited extent, solid-gas interfaces exert a marked influence on microbial metabolism, growth and interactions in these habitats. In considering the responses by microorganisms to these discontinuities in the aqueous phase, it is necessary to consider the conditions existing both within the bulk aqueous phase and at the interfaces. Heterotrophic microorganisms, in particular, are dependent upon adequate amounts of organic and inorganic nutrients for satisfactory metabolism and growth, although it is possible to distinguish between oligotrophic (autochthonous) microorganisms requiring low nutrient levels and zymogenous microorganisms requiring relatively high nutrient levels (Dahlem Group Report, 1979). The aqueous phase in most natural habitats, including many soils, oceans, lakes and streams, is oligotrophic and, thus, lacks sufficient nutrients to support vigorous (zymogenous) microbial activity. Under these conditions, interfaces become major sites of microbial growth and metabolism.

The physiocochemical properties of interfaces are different from those of the two phases involved. These interfaces serve as areas of nutrient accumulation in the form of inorganic ions, macromolecules, lipids, and other organic molecules. Consequently, an interface can be regarded as a haven for escape from an otherwise nutrient deficient aqueous phase (Marshall, 1976). The most obvious liquid-liquid interfaces are those resulting from the spreading of water-immiscible oils over the surface of a body of water. Air-water interfaces at the surfaces of oceans, lakes and streams are the most common gas-liquid interfaces although mixtures of H_2, CO_2, CH_4 and H_2S may be found in anaerobic soils and sediments and in gas bubbles originating from sediments. Solid-liquid interfaces are extremely varied and include large solid surfaces, such as ship hulls, rocks, aquatic plants and animals, and small surfaces such as plankton, biological debris and clay particles suspended in the aqueous phase.

ION AND MOLECULAR ACCUMULATION AT INTERFACES

Interfaces are charged and, hence, tend to attract ions of the opposite charge (gegen- or counter-ions), which serve to counterbalance the interfacial charge. Some of the counter-ions are held firmly at the interface (the Stern layer), but thermal agitation results in a loose association of most ions and this area is termed the diffuse electrical double layer of counter-ions (Fig.1). Most interfaces in nature possess

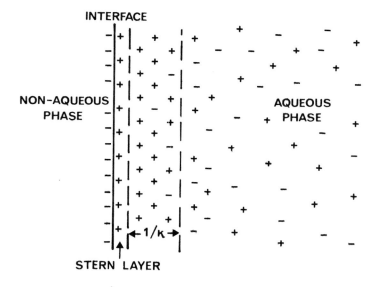

Fig. 1. *Schematic representation of a negatively-charged interface and counter-ions in the Stern and diffuse double layers. The theoretical thickness of the diffuse double layer is given by $1/\kappa$.*

a net-negative charge and, consequently, the majority of the
counter-ions in the diffuse double layer are cations. These
cations form at least part of the inorganic nutrient supply
for microorganisms accumulating at such interfaces. If the
surface charge properties of the interface are altered by the
accumulation of charged macromolecules, as detailed below,
then the types of counter-ions will vary accordingly.

Long chain amphipathic organic molecules, such as lipids
(Odham, Noren, Norkrans, Sodergren and Lofgren, 1978) readily
accumulate at interfaces. Adsorption at bubble gas-liquid
interfaces of amphipathic molecules present in low concentrations in the aqueous phase results in their selective removal
from and transport through the water column to the air-water
interface. This has been termed the adsubble (adsorptive
bubble) process by Lemlich (1972). The behaviour of polyelectrolytes at interfaces is dependent upon the charge density
of the chain and on salt concentration of the bulk solution
phase (Silberberg, 1978). Such polyelectrolytes generally
adsorb poorly from solutions of zero or low added salt and
adsorption increases with increasing salt concentration. The
double helical conformation of native DNA is opened and unwound
to a single stranded form at charged interfaces (Nurnberg
and Valenta, 1978). At strongly-charged interfaces the
adsorption and deconformation of DNA involves all base residues,
whereas at weakly-charged interfaces there is a preferential
adsorption and unwinding of regions rich in A-T pairs (Sequaris,
Valenta, Nurnberg and Malfoy, 1978).

The main factors involved in the adsorption of proteins at
interfaces have been summarized by Norde and Lyklema (1978)
as follows:

(i) Protein: The affinity of a protein for a given negatively
-charged interface increases with increasing hydrophobicity of
the protein, the greater the extent of structural rearrangement
of the protein on adsorption, and the less negatively- (or the
more positively-) charged groups of the protein become located
at the interface.

(ii) Interface: The affinity of a negatively-charged interface for a given protein increases with increasing hydrophobicity of the interface, the lower the interfacial charge, and
the better the surface charge is screened by specifically
adsorbed cations.

Norde and Lyklema (1978) have presented a model of a protein
adsorbed at a solid-liquid interface in which they recognise
three distinct regions in the adsorbed protein (Fig.2).
Region 1 (where 0 x m) contains positive groups of the
protein that are ion-paired to the negative groups at the

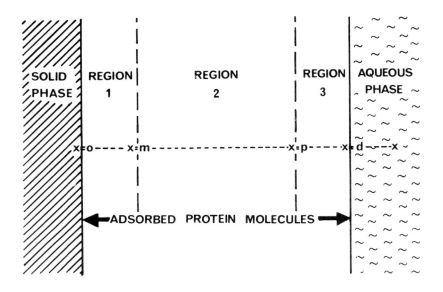

Fig.2. Model of a protein layer adsorbed at an interface showing the three regions described in the text (adapted from Norde and Lyklema, 1978).

solid surface (and is probably in the range of the extension
of the Stern layer). Region 2 (where $m < x \leqslant p$) is assumed
to be devoid of charged groups by analogy with the interior
of globular protein molecules. Region 3 (where $p < x \leqslant d$)
contains the charged groups exposed to the aqueous phase,
extends to the hydrodynamic shear plane, and accommodates
ions bound to the protein. Baier (1980) has shown that
glycoproteins tend to be selectively adsorbed to solid surfaces immersed in natural aqueous habitats, irrespective of
the original physicochemical properties of the surfaces, and
suggested that the configuration of the glycoproteins on the
different surfaces probably reflects the original solid surface properties. Adsorbed, surface-denatured protein at air-
water interfaces forms a protein gel, with the degree of
gelation being highly pH dependent (Warburton, 1978).
   The accumulation of organic molecules at interfaces
results in significant alterations to the interfacial charge,
free, energy, and wetting properties. From the model in Fig.

2, it is obvious that the solid surface will assume a net
surface charge characteristic of the outermost portion
(Region 3) of the protein because this region extends to the
hydrodynamic shear plane. For example, a comparison of
electrophoretic mobilities of various particulate materials
by Neihof and Loeb (1972, 1974) revealed that these solid
surfaces acquired different surface charge properties in
natural seawater from those in an artificial seawater (Table
1). The convergence of mobility values for particulates
suspended in natural seawater was the result of adsorption of
both high and low molecular weight materials to the surface
of the particulates. Destruction of the organics in natural
seawater by UV irradiation resulted in particle mobilities
comparable to those in artificial seawater. Adsorption of
organic and other colloidal materials to the surfaces of
soil bacteria also resulted in significant alterations to
their elctrophoretic properties (Lahav, 1962: Santoro and
Stotzky, 1967: Marshall, 1968).

TABLE 1

*Electrophoretic Mobilities of Particulates in Natural
and Artificial Seawater (adapted from Neihof and
Loeb, 1972)*

Particulate	Electrophoretic mobility ($ms^{-1}V^{-1}cm^{-1}$)	
	Artificial seawater	Natural seawater
Bentonite	− 1.49	− 1.11
Calcium carbonate	− 1.19	− 0.50
Chlorinated wax	− 1.13	− 0.68
Glass	− 0.75	− 0.87
Sephadex	0.00	− 0.15
Anion exchange resin	+ 1.19	− 0.38

The alteration in wettability of solid surfaces is conven-
iently determined by measurement of critical surface tension
(γc) values, using the contact angles (θ) made by a homologous
series of liquids on the surface and defining c as the intercept of the horizontal axis at cos θ = 1 with the extrapolated
straight line plot of cos θ against the surface tension of the
liquids (Baier, Shafrin and Zisman, 1968). Adsorption of
glycoproteins to solid surfaces of different original c
values results in generally more hydrophobic properties with
γc values of about 28-30 dynes cm^{-1} (Baier, 1979, 1980).
Lowering of the surface tension of water on spreading a mono-

layer of an amphipathic molecule is known as the surface
pressure and is determined by the use of a surface balance
(Kjelleberg, Norkrans, Lofgren and Larsson, 1976). Alterations
to the surface tension at natural air-water interfaces can be
determined by the method of Adam (1937) based on the behaviour
at the water surface of drops of a hydrocarbon oil, the
spreading coefficient of which is modified by dissolving
increasing amounts of a more hydrophilic liquid in the oil.

TRANSPORT OF MICROORGANISMS TO INTERFACES

Microorganisms suspended in the aqueous phase are transported over relatively long distances mainly by currents and
wave motion and, as a result, some microorganisms reach the
vicinity of various interfaces. Certain microorganisms possess
mechanisms for increasing their buoyancy, thereby increasing
the opportunities for passive movement upwards through a water
column towards the air-water (and oil-water) interface. These
mechanisms include the development of gas vacuoles (Walsby,
1974), prosthecae (Schmidt, 1971), and varying degrees of cell-
surface hydrophobicity (Mudd and Mudd, 1924; Marshall and
Cruickshank, 1973). Bubble formation, either by wave motion
or by gas production in sediments, provides an effective
means for the adsorption of bacteria and their transport to
air-water interfaces (Carlucci and Williams, 1965). The
preferential concentration of certain bacteria by this process,
probably results from the greater adsorption of relatively
hydrophobic bacteria at the bubble surface. Over shorter
distances, chemotactic responses to the nutrient gradient
established at interfaces will provide a selective advantage
to appropriate motile bacteria (Young and Mitchell, 1973).

MICROBIAL RESPONSES AT INTERFACES

When a negatively-charged microorganism reaches the vicinity
of an interface, it becomes subjected to a variety of short
range forces that tend to hold the microorganism at or near
the interface. Where the interface possesses or assumes a net
positive charge the microorganisms will be drawn to the interface by electrostatic attraction. Extremely hydrophobic
bacteria are attracted to an air-water interface or, more
correctly, tend to be rejected from the water phase. In an
oil-water system, such bacteria partition into the oil phase
(Mudd and Mudd, 1924). The possession of a relatively hydrophobic pole by some otherwise hydrophilic bacteria leads to
a common perpendicular mode of orientation by these bacteria
at all types of interfaces (Marshall and Cruickshank, 1973),
as illustrated for a budding, prosthecate hyphomicrobium in
Fig.3. This mode of orientation by bacteria at interfaces is
surprisingly frequent in natural habitats (Sieburth, 1975).

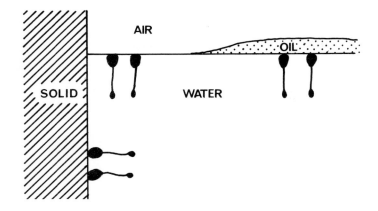

*Fig.3. Schematic representation of the perpendicular orient-
ation of a budding, prosthecate hyphomicrobium at solid-water,
air-water and oil-water interfaces.*

The advantages of this mode of orientation in the process of
manganese deposition by hyphomicrobia attached to the surface
of hydro-electric pipelines has been described by Tyler and
Marshall (1967).

At first glance, there appears to be a problem in explaining how a negatively-charged microorganism can become associated
with a negatively-charged interface. A diffuse double layer
of counter-ions (cations) will be associated with both the
microorganism and the appropriate interface (Fig.4) and, as
the microorganism approaches the interface, the cation clouds
will interact tending to produce a mutual repulsion effect
between the interacting surfaces. This is precisely what
happens at low salt concentrations (Fig.5, top) where the
diffuse double layer repulsion energies exceed the van der
Waals attraction energies at all levels of particle
separation except if both interfaces ever achieve contact
(primary attraction minimum). Under these conditions microorganisms are actively repelled from the interface. Marshall,
Stout and Mitchell (1971) reported that all cells of the non-
motile, marine *Achromobacter* R8 were repelled from a glass
surface at concentrations of $5 \times 10^{-4.5}$M for a uni-univalent
salt (NaCl) and $5 \times 10^{-3.4}$M for a di-divalent salt ($MgSO_4$). At
higher salt concentrations (Fig.5, bottom) the diffuse double
layer is more compressed, giving a resultant energy of interaction with a secondary attraction minimum at moderate particle
separations and a significant repulsion barrier at shorter
particle separations. Microorganisms attracted to the secondary

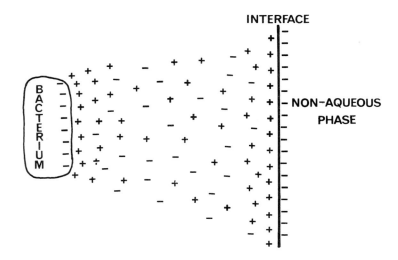

Fig.4. Schematic reperesentation of the interaction between potentially overlapping cation clouds accompanying a negatively-charged bacterium as it approaches a negatively-charged interface.

minimum are reversibly sorbed (Marshall *et al.*, 1971), and can escape from the interface on the application of an appropriate shear force such as that generated by flagellar activity. For instance, the kinetic energy of a marine pseudomonad (5.45 x 10^{-18} ergs) is sufficient to enable the bacteria to break away from the surface, but is insufficient to overcome the repulsion energy barrier (5.0 x 10^{-12} ergs) existing near a solid surface immersed in seawater (Marshall *et al.*, 1971). Gingell (1978) has observed the behaviour of red blood cells at a hexadecane-water interface and found that all cells attached to the inter-face at a concentration of 10^{-3} M NaCl, but most cells detached at 2 x 10^{-4} M NaCl. Using interference microscopy, Gingell estimated that the point of closest contact between red blood cells and a solid-water interface was about 1500°A at a concentration of 2 x 10^{-4} M NaCl, about 1000°A at 10^{-3} M NaCl (this may correspond to the secondary minimum), and complete contact (primary medium ?) at 0.2 M NaCl. Similar results were obtained at an oil-water interface.

Provided that the microorganisms are not subjected to

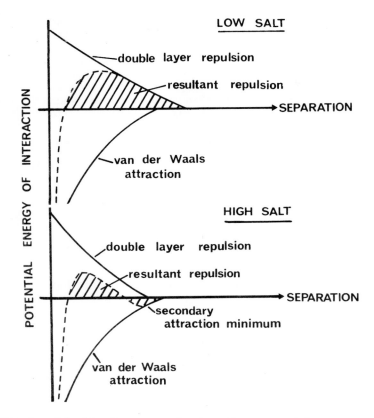

*Fig.5. Idealized curves showing the potential energy of
interaction (either repulsion or attraction) between a
bacterium and an interface of like charge at low (top) and
high (bottom) electrolyte concentrations.*

strong shear forces near an interface, attraction to and
reversible sorption at the point of the secondary minimum
should be sufficient for the organisms to benefit from the
enriched nutrient status of the interface. Consequently,
one can confidently expect significant metabolism and growth
of microorganisms at the interface leading to succession on a
temporal basis. Such behaviour may help explain the apparent
anomaly in the observed paucity of microorganisms in association with detritus in the open, nutrient-poor oceans
(Williams, 1970; Wiebe and Pomeroy, 1972; Hobbie, Holm-Hansen,
Packard, Pomeroy, Sheldon, Thomas and Wiebe 1972) The
filtration techniques used in these studies may provide

sufficient shear to dislodge reversibly sorbed microorganisms from the particulate material, thereby giving a possibly false representation of the natural system.

In natural conditions where consistent or spasmodic shear forces exist near interfaces, it is obviously an advantage for a microorganism to become more firmly associated with the interface. This must certainly occur in some instances where electrostatic and hydrophobic interactions are involved. In the case of the negatively-charged bacterium reversibly sorbed at the site of the secondary minimum, more permanent contact with the interface is achieved by the process of polymer bridging (Marshall *et al.*, 1971; Fletcher and Floodgate, 1973). This involves the production of extracellular polymers by the microorganisms and the form of attachment may be one of three possible types (Marshall, 1976).

Specific permanent adhesion

This occurs when microorganisms only attach to specific surfaces and must involve interactions between complementary molecular structures produced by the microorganism and the attachment surface (Jones, 1977). This form of adhesion has been described only in organism-organism interactions.

Non-specific permanent adhesion

Many microorganisms are capable of permanent adhesion to a wide variety of solid surfaces. Little is known of the chemical structure of the bridging polymers involved, although Corpe (1970) and Fletcher and Floodgate (1973) have identified acidic polysaccharides in the attachment of marine bacteria to solid surfaces.

Detailed descriptions of permanent adhesion by both specific and non-specific mechanisms are given elsewhere and will not be discussed further in this paper.

Temporary adhesion

This describes the process of adhesion of gliding bacteria at solid-water and air-water interfaces where the organisms attach to but are able to migrate across the interface. Humphrey, Dickson and Marshall (1979) have described adhesion by *Flexibacter* BH3 in terms of the Stefan adhesion principle. A slime produced by the bacterium functions as a Stefan adhesive by increasing the adhesiveness of the bacterium (high force preventing separation) but allowing translational movement across the interface (low horizontal drag). Obviously, the slime must have physicochemical properties different from those responsible for the permanent attachment of other bacteria to fixed sites on surfaces. The extracell-

ular slime produced by this *Flexibacter* BH3 is a glycoprotein
with viscous properties characteristic of a linear polymer.
The lack of a significant amount of cross-linkage between
polymer chains should eliminate any cementation between the
bacterium and the surface, but also should provide a suitable
lubrication effect to reduce the horizontal drag as the
organism moves over the surface. The presence of a viscous
fluid rather than water between the bacterium and the surface
results in a substantial increase in the force preventing
separation from the surface (Humphrey *et al.*, 1979).

CONCLUSIONS

In terms of the ecology of microorganisms, interfaces can
be regarded as aqueous phase discontinuities which, because
of their unusual properties, have a major influence on the
activity of and interactions between microorganisms in natural
habitats. The selective accumulation of ions, macomolecules
and other organic molecules at interfaces alters the physicochemical properties of the interfaces and plays a significant
role in the reactions of microorganisms at such interfaces.
Microorganisms are transported to interfaces by a variety of
physical, physicochemical and biological mechanisms. In
nutrient-deficient (oligotrophic) habitats, association with
the nutrients accumulated at interfaces provides the only
means for vigorous growth of zymogenous microorganisms.

REFERENCES

Adam, N.K. (1937). A rapid method for determining the
    lowering of tension of exposed water surfaces, with
    some observations on the surface tension of the sea
    and of inland waters, *Proceedings of the Royal Society,
    Series B* 122, 134-139.
Baier, R.E., Shafrin, E.G. and Zisman, W.A. (1968). Ad-
    hesion: mechanisms that assist or impede it, *Science*
    162, 1360-1368.
Baier, R.E. (1979). Substrate influences on adhesion of
    microorganisms and their resultant new surface properties.
    In "Adsorption of Microorganisms to Surfaces" (Ed. G
    Bitton and K.C. Marshall), Wiley, New York, in press.
Baier, R.E. (1980). Early events of micro-biofouling of all
    heat transfer equipment. *In* "Proceedings of the Inter-
    national Conference on the Fouling of Heat Transfer
    Equipment" (Ed. J.C. Knudson and E.F.C. Somerscales)
    Hemisphere Publishing Corp., Washington, in press.
Carlucci, A.F. and Williams, P.M. (1965). Concentration of
    bacteria from sea water by bubble scavenging, *Journal du
    Conseil Permanent International pour l'Exploration de la
    Mer* 30, 28-33.

Corpe, W.A. (1970). An acid polysaccharide produced by a
    primary film-forming marine bacterium. *Developments in
    Industrial Microbiology* 11, 402-412.
Dahlem Group Report (1979). Life under conditions of low
    nutrient concentrations. *In* "Proceedings of the Dahlem
    Workshop on Strategy of Life in Extreme Environments"
    (Ed. M. Shilo), Dahlem Konferenzen, Berlin, in press.
Fletcher, M. and Floodgate, G.D. (1973). An electron-
    microscopic demonstration of an acidic polysaccharide
    involved in the adhesion of a marine bacterium to a solid
    surface, *Journal of General Microbiology* 74, 325-334.
Gingell, D. (1978). Experimental study of cell adhesion.
    In "Ions in Macromolecular and Biological Systems" (Ed.
    D.H. Everett and B. Vincent), Colston Papers No. 29,
    pp. 237-241, Scientechnica, Bristol.
Hobbie, J.E., Holm-Hansen, O., Packard, T.T., Pomeroy, L.R.,
    Sheldon, R.W., Thomas, J.P. and Wiebe W.J. (1972). A
    study of the distribution and activity of microorganisms
    in ocean water, *Limnology and Oceanography* 17, 544-555.
Humphrey, B.A., Dickson, M.R. and Marshall, K.C. (1979).
    Physicochemical and in situ observations on the adhesion
    of gliding bacteria to surfaces, *Archives of Microbiology*
    120, 231-238.
Jones, G.W. (1977). The attachment of bacteria to the
    surfaces of animal cells. *In* "Microbial Interactions"
    (Ed. J.C. Reissig), Receptors and Recognition, Series B,
    Vol. 3, pp. 140-176, Chapman and Hall, London.
Kjelleberg, S., Norkrans, B., Lofgren, H. and Larsson, K.
    (1976). A surface balance study of the interaction
    between microorganisms and lipid monolayer at the air/
    water interface, *Applied and Environmental Microbiology*
    31, 609-611.
Lahav, N. (1962). Adsorption of sodium bentonite particles
    on *Bacillus subtilis, Plant and Soil* 17, 191-208.
Lemlich, R. (1972). Adsubble processes: foam fractionation
    and bubble fractionation, *Journal of Geophysical Research*
    77, 5204-5210.
Marshall, K.C. (1968). Interaction between colloidal mont-
    morillonite and cells of Rhizobium species with different
    ionogenic surfaces, *Biochimica et Biophysica Acta* 156,
    179-186.
Marshall, K.C., Stout, R. and Mitchell, R. (1971). Mech-
    anism of the initial events in the sorption of marine
    bacteria to surfaces, *Journal of General Microbiology*
    68, 337-348.
Marshall, K.C. and Cruickshank, R.H. (1973). Cell surface
    hydrophobicity and the orientation of certain bacteria at

interfaces, *Archiv für Mikrobiologie* 91, 29-40.

Marshall, K.C. (1976). "Interfaces in Microbial Ecology" Harvard Univ. Press, Cambridge, Mass.

Mudd, S. and Mudd, E.B.H. (1924). The penetration of bacteria through capillary spaces. IV. A kinetic mechanism in interfaces, *Journal of Experimental Medicine* 40, 633-645.

Neihof, R.A. and Loeb, G.I. (1972). The surface charge of particulate matter in seawater, *Limnology and Oceanography* 17, 7-16.

Neihof, R. and Loeb, G. (1974). Dissolved organic matter in seawater and the electric charge of immersed surfaces, *Journal of Marine Research* 32, 5-12.

Norde, W. and Lyklema, J. (1978). Adsorption of proteins from aqueous solution on negatively charged polystyrene surfaces. *In* "Ions in Macromolecular and Biological Systems" (Ed. D.H. Everett and B. Vincent), Colston Papers No.29, pp. 11-35, Scientechnica, Bristol.

Nurnberg, H.W. and Valenta, P. (1978). Bioelectrochemical behaviour and deconformation of native DNA at charged interfaces. *In* "Ions in Macromolecular and Biological Systems" (Ed. D.H. Everett and B. Vincent) Colston Papers No.29, pp. 201-229, Scientechnica, Bristol.

Odham, G., Noren, B., Norkrans, B., Sodergren, A. and Lofgren, H. (1978). Biological and chemical aspects of the aquatic lipid surface microlayer, *Progress in the Chemistry of Fats and other Lipids* 16, 31-44.

Santoro, T. and Stotzky, O. (1967). Effect of cations and pH on the electrophoretic mobility of microbial cells and clay minerals, *Bacteriological Proceedings* A24.

Schmidt, J.M. (1971). Prosthecate bacteria, *Annual Review of Microbiology* 25, 93-110.

Sequaris, J.M., Valenta, P., Nurnberg, H.W. and Malfoy, B. (1978). ON the interfacial behaviour of double stranded polynucleotides in alkaline solution. *In* "Ions in Macromolecular and Biological Systems" (Ed. D.H. Everett and B. Vincent), Colston Papers No.29, pp. 230-234. Scientechnica, Bristol.

Sieburth, J.McN. (1975). "Microbial Seascapes". University Park Press, Baltimore.

Silberberg, A. (1978). Polyelectrolyte behaviour at interfaces. *In* "Ions in Macromolecular and Biological Systems" (Ed. D.H. Everett and B. Vincent), Colston Papers No.29, pp. 1-8, Scientechnica, Bristol.

Tyler, P.A. and Marshall, K.C. (1967). Form and function in manganese-oxidizing bacteria, *Archiv für Mikrobiologie* 56, 344-353.

Walsby, A.E. (1974). The identification of gas vacuoles and their abundance in the hypolimnetic bacteria of Arco Lake, Minnesota, *Microbial Ecology* 1, 51-61.

Warburton, B. (1978). The effect of hydrogen ion concentration on the rate of gelation of thin collagen films at the air-water interface. *In* "Ions in Macromolecular and Biological Systems" (Ed. D.H. Everett and B. Vincent), Colston Papers No.29, pp. 273-281, Scientechnica, Bristol.

Wiebe, W.J. and Pomeroy, L.R. (1972). Microorganisms and their association with aggregates and detritus in the sea: a microscopic study, *Memorie Dell 'Instituto Italiano di Idorbiologie Dott Marco de Marchi 29 Supplement* 225-352..

Williams, P.J. LeB. (1970). Heterotrophic utilization of dissolved organic compounds in the sea. I. Size distribution of population and relationshiop between respiration and incorporation of growth substances, *Journal of the Marine Biological Association, United Kingdom* 50, 859-870.

Young, L.Y. and Mitchell, R. (1973). The role of chemotactic responses in primary film formation. *In* "Proceedings of the Third International Congress on Marine Corrosion and Fouling" (Ed. R.F. Acker, B.F. Brown, J.R. DePalma and W.P. Iverson), pp. 617-624, Northwestern Univ. Press, Evanston, Illinois.

BIODEGRADATION: SOME ATTITUDES AND STRATEGIES OF MICROORGANISMS AND MICROBIOLOGISTS

Alan T. Bull

Department of Applied Biology
University of Wales Institute of Science and Technology
Cardiff CF1 3NU, Wales

PREAMBLE

Biodegradation means the biological - usually microbiological - transformation of chemicals. The process can result in complete mineralization of the substance or lead to the loss of some or all of its characteristic properties. Biodegradation *may* be accompanied by utilisation of the chemical as a source of energy or nutrient; likewise it *may* result in detoxification with reference either to the transforming population, a specific target organism or organisms in general. The number of terms used to describe chemical transformations, particularly of xenobiotic chemicals (environmental foreign compounds) such as pesticides and pollutants of industrial origin, is large and frequently confusing and the reader might refer to Stafford and Callely (1977) and Hill (1978) for additional guidance. The antithesis of biodegradation is persistence: use of this term is best made in a pragmatic way such that a chemical is defined as being persistent in a given situation if it retains its identity for an arbitrary time. Such usage, as Giger and Roberts (1978) have commented, avoids the conceptual difficulty of differentiating between inherent resistance of a chemical to degradation (recalcitrance) and persistence occasioned by an environment that creates metabolic impotence in an otherwise competent population. Of course, there are several basic biochemical reasons, such as deficiencies of appropriate uptake, catabolism and control mechanisms, that may make certain chemicals resistant to biodegradation and these have been appraised recently by Slater and Somerville (1979). Nevertheless, the terms persistence and recalcitrance will be used synonymously in this essay. Unfortunately, the meanings of biodegradation and persistence themselves are blurred by considerations of

the extent and rate of transformation. Thus, not infrequently
are parent chemicals metabolized to more persistent products,
while persistence itself can only be related to an arbitrary
time scale on which chemicals like DDT mark one end of a
continum. In one context, however, a quantative definition
of persistence is necessary, namely for official regulatory
purposes and such a definition is best founded on the test
method being applied (Tiedje, 1979).

Chemicals that are considered to be persistent in most
environments are of two types. First there are natural organic
compounds including paleobiochemicals such as humus, lignin,
tannin, melanin and complexes of inherently biodegradable
materials with polyaromatics, for example, lignified wood,
melanized fungal walls, tanned proteins. Secondly there is
the enormous variety of synthetic organic chemicals (xenobiotics)
that enter the environment either deliberately, via agricultural
practice for example, or less so as a consequence of industrial-
isation. Xenobiotics that have caused most concern include
pesticides, detergents, coolants, polymers, resins and solvents
but the number of such chemicals is ever expanding as also are
the purposes for which they are being used. This latter group
of chemicals will be the major concern of the present discussion.

It is perhaps facile to question the reasons for studying
biodegradation but it is worth brief mention. The fate and
effect of xenobiotics in the biosphere has generated massive
world-wide research programmes focussed directly or otherwise
on environmental quality. Similarly, biodegradative activities
in natural material and energy cycling constitute one of the
most important processes in water, sediment, soil and other
ecosystems. Moreover, these two approaches are usefully
integrated in analyses of the effect of herbicides on organic
matter turnover in soil, for example, or in investigations of
the effect of readily assimilable organic substrates on the
persistence of xenobiotics. Relatively little information is
available on the interaction of xenobiotics, organic matter
turnover and soil factors but the effects can be dramatic and
unexpected. We have observed that the herbicide dalapon and
the fungicide triphenyltin acetate (TPTA) when applied at
100ppm to aerated pasture soil notably enhanced the rate of
humus fraction turnover. The half life of this fraction was
reduced from 430 to 150 days on application of the pesticides
and to 53 days (dalapon) and 75 days (TPTA) when the soil was
amended with fungal mycelia (0.5% w/w) (H.S.N. Hussain and
A.T. Bull, unpublished observations). Present impetus for
the unflagging concern with biodegradation is at least two-
fold. On the one hand stricter national and international
regulations relating to public health and safety, as typified
by the U.S. Toxic Substances Control Act of 1976, necessitate

the development of tests and guidelines for industry and such tests need to be based on modern ideas and information. In this regard it is prudent to focus effort onto major classes of widely disseminated chemicals, especially those where evidence or suspicion of toxigenicity and bioaccumulation exists. On the other hand, exploitation of photosynthetically renewable raw materials as feedstocks for biotechnological processes is demanding a much greater understanding of cellulose, hemicellulose and lignin deploymerization (Bull, Ellwood and Ratledge, 1979) by microorganisms. Finally, waste and effluent processing can have implications both for environmental management and biotechnology; in both cases basic understanding of biodegradation is essential for the design of effective processes.

APOLOGIA

Assuming that we know what biodegradation is and that it is a subject warranting study, we need to address the question of how it should be investigated. An attempt to answer this question is the chief intent of this essay; major concern will be for the establishment of principles, not the cataloguing of data and the hope is that it will catalyze and not constrain thought.

It is now exactly a century ago since Robert Koch published his pioneering papers on the aetiology of infectious diseases and the technology for isolating and growing pure cultures of the causative organisms. It is my contention that most contempory attitudes and experimental strategies found in the study of biodegradation stem from the Koch tradition of microbiology. Boldy stated the approach is to study a single organism which has been selected on the basis of an ability to utilise a given organic compound as a sole source of carbon and/or energy. Although much has been accomplished via this approach, it contains an unfortunate element of opportunism and perforce engenders a rather simplistic and sometimes false view of biodegradation and, at the same time, has tended to obscure many areas of necessary enquiry. Only very exceptionally do monospecies populations of microorganisms growing on a single substrate develop in nature. If any or all of the above are considered to be overstatements, the intention has been deliberate in order to highlight problems and encourage new thinking.

Biodegradation is a field of activity that can be examined from several viewpoints, among them being microbiological, biochemical, engineering and ecological. Current needs suggest that a multidisciplinary approach will be rewarding. But whatever approach is taken, work on biodegradation normally assumes some relationship to natural or man-made environments.

Unfortunately such environments are difficult milieu in which
to study chemical transformations and a large proportion of
investigations have been of the *in vitro* type alluded to
above. Clearly there are inherent problems both in field
and laboratory studies and this matter will be discussed
briefly later in this essay. The position adopted here will
be one of environmentally relevant laboratory modelling from
which cautious extrapolation to the environment can be made
and in which the generation of testable hypotheses is a major
objective.

CONCEPTUAL AND EXPERIMENTAL CONSTRAINTS

This discussion begins with the premise that most investigations
of biodegradation have been based on the Koch pure culture
tradition and allied to the metabolism of single substrates in
aerobic, batch systems in which environment conditions have
been made constant and the physical state of the experimental
system has been kept as homogeneous as possible.

Pure Culture Analyses

Undoubtedly pure culture studies have provided many data
that are relevant to the understanding of biodegradation in
the environment. Thus a knowledge of the metabolism of
different microorganisms can give clues to:
    i)   the alternative strategies available for biodegradation,
    ii)   the identification of degradation products,
    iii) the biochemical interdependency within microbial
communities effecting degradation and
    iv)   the mechanisms by which microorganisms evolve in order
to metabolise and/or to resist xenobiotic chemicals (Gibson,
1979). However, there are several serious drawbacks to the
pure culture approach: the biodegradative capacity of microbial
communities (consortia; mixed cultures) may be greater, both
quantatively and qualitatively, than pure cultures, while resort
to pure cultures denies analysis of genetic exchange between
populations that may lead to altered biodegradative activity.
The herbicide dalapon (2,2-dichloropropionic acid) is
readily biodegradable. A seven-membered microbial community
isolated by Senior, Bull and Slater (1976) contained two
bacteria and one fungus that could grow independently on
dalapon. The maximum rates of dalapon degradation by the
bacteria alone were 11.5 and 7.7 gl^{-1} d^{-1} whereas the
community rate was 14 gl^{-1} day^{-1} which was the highest reported
rate of pesticide degradation. Similarly, the degree of
resistance to toxic substances may be significantly higher in
microbial communities than in the individual species comprising
them. A recent illustration of this phenomenon has been
recorded by Charley and Bull (1979) in relation to bacterial

resistance to the bioaccumulation of silver.

An increasing number of cases are known where widely
different xenobiotics are degraded by communities of organisms.
The list includes herbicides: trichloroacetic acid (Jensen,
1957) and propham (isopropylphenylcarbamate) (McClure, 1973);
insecticides: diazinon (O,O-diethyl O-2-isopropyl-4-methyl-
6-pyrimidinyl thiophosphate (Gunner and Zuckerman, 1968) and
parathion (O,O-diethyl O-p-nitrophenyl phosphorothionate)
(Munnecke and Hsieh, 1974); ixodicides: amitraz (1,5-di-(2,
4-dimethylphenyl)-3-methyl-1,3,5-triazapenta-1,4-diene) (Baker
and Woods, 1977); detergents (linear alkylbenzenesulphonates)
(Johanides and Hršak, 1976); and cycloparaffins (cyclohexane)
(Beam and Perry, 1974; Stirling, Watkinson and Higgins, 1976).
In some cases obligatory association of organisms is not
required for mineralization of the compound: here the
association is based on nutrient cross-feeding that can be
satisfied in pure cultures by medium supplementation, such
is the case in TCA and dalapon degradation referred to above.
However, biodegradation of diazinon, parathion, amitraz and
alkylbenzenesulphonates necessitated the synergistic metabolic
attack of two or more organisms. Most of the biodegrading
communities studied to date are composed of or dominated by
bacteria (this may simply reflect isolation procedures!) but
a few have greater complexity and contain bacteria and fungi
(dalapon; Senior *et al.*, 1976) or bacteria, actinomycetes
and fungi (propham; McClure, 1973). Such facts prompt another
general comment on biodegradation research; there has been a
disproportionate emphasis on bacteria. Certainly bacteria
are the most abundant microorganisms in the ecosystems of
interest but, for example, filamentous fungi normally constitute
the largest fraction of the total microbial biomass in soils.
Again, most studies have been made with organisms isolated
from terrestrial or freshwater locations and biodegradation
mechanisms, rates and populations operating in estuarine,
brackish and marine environments are very much less well
documented. Yet 99% of the biosphere is made up of marine
environments (Sverdrup, Johnson and Fleming, 1942) which in
turn contain the greatest range and concentration of natural
organohalides (Gibson, 1979).

A priori the acquisition of catabolic plasmids appears to
be an attractive hypothesis for explaining, at least in part,
the evolution of metabolic capabilities and, moreover, the
greater the gene pool in a microbial community, the more
likely it is that capabilities can evolve for the degradation
of unusual or novel substrates. Chromosomal gene mutations
coupled to the interchange of genes between chromosomes and
plasmids could provide a rapid means of spreading new metabolic
activities once they had arisen in an individual and ensure

the retention of such genes in a population despite possible
loss of the original activity (Clarke, 1978; Slater and Godwin,
this volume, pp. 137). Within the last two or three years
several reports give credence to these suggestions with respect
to the biodegradation of xenobiotics. The ability of *Alcaligenes
paradoxus* to degrade the herbicide 2,4-D (2,4-dichlorophenoxy-
acetic acid) is encoded by the conjugal plasmid pJP1 (Pemberton
and Fisher, 1977; Fisher, Appleton and Pembleton, 1978) thereby
enabling the conversion of 2,4-D to 2,4-dichlorophenol. Catabolism
of the latter is probably encoded on chromosomal genes. Knackmuss
has constructed strains of *Pseudomonas* that can mineralise
haloaromatics. Thus, the TOL plasmid of *Pseudomonas putida*,
which encodes benzoate-1,2-dioxygenase, was transferred to
Pseudomonas sp. B13 and the exconjugant was able to utilise
3-chloroaniline (Reineke and Knackmuss, 1979). Neither of the
wild-type strains alone was able to utilise haloaromatics:
Pseudomonas sp. B13 could not hydroxylate chlorobenzoates while
P. putida could not catabolise chlorocatechols. The significance
of Knackmuss' work is greatly extended by subsequent demonstration
of this phenomenon in continuous cultures in the presence of a
natural soil microflora. The conversion of alkylbenenesulphonates
to the corresponding alkylcatechols and their catabolism via *meta*
aromatic ring cleavage in *P. testosteroni* is specified by the ASL
plasmid (Sagoo and Cain, 1979). Plasmid transfer occurred over a
wide range of environmental conditions and exconjugants were
formed with *P. arvilla* strains unable to degrade the surfactants.
Once the ASL plasmid was in *P. arvilla* it could be transferred to
other *Pseudomonas* species. It seems likely that the ability to
degrade persistent biogenic substances such as lignin might also
be specified by plasmid genes. Recently Salkinoja-Salonen and
her colleagues have proved the involvement of plasmids in the
degradation of lignin-derived compounds such as vanilic, veratic,
ferulic, syringic and protocatechuic acids (Salkinoja-Salonen,
Väisänen and Paterson, 1979). At least two plasmid encoded pathways have been revealed in bacterium K17 (proposed genus *Ligno-
bacter*), one for the *ortho*-hydroxylation of *p*-hydroxybenzoate,
the other for salicylate metabolism (the latter plasmid is
distinct from the SAL plasmid of *P. putida*).

Single Substrate Systems

There are peremptory reasons for analysing mixed substrate
systems in the context of biodegradation. By mixed substrate
I mean multiple sources of a given nutrient (usually but not
invariably carbon) which, in chemical terms, may be homolgous
or nonhomolgous compounds. A significant proportion of total
biodegradative activity, particularly of xenobiotics, involves
cometabolism (Jensen, 1963; Horvath, 1972), a process in which

metabolism of a substrate is not accompanied by nutrient
assimilation or energy conservation. The substrate being
cometabolised is thought to be acted upon by an enzyme having
some other physiological role but the product is not further
metabolised by the organism in question. The significance
of cometabolism in the degradation of xenobiotics in natural
environments has rarely been determined in quantitative terms.
However, the significance may be large as Bourquin (1975)
has observed for malathion (diethyl mercaptosuccinyl O,O-dimethyl
dithiophosphate) in artificial salt marsh ecosystems. In both
water column and sediment bacteria that utilised malathion as
a sole carbon source comprised only about 10% of the bacterial
population that could degrade the insecticide.

Generation of cometabolic products can have several consequences:

i)   They may be more toxic to microorganisms than the parent
chemical and some examples are: 1-naphthol from the insecticide
carbaryl (1-naphthyl methylcarbamate) (Bollag and Liu, 1971),
3,5-dichlorocatechol from 2,3,6-trichlorobenzoate, (Horvath,
1971) and p-nitrophenol from parathion (Munnecke and Hsieh,
1974).

ii) Cometabolic transformations may lead to increased
recalcitrance, for example, the herbicide propanil (3,4-
dichloropropionamilide) degrades in soil to 3,4-dichloroaniline
and the latter undergoes peroxidase condensation to the
corresponding azo compound (Bartha and Pramer, 1967). The
formation of recalcitrant polymers via cometabolism, such as
occurs in the transformation of 2,4,6-trinitrotoluene (Carpenter
McCormick, Cornell and Kaplan, 1978), may be relatively common
in the biodegradation of aromatic compounds and merits further
examination.

iii) The cometabolic products of one organism may provide a
growth substrate for one or more other organisms, for example,
p-chlorophenyl acetic acid from DDT (1,1,1-trichloro-2,2-bis
(p-chlorophenyl) ethane (Pfaender and Alexander, 1972) and
p-nitrophenol from parathion (Daughton and Hsieh, 1977).

The pattern of biodegradation can be modified by the
presence of readily metabolised substrates. The insecticide
mexacarbate (4-dimethylamino-3,5-xylyl methylcarbamate) is
cometabolised to a decarbamylated product in the presence of
carbohydrates but in their absence is dealkylated and
catabolised at an increased rate (Benezet and Matsumura, 1974).
Chou and Bohonos (1979) also observed a depressing effect of
organic substrates on the biodegradation of heterocyclic aromatics
by mixed microbial populations. The implications of these
studies for biodegradation in waste treatment plants that
receive industrial and domestic loads and ecosystems that are
supplied with organic nutrients need no emphasis. However,
caution is necessary before extrapolating too widely from such

observations. Thus, a wide range of phenylurea, carbamate, phenoxy acid and triazene herbicides showed accelerated degradation in the soil following application of microbial nutrients (McClure, 1970). The presence of a readily assimilated substrate may enable an organism to degrade an otherwise recalcitrant compound by furnishing energy or reductant necessary for enzyme synthesis or enzyme activity. Work on aromatic catabolism from our laboratory illustrates this latter point. Under conditions of carbon-limited growth in a chemostat, the addition of glucose (a catabolite repressor) to a p-hydroxybenzoate medium stimulated the synthesis of all enzymes involved in metabolising 3-oxoadipate in *Rhodotorula mucilaginosa* from between 20% (hydrobenzoate mono-oxygenase) to 74% (carboxymuconate cyclase) (Huber, Street, Bull, Cook and Cain, 1975). Similar but even greater enhancement of synthesis was observed in *Aspergillus nidulans* growing under identical conditions. Furthermore, whereas glucose concentrations above 0.3 to 0.5 mM caused progressively greater catabolite repression of synthesis, 10-fold higher concentrations of pyruvate sustained the enhanced rates of synthesis (M. Karimian, R.B. Cain and A.T. Bull, unpublished observations). We have interpreted these data in terms of an increased availability of NAD(P)H which is required in the initial hydroxylation step. The relevance of these findings to lignin biodegradation is obvious and subsequent research by Kirk's group supports this view. White rot fungi such as *Phanerochaete chrysosporium* could not utilise or degrade lignin unless a growth substrate (cellulose, glucose) was provided; moreover, lignin-related aromatics did not serve as sole carbon and energy sources for growth (Kirk, Connors and Zeikus, 1976; Kirk, Yang and Keyser, 1978). It can be argued that such is the stability of lignin that there will be a net consumption of energy associated with its biodegradation and for this reason an auxiliary substrate is required to be provided.

Formate is oxidised constitutively by the marine bacterium *Beneckea natriegens* when it is fed to glucose or glycerol limited chemostat cultures. However, formate cannot serve as a sole source of carbon and energy for the growth of this species. The growth yield on glucose increased from 83 to 92 g mole^{-1} when formate was present and formate-carbon was not incorporated into biomass (Linton, Griffiths, Harrison and Bull, 1979). During growth on glucose-formate mixtures, energy (reductant) generated from formate oxidation has a sparing effect on glucose catabolism such that a greater proportion is diverted to biomass synthesis. Consequently a switch from energy limited to carbon limited growth may occur on the addition of formate (Linton and Stephenson, 1978). Because both chemostats and turbidostats select for organisms

having higher growth yields, considerations of this type of
data are relevant to the enrichment isolation of biodegrading
populations (see below).

A final reason for examining mixed substrate systems is the
opportunity they afford for studying adaptation of organisms
to novel substrates. The dalapon community isolated in our
laboratory is a good example of this point (Senior *et al.*,
1976). Within the community was a strain of *Pseudomonas
putida* that could not grow on dalapon. Following nearly 3000h
of growth in a chemostat a mutant of this bacterium that could
be grown on dalapon was selected. The wild-type parent was
then grown as a monoculture in a carbon limited chemostat on
a propionate-dalapon (1:5) mixture and in several such
experiments dalapon utilising mutants were selected. Subsequent
analyses (Slater, Lovatt, Weightman, Senior and Bull, 1979)
revealed that the ability of the mutant to grow on dalapon
was due to a 16-fold increase in dehalogenase activity
compared with the wild-type. The significance of these
findings lies in the fact that a *carrying substrate* (propionate
in the monoculture experiments, some undefined metabolite
or lysis product in the community) permits the growth of a
nonbiodegrader while continuously being exposed to the non-
utilisable chemical; the carbon limited nature of the system
ensures a strong and unremitting selective pressure for
dalapon utilising mutants.

Batch cultures

The vast majority of laboratory based studies of biodegradation
have been made in batch (closed) culture systems. On several
occasions I have advocated the deployment of continuous (open)
cultures for ecological research and particularly for analyses
of pollution effects and biodegradation (Bull, 1974; Bull and
Slater, 1976; Slater and Bull, 1978; Bull and Brown, 1979).
Consequently detailed arguments will not be reiterated here.
It will suffice for the present discussion to emphasise the
infinitely variable and unique environments that can be
established in continuous systems, the ease of control of
such variables, the opportunity to observe long term experiments
(*inter alia* adaptation to xenobiotics) and the chronic effects
of xenobiotics. Among the most enlightening applications in
the latter context are the elegant investigations of Wurster
and his colleagues on the effects of polychlorinated biphenyls
on marine phytoplankton communities (Bull and Brown, 1979,
for review). Continuous culture systems also have special
value in the isolation of biodegrading communites (see
below).

Aerobiosis

Anaerobic ecosystems are of considerable significance and
are widespread on our planet, the best known being estuarine
and marine sediments, water logged soils, gastrointestinal
tracts including the rumen, and various anaerobic digester
systems. Additionally anaerobic niches are commonly found in
well aerated ecosystems, for example, soil crumbs in aerated
soil. Anaerobic ecosystems, particularly benthic sediments,
are frequently cited as repositories for xenobiotics and the
concept of the ultimate sink is familiar to scientists and
laymen alike. It is surprising, therefore, that studies of
biodegradation and persistence under anaerobic conditions
have been neglected. One reason for this situation is the
greater opportunistic appeal of aerobic microbiology and
the general belief that anaerobes are difficult to grow. Two
recent reviews of biodegradation in anaerobic environments
(Evans, 1977; Williams, 1977) should help to dispel this view.
We now know that biodegradation of several xenobiotics can
proceed anaerobically via photometabolism, fermentation and
anaerobic respiration and that a wide array of reaction
mechanisms - reduction, hydrolysis, dealkylation, dehalogenation -
are possible in the absence of molecular oxygen. Evans'
group have made a special study of the anaerobic degradation
of benzoate by *Rhodopseudomonas palustris* and by a *Moraxella*
species. Photometabolism proceeds via reduction and hydration
of the coenzyme A thioester of benzoate to cyclohexanol-2-
carboxylic acid and subsequent cleavage of the alicyclic ring
to yield pimelate that feeds into intermediary metabolism.
Nitrate respiration of benzoate by *Moraxella* sp. results in
the production of adipate; reduction is made by a ferrodoxin-
type agent and is followed by ring cleavage. Growing attention
also is being given to the degradation of aromatics by methanogenic bacteria. Communities of methanogenic bacteria ferment
benzoate, phenol catechol to methane and carbon dioxide,
reduction of the benzene nucleus being following by ring
cleavage. The carbon 1 of benzoate is converted to methane
and carbons 4 and 7 predominantly to CO_2 (Fina, Bridges,
Coblentz and Roberts (1978). The range of aromatics that
serve as methanogenic substrates include several lignin-
derived compounds (vanillin, syringealdehyde, ferulic acid
and cinnamic acid; Healy and Young, 1979).

The relative importance of aerobic and anaerobic routes
of biodegradation in the biosphere remains to be quantified:
what is clear is that concentration on aerobic processes will
give a grossly distorted view of the total picture. Some
observations of Bollag and Russel (1976) are interesting
in this context. They found that a *Paracoccus* sp. could

transform 4-chloroaniline under aerobic and anaerobic
conditions and the rate and extent of transformation were
much greater in anaerobic cultures. It is also known that
different pathways for xenobiotic degradation operate under
aerobic and anaerobic (or microaerophilic) conditions.
Aerobically parathion is degraded by a bacterial community
to p-aminophenol and diethyl thiophosphate (Munnecke and
Hsieh, 1975). Finally, just as aerobic cometabolism can
give rise to products having greater toxicity and relcalcitrance
than the parent compound, anaerobic transformation of xenobiotics
can produce analogous results. Thus, one pesticide, DDT, may
be converted to another, dicofol (4,4'-dichloro-α-(trichloromethyl)
-benzhydrol), in anaerobic soils (Guenzi and Beard, 1968), while
the growth regulator disugran (methyl 3,6-dichloro-O-anisate)
is converted to a non-metabolised herbicide dicamba (2-methoxy-
3,6-dichlorobenzoic acid) by sheep rumen fluid (Ivie, Clark
and Rushing, 1974).

The Constant Environment Syndrome

    The control of environmental variables at constant values
has been an objective of countless microbiological investigations and has been necessary for much physiological and
biochemical work. However, biodegrading populations in nature
are frequently subjected to environmental perturbations of
natural or man-made origin, perturbations that are regular
(temperature and salinity changes in estuarine ecosystems)
or irregular (point organic and pollutant loading) in time.
It is quite conceivable that fluctuating environments might
be necessary for certain biodegradative activity or influence
rates of transformation. For example, DDT is converted,
under anaerobic conditions to chlorinated diphenlyethanes
and dichlorobenzophenone by *Hydrogenomonas* and by freshwater
sediment and sewage sludge communities (Pfaender and
Alexander, 1972). Subsequent aeration of the *Hydrogenomonas*
cultures lead to ring cleavage and appearance of p-chloro-
phenylacetic acid. The latter was utilised as a growth
substrate by an *Arthrobacter* sp. isolated from sewage.
Eglinton's group has provided good evidence that the conversion
of DDT to DDD (1,1-dichloro-2,2-bis (p-chlorophenyl ethane)
can occur abiotically in reducing environments via the action
of reduced iron porphyrins; the latter were believed to be
released on the decay of the biomass in the environment (Zoro
Hunter, Eglington and Ware, 1972). One possible route for
DDT mineralisation, therefore, could require sequential
anaerobiosis and aerobiosis, conditions readily established
in many ecosystems.
    The knowledge of alternative routes by which compounds can
be metabolised by microorganisms (Bull and Brown, 1979) has

yet to be fully exploited by those involved with biodegradation but is evidenced by the studies of mexacarbate, chloroaniline and parathion referred to above. There are difficulties in predicting biodegradative patterns and considerably more information is needed on the qualitative and quantitative effects of fluctuating environments: such conditions are readily established in laboratory models.

There is a notable dearth of information on the effects of stress situations on biodegradation; stress can be taken to include significant changes in physico-chemical variables, such as salinity and pH, and the discharge of multiple xenobiotics into the ecosystem of interest. Combinations of pesticides, for example, often occur in the environment as a consequence of agricultural practice, whilst rivers, estuaries and lakes may receive an *olla potridia* of pollutant toxicants if located near industrial centres. Some work has been done on the biodegradation of pesticide mixtures, particularly by Kaufman and his colleagues (Kaufman, 1972), from which a few generalisations can be drawn:

i)   increased recalcitrance: for example, dalapon degradation inhibited by a second herbicide, amitrole (3-amino-1,2,4-triazole) organophosphorus insecticide persistence in soil increased by addition of alkylbenzenesulphonates (Lichtenstein, 1966).

ii)   decreased recalcitrance: for example, persistence of dicamba in soil when combined with 2,4-D.

iii) complex formation: for example, asymmetric azobenzenes from the condensation of aniline residues from different herbicides (Bartha, 1969).

Culture and System Homogeneity

Any discontinuity in the aqueous phase produces an interface or surface. Such interfaces can be defined as phase or density gradients that provide unique environments not found in either of the adjoining phases and in which microbial activity may be significantly altered. Clearly interfaces are usual features of natural environments and include air/water (for example, marine microlayers); water/water (for example, coastal fronts, upwellings, freshwater-brackish water-seawater); oil/water, oil/air; sediment and soil/water. However, most biodegradation studies have not, or only incidently have, considered interface effects and the importance of surfaces, films and aggregations remains poorly understood for the main part. The relationship of surface phenomena to pesticide breakdown has been discussed recently by Burns (1979). There are at least five interface phenomena that require further analysis in this context.

Rate Effects   Is the availability and rate of transformation
of a compound enhanced or reduced at an interface? Generalised
statements on the advantages gained by attached microorganisms
in the assimilation of nutrients from very low concentrations
and increased rates of metabolism are common in the literature
and are usually based on the pioneering research of ZoBell.
Unfortunately critical data are few and hypothesis has tended
to become theory simply by restatement. A number of studies
have revealed enhanced rates of xenobiotic degradation when
the population is associated with an interface. Lee and
Ryan (1979), for example, recently reported greatly increased
rates of organochlorine degradation in estuarine sediment
compared with the water column. Some half-life values in days
are as follows with the water column data in parentheses:
1,4,5-trichlorophenoxyacetic acid, 35 (1400); 2,4,5-trichloro-
phenol, 23 (690); *p*-chlorophenol, 3 (20); chlorobenzene,
75 (150). In another context, Pawlowsky and Howell (1973)
observed a 3-fold increase in the critical dilution rate of
a phenol-metabolising community when the biomass was attached
to the reactor; the implications for rates of substrate
transformation was obvious. Similarly, the rates of denitrification in fixed films (Compare and Griffiths, 1977) and
degradation of coal conversion effluents in fluidized-beds
(Lee, Scott and Hancher, 1979) are reported to be 100- and
50- times greater respectively than continuous stirred tank
reactors.

Retention of Biomass   The retention of organisms at a stable
interface can endow the population with greater tolerance of
environmental perturbations. The only present illustrations
of this phenomenon come from waste treatment systems but the
principle also has considerable relevance for biodegradation
in natural ecosystems. An interesting example is the upward
flow, anaerobic fixed-film digester developed at Oak Ridge
National Laboratory which showed remarkable operating tolerance
compared to conventional dispersed biomass reactors. Thus,
the methanogenic community remained stable over wide ranges
of pH (3 to 10), temperature (10 to 25°C), total suspended
solids (60 to 210ppm) and BOD (60-180ppm) (Genung, Millian,
Hancher and Pitt, 1979).

Partitioning of Chemicals   It is well known that partitioning
effects can occur at interfaces and in films and the consequences
for microbial degradation are important and little studied.
Sayler and Colwell (1976) have analysed the partitioning of
highly insoluble polyaromatic hydrocarbons and polychlorinated
biphenyls (PCB) in crude oil and suspended sediments and found
partitioning effects of several hundred or thousand percent

in each system. Moreover, both oil and sediment strongly
partitioned mercury with the result that a toxic environment
developed. Similar partitioning occurs in surface films.
Such multipollutant environments are likely to inhibit
microorganisms capable of degrading the individual chemicals
thus rendering them all more recalcitrant. Further examination
of these effects is urgently required. In an analogous way,
chemical pollutants can be sequestered from the environment,
and thereby made less readily degradable, by bioaccumulation.
Bacteria and phytoplankton are known to bioaccumulate certain
toxic xenobiotics and concentration ratios as high as 56,000
have been observed for the insecticide chlordane in *Caulobacter*
sp. (Grimes and Morrison, 1975).

Genetic Exchange Information is almost totally lacking on
the roles played by attachment to solid surfaces, or
concentration at interfaces, in influencing genetic exchange
within and between different microbial populations and the
bearing that they may have on the evolution of novel catabolic
capabilities or the dissemination of extant properties
throughout an ecosystem. It is known that sorption of or
onto sediments can modify microbial interactions (for example,
Escherichia coli-bacteriophage; Roper and Marshall, 1974)
and from such evidence it is tempting to infer that the
frequency of genetic interactions may also be conditioned
by sorption phenomena.

INTERLUDE ONE: ABIOTIC DEGRADATION

Non-biological transformations of xenobiotics are well known
and such transformations of pesticides have been reviewed
by Crosby (1970; 1976). Abiotic transformations are induced
by soil organic matter, in which nucleophilic groups are
abundant and reactive; soil minerals; water; and light,
particularly ultraviolet (oxidation, reduction, elimination,
hydrolysis, isomerisation). Photochemical transformation
probably is the most important abiotic activity and very many
pesticides are susceptible under experimental and field
conditions. The purpose of introducing abiotic degradation
into this discussion is to reiterate that such transformations
can influence subsequent microbiological attack. The herbicide
paraquat (1,1-dimethyl-4,4-bipyridilium dichloride) was not
degraded by an *Achromobacter* sp. but the photolytic product
1-methyl-4-pyridinium chloride could be used as a sole source
of carbon (Cain, Wright and Houghton, 1970). Photo-oxidations
may generate a family of products each of which may vary in
its susceptibility to metabolism. Thus monuron (3-(p-chlorophenyl)
-1,1-dimethylurea) can be hydroxylated, dehalogenated or have
one or both N-methyl groups eliminated by phototransformation

in aqueous environments. Certain synthetic polymers also
are sensitive to partial photodegradation; free radical
reactions caused by sunlight generate short chains containing
polar groups that are susceptible to microbial degradation.
Polyethylene (Kaplan, 1975) and polybutene (Reich and Bartha,
1977) are attacked in this way but in both cases the residual
long chain polymers resist further decomposition.

BIODEGRADATION: INVESTIGATIVE STRATEGIES

Old Problems

Numerous debates have been held on the merits of different
investigative strategies for studying biodegradation and, in
particular, resolution of the *in vitro* (laboratory) versus
in situ (field) conflict remains remote. It would be pointless
to rehearse the arguments yet again, instead I shall adopt a
position and attempt to defend it. Useful background reading
in this context can be found in Hill and Arnold (1978) and
in a lucid essay by Kaplan (1977).

The complexity of natural environments presents the
researcher with an array of daunting methodological challenges.
It is difficult to follow transformations in such environments
and great caution needs to be exercised if biodegradative
routes are wished to be defined *in situ*. Consequently much
of the work on biodegradation has been done *in vitro* and
done, moreover, hide-bound by traditional microbiology and
biochemistry and without reference to many of the issues
raised in the foregoing discussion. Most people would now
agree that there are grave inadequacies in such a strategy.
In recent years much attention has been given to the development
of environmentally relevant laboratory models in order to
generate information, answer complex questions and to probe
new areas. It is this latter type of approach that my
colleagues and I have favoured and will be focus of the
remaining discussion.

New Options

Environmentally relevant laboratory models have gradually
become known as *microcosms* and they are simply experimental
designs for studying or simulating activities that occur in
nature, in the present instance, biodegradation. Microcosm
studies may have screening or predictive objectives. Microcosms frequently mean all things to all men but some useful
definitions have been provided by Matsumura, (1979).

   i)   Benchmark approaches: determination of physicochemical
parameters;

   ii) Single reaction approaches: determination of simple
cause and effect relationships;

iii) Multicomponent system approach (microcosm *sensu stricto*):
closer approximation to environment, determination of ecological
interactions;

iv) System analysis approach: mathematical modelling of
rates at which compounds are biodegraded or transported
through an ecosystem (Blau and Neely, 1975 for review). I
shall concentrate on microcosms of the third category: they
can be constructed either by taking a sample of the environment and analysing it under defined laboratory conditions, (for
example, sediment core), or by assembling a limited number of
species together in an artificial laboratory system (for example,
chemostat). Whatever type of microcosm is deployed it must be
appreciated that the information generated will be a function
of the design features of that microcosm. Some microcosms
will enable biodegradative capacity to be assessed, others
will be limited to revealing biodegradative potential.

The physical range of microcosms is large and encompasses
simple flask (closed) systems, chemostats of varying degrees
of complexity and instrumented terrestrial chambers and
channels (Metcalf, 1977). In addition it is possible to
model gradient systems with multistage chemostats in which
there is a counterflow of nutrients, dissolved oxygen,
xenobiotics etc. (the gradostat: Lovitt and Wimpenny, 1979);
and interphase phenomena with systems such as fluidised-
bed and microbial film reactors. The choice of physical model
is often made intuitively without a critical comparison of
alternatives but one valuable assessment was made by Falco,
Sampson and Carsel, (1977) in the context of malathion
degradation. These authors compared degradation by a bacterial
community (Paris. Lewis and Wolfe, 1975) in flasks, chemostats
and an instrumented channel with a mathematical model.
Comparable data were obtained from all three physical models
when the insecticide concentration was held below 1 mg l^{-1}
but at higher concentrations degradation rates in flasks and
chemostats were significantly different: these differences
were attributed to large changes in degradation rate over
narrow concentration ranges and the averaging out of such
effects in the closed system. Falco and his colleagues concluded
that chemostat behaviour more accurately described observed
field phenomena over a malathion range 1.0 to 4.0 mg l^{-1}.
It was emphasised that the application of mathematical models
must be restricted to conditions that have been tested experimentally. It will be obvious that, depending on the nature of
the system being modelled, special features may need to be
incorporated; evaluation of plant root effects will serve
to illustrate this point. For example, root exudates may be
an important source of growth substrates for cometabolic
transformation of xenobiotics in soil. Recently Hsu and

Bartha (1979) have shown that diazinon and parathion are mineralised more rapidly in rhizosphere microcosms and that in the case of parathion the enhancement was also included by applying root exudate to root-free systems. Population selection and/or increased cometabolism were considered to be the most likely causes of these stimulations. Studies of this sort are extremely important in demonstrating that persistence of xenobiotics in the environment may be reduced by rhizosphere and similar effects.

It would be facile indeed to conclude that microcosms are the systems of choice for all biodegradation research; there are, as Matsumura (1979) has indicated, problems attending their use. It cannot be stated too strongly that *Microcosms should be viewed as analytical tools and not reproductions of ecosystems.* Furthermore, microcosms studies do not replace environment studies but they can provide considerable insight and guidance for the latter. Nevertheless, and irrespective of type, microcosms have the advantages of standardisation, replicability, ease of control and manipulation and with the proviso that they are made functionally similar to the ecosystem being mimicked (for example, providing appropriate electron acceptors in anaerobic systems), they are sound investigative systems for biodegradation work.

INTERLUDE TWO: ENRICHMENT ISOLATION

Attempts to isolate biodegradative microorganisms by enrichment have, with few exceptions, relied on traditional dilution techniques and batch systems. There are several inherent drawbacks to such approaches - inappropriate for the isolation of interacting communities, selection largely on the basis of maximum specific growth rate because of substrate excess conditions. Indeed persistence with these methods may well be the cause of failure to isolate effective organisms. In contrast, continuous-flow enrichment procedures enable enrichment isolation to be made on a variety of criteria, for example, organisms having high substrate affinities and/or low specific growth rates, both of which properties may be of high selective advantage in natural environments. Continuous-flow systems also are ideal for isolating microbial communities. Various types of continuous-flow enrichment culture can be used, depending on requirements and some options are:

   i)   chemostat operating at a low dilution rate.
   ii)   2-stage chemostat, the second stage operating at a high dilution rate and the first at a low dilution rate in order to provide a continuous inoculum,
   iii) turbidostat; here selection will be on the basis of highest maximum specific growth rate. The nature of the growth

rate limiting substrate can also be manipulated at will and may have important repercussions. Thus, Campacci, New and Tchan (1977) could isolate amitrole degrading bacteria only if the herbicide was made the sole source of nitrogen; if it functioned as a sole carbon source the enrichment was abortive. Likewise, the judicious provision of mixed carbon (or other) nutrients in continuous flow enrichments could be revealing (*Single Substrate Systems*) and also be appropriate for the isolation of cometabolising communities.

The basic chemostat enrichment procedure is easily adapted for special purposes, for the isolation organisms developing at interfaces, for example. In our laboratory we have used fluidised-beds to enrich the attached estuarine communities capable of degrading phenylurea herbicides while Liu (1979) has stabilised emulsion preparations of lipophilic compounds on finely divided lignin sulphonate and successfully enriched n-alkane, aromatic hydrocarbon, phenol and PCB degrading organisms.

One final and seemingly trivial point should be added on enrichment isolation. This is the question of environment sampling. Reference has been made previously to the much more active organochlorine degrading populations in sediment rather than water column in an estuarine ecosystem (Lee and Ryan, 1979). Our own work on dicamba and fenuron biodegradation in estuarine environments has revealed considerable variations in herbicide toxicity as a function of sediment depth (increased resistance with depth) and a marked seasonal variation in toxicity (greatest resistance in mid-summer) (S.F. Minney, R.J. Parkes and A.T. Bull, unpublished observations). Watson (1977) also has reported seasonal variation of 2,4-D biodegradation in river water; maximum activity occurred during winter flood conditions and may be related to increased sediment, nutrient and/or organism loads in the water. Observations such as these highlight the importance of integrated laboratory-field analyses and having as detailed an understanding of the receiving environment prior to setting up microcosm models.

SOME FEATURES OF MICROCOSM BEHAVIOUR

Relation to Natural Environments

One advantage of microcosms, so far assumed and not proved, is their replicability. In fact microcosms, be they of the multicomponent food chain developed by Metcalf or the simpler microbial community types, appear to have good replicability and certainly comparable to that found in other types of statistical trials. Two early and detailed experimental confirmations of this assumption are to be found in the work

of Beyers (1963) and Abbott (1966). Good agreement between
microcosm and environment has also been reported for pesticide
biodegradation and bioaccumulation. Several studies in Metcalf's
laboratory (Metcalf, Shanga and Kapoor, 1971; Cole, Metcalf
and Sanborn, 1976) have shown that the fates of insecticides
such as DDT, methoxychlor and aldrin in complex, multitrophic
microcosms are similar to those observed under field conditions.
The biodegradation of natural substrates such as leaf litter
(Bott, Preslan, Finlay and Brunker, 1977) in microcosms also
approximates field behaviour quite well.
   Microbial microcosms also exhibit those basic properties
common to all ecosystems. Gorden, Beyers, Odum and Eagon,
(1969), for example, compared the attributes expected in
developing and mature ecosystems with a simple batch laboratory
model, that is, a comparison of natural and microcosm bacterial
succession. Of the attributes monitored (biomass production,
photosynthesis/respiration ratio, species diversity and inter-
dependence, stability, life cycles and stratification) only
species diversity showed discrepancies between the ecosystem
and the microcosm and this was ascribed to the use of a point
inoculum from a mature system. A more complex microbial
microcosm was studied by Cooper and Copeland (1973) who
developed a 5-stage counter flow chemostat (*cf*. gradostat)
with which to model a shallow estuary. Salinity patterns
very similar to those of the estuary were established and
drought and flood conditions could be simulated by varying
the flow from the fresh and salt water medium reservoirs
feeding the vessels from opposite ends of the chain. Metabolism
in the predominantly freshwater vessels was heterotrophic but
shifted to autotrophic metabolism when drought was simulated.
The addition of industrial effluent to freshwater vessels
reinforced their heterotrophic character but imposition of a
primary stress (that is, drought) rendered the receiving
communities more susceptible to the effluent. The microflora
of the microcosm was qualitatively but not quantatively
similar to that of the estuary water column. In our laboratory
we have observed good replicability of chemostat microcosms
developed to study dalapon degradation (E. Senior, J.H. Slater
and A.T. Bull, unpublished observations) while, in quantative
terms, the rates of fenuron degradation by microbial communities
growing in chemostats accord well with reports of field
persistence (S.F. Minney, R.J. Parkes and A.T. Bull, unpublished
observations).

Structure and Stability

   If sufficient time is allowed, theory predicts (Slater and
Bull, 1978) that following inoculation of an enrichment
chemostat one organism will competitively displace all others

because the chance of two organisms having identical maximum
specific growth rates, substrate affinities, maintenance
energy requirements and growth yields is exceedingly small.
However, in practice one invariably selects a multispecies
community; I know of no exception to this statement. It is
not uncommon, especially where xenobiotics are being studied,
to find two types of population in these communities, namely
primary and secondary (Slater and Bull, 1978). Primary
species are those that can grow as monocultures on the substrate
of interest; in contrast, secondary species cannot metabolise
the substrate and rely on metabolites of primary species and/or
lytic products to sustain their growth. It has been stated,
mistakenly, in the past that mixed cultures are inherently
unstable. This could be true of synthesised mixtures but with
respect to communities isolated by continuous-flow enrichment
it is without foundation. On the contrary, mixed cultures
may be considerably more stable than monocultures under
circumstances where a substrate, product or medium component
is growth inhibitory (Osman, Bull and Slater, 1976; Harrison,
1978; Charley and Bull, 1979). We have analysed some of our
herbicide degrading communities over several thousand hours,
in excess of 20,000 h in the case of dalapon, and stability
has been maintained. In parenthesis, it is preferable to
refer to stable rather than to steady states when describing
chemostat microcosms. Such systems exhibit oscillations in
population sizes of the component organisms and residual
limiting substrate concentration, the amplitude and period of
which are system dependent. On perturbation microcosm
parameters return to the original equilibrium position,
establish new equilibrium positions or remain unstable.
That population changes can occur in these communities is
evident from the work of Senior *et al.*, 1976) (see *Single
Substrate Systems*) and chemostat microcosms of the sort
described should be exploited further for studies of
adaptation mechanisms.

Interactions

The selection of communities in enrichment chemostats (see
Structure and Stability) is due to a variety of species interactions that can eliminate free competition between primaries
for the growth limiting substrate and enable secondaries to
coexist with them. Among these interactions, many of which
are elements of biodegrading communities, are:
    i)   commensalism (for example, cyclohexane, Stirling *et al.*,
1976);
    ii)   synergism (for example, diazinon, Gunner and Zuckerman,
1968; alkylbenzenesulphonates, Johanides and Hrsak, 1976;
    iii) relief of substrate inhibition;

iv)   cometabolism (for example, parathion, Munnecke and
Hsieh, 1974; amitraz, Baker and Woods 1977).

In the degradation of parathion *Pseudomonas stutzeri* effected
an initial cometabilism to *p*-nitrophenol and diethyl thiophosphate; the *p*-nitrophenol is growth inhibitory but serves as
a growth substrate for a second component of the community,
P. aeruginosa. Thus the two bacteria establish a mutualistic
relationship in which the former grows at the expense of
metabolites (or lytic products) of the latter. An interesting
example of alleviated inhibition is provided by the orcinol
(3-methyl resorcinol) degrading community isolated by Osman
(Bull and Brown, 1979). This stable community comprised three
bacteria only one of which, *P. stutzeri*, utilised the phenol.
However, typical substrate-inhibited growth occurred when this
bacterium was grown in pure culture on orcinol. Kinetic
analyses of this system revealed that the selective advantage
possessed by the community resided in the fact that S_{max}, that
concentration of orcinol at which specific growth rate is
maximum was increased 33% by the presence of the two secondary
organisms.

It would be surprising if new categories of factors
implicated in the maintenance of microbial communities and
affecting biodegradation activities are not identified in the
near future. de Freitas and Fredrickson (1979) have recently
discussed the production of autoinhibitors in this light
while the production of microbial surfactants (Margaritis,
Zajic and Gerson, 1979) might have interesting repercussions
for the degradation of lipophilic compounds.

INTERLUDE THREE: QUESTIONS OF CONCENTRATION

The final topic which I wish to touch upon hinges on
concentration effects. One obvious effect of concentration
can be related to toxicity of the compound under study, a
self-evident matter but one that is relevant to the isolation
of organisms and analysis of degradation routes and not always
sufficiently well appreciated. Much more important, however,
is the degradation of compounds present at very low concentrations and for xenobiotics there is a dearth of information.
Recent work from Alexander's laboratory (DiGeronimo, Boethling
and Alexander, 1979; Boethling and Alexander, 1979) has
sought to redress this situation. This group studied the
degradation rates of secondary amines, organohalogens and
Sevin (1-naphthyl-N-methylcarbamate) by freshwater communities.
Some compounds (dimethylamine, *p*-chlorobenzoate) were
degraded at rates directly proportional to their concentrations;
thus rates increased 100- and 10-fold respectively as
concentration was *reduced* over the µg ml^{-1} to pg ml^{-1} range.
The rate of degradation of other compounds (2,4-D, Sevin) was

very low or zero at concentrations in the ng ml^{-1} and pg ml^{-1} range and a *threshhold* existed below which mineralisation was insignificant. Xenobiotics persisting in the environment at such low concentrations could cause problems either directly by toxicity (PCBs, for example can be toxic at μg ml^{-1} levels) or via bioaccumulation and transfer to higher trophic levels. Such data clearly have serious implications for the design of biodegradability tests.

PROSPECT

Undoubtedly much remains to be done and comprehended in the area of microbial degradation and of the areas warranting further inquiry, the following have commended themselves to me.

i) *Effects of environmental stresses* Large areas of ignorance exist about the constraints of natural stresses such as salinity and of pollutants, particularly multiple pollutants, on biodegradation of chemicals. Techniques for recognising stress need to be refined and Goulder's recent appraisal of mineralisation kinetics in this context is noteworthy (Goulder, Blanchard, Sanderson and Wright, 1979). Behaviour of degradative microorganisms at very low specific growth rates and substrate concentrations similarly is poorly researched.

ii) *Evolution of catabolic capabilities* The early discussion suggests that there is a growing need to investigate mutation and recombination in mixed communities with respect to degradation and resistance. The competitiveness of strains engineered for particular biodegradative purposes has received little or no critical evaluation.

iii) *Monitoring technology* New technology probably is not required but continual refinement of exising protocols is desirable as new basic information on biodegradation appears. Quantitative information generally needs to be collected on occurrence, pathways, fates and effects of xenobiotics in natural environments.

iv) *Applications* One of the more obvious problems is the development of methods for detoxifying harmful xenobiotics both at their sites of intentional and polluting discharge into the environment and in more contained situations at sites of manufacture. Treatment of such materials when they are dispersed in the environment is difficult but some success has been achieved by seeding degrading populations into polluted sites, encouraging growth of natural populations by supplying nutrients such as N and P, and encouraging cometabolic degradation by introducing appropriate substrates. Golovleva and Skryabin (1978), for example, have reported accelerated removal of 2,4-D from paddy field waste waters and ordram from reservoirs by the addition of propionate and ethanol respectively as cosubstrates. The preparation of potent detoxifying enzymes from mixed cultures has been made by Munnecke (1978) for treating effluents, equipment and containers contaminated by parathion and one can expect further developments of this in order to deal with other hazardous materials.

But perhaps the final comment should be Sam Weller's *"Yes I have a pair of eyes," replied Sam, "and that's just it. If they wos a pair o' patent double million magnifyin' gas microscopes of hextra power, p'raps I might be able to see through a flight o' stairs and a deal door; but bein' only eyes, you see, my wision's limited."*

ACKNOWLEDGEMENTS

It is a pleasure to thank all those colleagues who, over the years, have contributed to my thinking on microbial degradation and particularly those who have been part of the research programmes. I thank Nisar Hussain, Mahmoud Karimian, Susan Minney, Aslan Osman and John Parkes for permission to report previously unpublished information and the Natural Environment Research Council, Science Research Council and Agricultural Research Council for sustaining the work of my laboratory.

REFERENCES

Abbott, W. (1966). Microcosm studies on estuarine waters. 1. The replicability of microcosms, *Journal of the Water Pollution Control Federation* 38, 258-270.

Baker, P.B. and Woods, D.R. (1977). Cometabolism of the ixodicide amitraz, *Journal of Applied Bacteriology* 42, 187-196.

Bartha, R. (1969). Pesticide interaction creates hybrid residue, *Science, N.Y.* 166, 1299-1300.

Bartha, R. and Pramer, D. (1967). Pesticide transformation to aniline and azo compounds in soil, *Science, N.Y.* 156, 1617-1618.

Beam, H.W. and Perry, J.J. (1974). Microbial degradation of cycloparaffinic hydrocarbons via cometabolism and commensalism, *Journal of General Microbiology* 82, 163-169.

Benezet, H.J. and Matsumura, F. (1974). Factors influencing the metabolism of mexacarbarbate by microorganisms, *Journal of Agricultural and Food Chemistry* 22, 427-430.

Beyers, R.J. (1963). The metabolism of 12 aquatic laboratory microecosystems, *Ecological Monographs* 33, 281-306.

Blau, G.E. and Neely, W.B. (1975). Mathematical model building with an application to determine the distribution of Dursban insecticide added to a simulated ecosystem, *Advances in Ecological Research* 9, 133-163.

130   A.T. BULL

Boethling, R.S. and Alexander, M. (1979). Effect of
  concentration of organic chemicals on their biodegradation
  by natural microbial communities, *Applied and Environmental
  Microbiology* 37, 1211-1216.
Bollag, J.-M, and Liu, S.-Y. (1971). Degradation of Sevin
  by soil microorganisms, *Soil Biology and Biochemistry* 3,
  337-345.
Bollag, J-M. and Russel, S. (1976). Aerobic versus anaerobic
  metabolism of halogenated anilines by a Paracoccus sp.,
  Microbial Ecology 3, 65-73.
Bott, T.L., Preslan, J., Finlay, J. and Brunker, R. (1977).
  The use of flowing-water microcosms and ecosystem streams
  to study microbial degradation of leaf litter and nitrilo-
  triacetic acid (NTA), *Development in Industrial Microbiology*
  18, 171-184.
Bourquin, A.W. (1975). Microbial-malathion interaction in
  artificial salt marsh ecosystems, U.S. Environmental
  Protection Agency Report No. EPA-660 3-75-035, pp. 1-41.
Bull, A.T. (1974). Microbial growth. *In* "Companion to
  Biochemistry" (Ed. A.T. Bull, J.R. Lagnado, J.O. Thomas
  and K.F. Tipton), pp. 415-442. Longman, London.
Bull, A.T. and Brown, C.M. (1979). Continuous culture
  applications to microbial biochemistry, *In* "Microbial
  Biochemistry" International Review of Biochemistry 21,
  (Ed. J.R. Quayle), pp. 177-226. University Park Press,
  Baltimore.
Bull, A.T., Ellwood, D.C. and Ratledge, C. (1979). The
  changing scene in microbial technology, *In* "Microbial
  Technology: Present State, Future Prospects" (Ed. A.T. Bull,
  D.C. Ellwood and C. Ratledge), pp. 1-28. Cambridge University
  Press, Cambridge.
Bull, A.T. and Slater, J.H. (1976). The teaching of
  continuous culture, *In* "Continuous Culture 6. Applications
  and New Fields" (Ed. A.C.R. Dean, D.C. Ellwood, C.G.T. Evans
  and J. Melling), pp. 49-68. Ellis Horwood, Chichester.
Burns, R.G. (1979). Interaction of microorganisms, their
  substrates and their products with soil surfaces. *In*
  "Adhesion of Microorganisms to Surfaces" (Ed. D.C. Ellwood,
  J. Melling and P. Rutter), pp. 109-138. Academic Press,
  London.
Cain, R.B., Wright, K.A. and Houghton, C. (1970). Microbial
  metabolism of the pyridine ring: bacterial degradation of
  1-methyl-4-carboxypyridinium chloride a photolytic product
  of paraquat, *Mededelingen Faculteit Landbouw Wetenschappen
  Gent* 35, 785-798.
Campacci, E.F., New, P.B. and Tchan, Y.T. (1977). Isolation
  of amitrole-degrading bacteria, *Nature London* 266, 164-165.
Carpenter, D.F., McCormick, N.G., Cornell, J.H. and Kaplan,

A.M. (1978). Microbial transformation of ^{14}C-labelled
2,4,6-trinitrotoluene in an activated-sludge system,
Applied and Environmental Microbiology 35, 949-954.

Charley, R.B. and Bull, A.T. (1979). Bioaccumulation of
silver by a multispecies community of bacteria, *Archives
of Microbiology* 123, 239-244.

Chou, T.-W. and Bohonos, N. (1979). Diauxic and cometabolic
phenomena in biodegradation evaluations. *In* "Microbial
Degradation of Pollutants in Marine Environments" (Ed. A.W.
Bourquin and P.H. Pritchard), pp. 76-88. U.S. Environmental
Protection Agency, Gulf Breeze, Florida.

Clarke, P.H. (1978). Experiments in microbial evolution.
In "The Bacteria VI" (Ed. L.N. Ornston and J.R. Sokatch),
pp. 137-218. Academic Press, New York.

Cole, L.J., Metcalf, R.L. and Sanborn, J.R. (1976).
Environmental fate of insecticides in terrestrial model
ecosystems, *International Journal of Environmental Studies*
10, 7-14.

Compare, A.L. and Griffiths, W.L. (1977). Continuous fixed-
film denitrification of high-strength industrial nitrate
wastes, *Development in Industrial Microbiology* 18, 717-722.

Cooper, D.C. and Copeland, B.J. (1973). Responses of
continuous-series estuarine micro ecosystems to point-
source input variations, *Ecological Monographs* 43, 213-236.

Crosby, D.G. (1970). Non biological degradation of pesticides
in soils. *In* "Pesticides in the Soil", pp. 86-94.
Michigan State University, East Lansing.

Crosby, D.C. (1976). Non biological degradation of herbicides
in soil, *In* "Herbicides: Physiology, Biochemistry, Ecology,
Volume 2" (Ed. L.J. Audus). Academic Press, London.

Daughton, C.G. and Hsieh, D.P. (1977). Parathion utilisation
by bacterial symbionts in a chemostat, *Applied and Environmental Microbiology* 34, 175-184.

DiGeronimo, M.J., Boethling, R.S. and Alexander, M. (1979).
Effect of chemical structure and concentration on microbial
degradation in model ecosystems, *In* "Microbial Degradation
of Pollutants in Marine Environments" (Ed. A.W. Bourquin
and P.H. Pritchard), pp. 154-166. U.S. Environmental
Protection Agency, Gulf Breeze, Florida.

Evans, W.C. (1977). Biochemistry of the bacterial metabolism
of aromatic compounds in anaerobic environments, *Nature,
London* 270, 17-22.

Falco, J.W., Sampson, K.T. and Carsel, R.F. (1977). Physical
modelling of pesticide degradation, *Development in Industrial
Microbiology* 18, 193-202.

Fina, L.R., Bridges, R.L., Coblentz, T.H. and Roberts, F.F.
(1978). The anaerobic decomposition of benzoic acid during
methane formation III. The fate of carbon four and the

identification of proponoic acid, *Archives of Microbiology* 118, 169-179.

Fisher, P.R., Appleton, J. and Pemberton, J.M. (1978). Isolation and characterisation of the pesticide-degrading plasmic pJP1 from *Alcaligenes paradoxus*, *Journal of Bacteriology* 135, 798-804.

de Freitas, M.J. and Fredrickson, A.G. (1978). Inhibition as a factor in the maintenance of diversity of microbial ecosystems, *Journal of General Microbiology* 106, 307-320.

Genung, R.K., Million, D.L., Hancher, C.W. and Pitt, W.W. (1979). Pilot plant demonstration of an anaerobic fixed-film bioreactor for wastewater treatment, *Biotechnology and Bioengineering Symposium* No.8 329-344.

Gibson, D.T. (1979). Report of Task Group II: Biochemistry of microbial degradation. *In* "Microbial Degradation of Pollutants in Marine Environments" (Ed. A.W. Bourquin and P.H. Pritchard) pp. 514-518. U.S. Environmental Protection Agency, Gulf Breeze, Florida.

Giger, W. and Roberts, P.V. (1978). Characterisation of persistent organic carbon. *In* "Water Pollution Microbiology Volume 2" (Ed. R. Mitchell), pp. 135-175. John Wiley & Sons New York.

Golovleva, L.D. and Skryabin, G.K. (1978). Cometabolism of foreign compounds. *In* "Environmental Transport and Transformation of Pesticides", pp. 73-85. U.S. Environmental Protection Agency, Athens, Georgia.

Gorden, R.W., Beyers, R.J., Odum, E.P. and Eagon, R.G. (1969). Studies of a simple laboratory micro ecosystem: bacterial activities in a heterotrophic succession, *Ecology* 50, 86-100.

Goulder, R., Blanchard, A.S.. Sanderson, P.L. and Wright, B. (1979). A note on the recognition of pollution stress in populations of estuarine bacteria, *Journal of Applied Bacteriology* 46, 285-289.

Grimes, D.J. and Morrison, S.M. (1975). Bacterial bioconcentration of chlorinated hydrocarbon insecticides from aqueous systems, *Microbial Ecology* 2, 43-59.

Guenzi, W.D. and Beard, W.E. (1968). Anaerobic conversion of DDT to DDD and aerobic stability of DDT in soil, *Soil Science Society of American Proceedings* 32, 522-527.

Gunner, H.B. and Zuckerman, B.M. (1968). Degradation of diazinon by synergistic microbial action, *Nature, London* 217, 1183-1184.

Harrison, D.E.F. (1978). Mixed cultures in industrial fermentation processes, *Advances in Applied Microbiology* 24, 129-164.

Healy, J.B. and Young, L.Y. (1979). Methanogenic biodegradation of aromatic compounds, *In* "Microbial Degradation of Pollutants in Marine Environments" (Ed. A.W. Bourquin and

P.H. Pritchard), pp. 348-359. U.S. Environmental Protection
Agency, Gulf Breeze, Florida.
Hill, I.R. (1978). Microbial transformation of pesticides,
    In "Pesticide Microbiology" (Ed. I.R. Hill and S.J.L.
    Wright), pp. 137-202. Academic Press, London.
Hill, I.R. and Arnold, D.J. (1978). Transformations of
    pesticides in the environment − the experimental approach.
    In " Pesticide Microbiology" (Ed. I.R. Hill and S.J.L.
    Wright), pp. 203-245. Academic Press, London.
Horvath, R.S. (1971). Cometabolism of the herbicide, 2,3,6-
    trichlorobenzoate, *Journal of Agricultural and Food
    Chemistry* 19, 291-293.
Horvath, R.S. (1972). Microbial cometabolism and the
    degradation of organic compounds in nature, *Bacteriological
    Reviews* 36, 146-155.
Hsu, T.-S. and Bartha, R. (1979). Accelerated mineralisation
    of two organophosphate insecticides in the rhizosphere,
    Applied and Environmental Microbiology 37, 36-41.
Huber, T.J., Street, J.R., Bull, A.T., Cook, K.A. and Cain,
    R.B. (1975). Aromatic metabolism in the fungi. Growth
    of *Rhodotorula mucilaginosa* in *p*-hydroxybenzoate-limited
    chemostats and the effect of growth rate on the synthesis
    of enzymes of the 3-oxoadipate pathway, *Archives of
    Microbiology* 102, 139-144.
Ivie, G.W.. Clark, D.E. and Rushing, D.D. (1974). Metabolic
    transformations of disurgran by rumen fluid of sheep
    maintained on dissimilar diets, *Journal of Agricultural
    and Food Chemistry* 22, 632-644.
Jensen, H.L. (1957). Decomposition of chloro-substituted
    aliphatic acids by soil bacteria, *Canadian Journal of
    Microbiology* 3, 151-164.
Jensen, H.L. (1963). Carbon nutrition of some microorganisms
    decomposing halogen-substituted aliphatic acids, *Acta Agric.
    Scand.* 13, 404-412.
Johanides, V. and Hršak, D. (1976). Changes in mixed
    bacterial culture during linear alkylbenzenesulfonate (LAS)
    biodegradation, *Abstracts of Papers, 5th International
    Fermentation Symposium,* p. 426. Verlag Versuchs-und
    Lehranstalt fur Spiritusfabrikation und Fermentationstech-
    nologie, Berlin.
Kaplan, A.M. (1975). Microbial decomposition of synthetic
    polymeric materials. *Proceedings of the 1st Intersectional
    Congress of IAMS,* Volume 2, 535-545. Science Council of
    Japan, Tokyo.
Kaplan, A.M. (1977). Microbial degradation of materials in
    laboratory and natural environments, *Development in
    Industrial Microbiology* 18, 203-211.
Kaufman, D.D. (1972). Pesticide metabolism. *In* "Pesticides

in the soil, pp. 72-86. Michigan State University, East
Lansing.

Kirk, T.K., Connors, W.J. and Zeikus, J.G. (1976). Requirement
for a growth substrate during lignin decomposition by two
wood-rotting fungi, *Applied and Environmental Microbiology*
32, 192-194.

Kirk, T.K., Yang, H.H. and Keyser, P. (1978). The chemistry
and physiology of the fungal degradation of lignin,
Development in Industrial Microbiology 18. 51-61.

Lee, R.F. and Ryan, C. (1979). Microbial degradation of
organochlorine compounds in estuarine waters and sediments.
In "Microbial Degradation of Pollutants in Marine Environments" (Ed. A.W. Bourquin and P.H. Pritchard), pp. 443-450.
U.S. Environmental Protection Agency, Gulf Breeze, Florida.

Lee, D.D., Scott, C.D. and Hancher, C.W. (1975). Fluidised-
bed bioreactor for coal-conversion effluents, *Journal of
Water Pollution Control* 51, 974-984.

Lichtenstein, E.P. (1966). Increase of persistence and
toxicity of parathion and diazinon in soils with detergents,
Journal of Economic Entomology 59, 985-993.

Linton, J.D., Griffiths, K., Harrison, D.E.F. and Bull, A.T.
(1979). Growth of *Beneckea natriegens* on mixtures of
glucose and formate in chemostat culture, *Society for
General Microbiology Quarterly* 6, 91.

Linton, J.D. and Stephenson, R.J. (1978). A preliminary
study on growth yields in relation to the carbon and energy
content of various organic growth substrates, *FEMS
Microbiology Letters* 3, 95-98.

Liu, D. (1979). A novel selective enrichment technique for
use in biodegradation studies. *In* "Microbial Degradation
of Pollutants in Marine Environments" (Ed. A.W. Bourquin
and P.H. Pritchard), pp. 370-379.

Lovitt, R.J. and Wimpenny, J.W.T. (1979). The gradostat:
a tool for investigating microbial growth and interactions
in solute gradients, *Society for General Microbiology
Quarterly* 6, 80.

Margaritis, A., Zajic, J.E. and Gerson, D.F. (1979).
Production and surface-active properties of microbial
surfactants, *Biotechnology and Bioengineering* 21, 1151-1162.

Matsumura, F. (1979). Report of Task Group IV: Microcosms.
In "Microbial Degradation of Pollutants in Marine
Environments" (Ed. A.W. Bourquin and P.H. Pritchard), pp.
520-524. U.S. Environmental Protection Agency, Gulf Breeze,
Florida.

McClure, G.W. (1970). Accelerated degradation of herbicides
in soil by the application of microbial nutrient broths,
Contributions of the Boyce Thompson Institute 24, 235-244.

McClure, G.W. (1973). Membrane biological filter device for

reducing waterborne biodegradable pollutants, *Water Research* 7, 1683-1690.

Metcalf, R.L. (1977). Model ecosystem approach to insecticide degradation: a critique, *Annual Reviews of Entomology* 22, 241-261.

Metcalf, R.L., Shanga, G.K. and Kapoor, I.P. (1971). Model ecosystem for the evaluation of pesticide biodegradability and ecological magnification, *Environmental Science and Technology* 5, 709-713.

Munnecke, D.M. (1978). Detoxification of pesticides using soluble or immobilised enzymes, *Process Biochemistry* 13, 14-16, 31.

Munnecke, D.M. and Hsieh, D.P.H. (1974). Microbial decontamination of parathion and *p*-nitrophenol in aqueous media, *Applied Microbiology* 28, 212-217.

Munnecke, D.M. and Hsieh, D.P.H. (1975). Microbial metabolism of a parathion-xylene pesticide formulation, *Applied Microbiology* 30, 575-580.

Osman, A., Bull, A.T. and Slater, J.H. (1976). Growth of mixed microbial populations on orcinol in continuous culture. *Abstracts of Papers, 5th International Fermentation Symposium* p. 124. Verlag Versuchs-und Lehranstalt fur Spiritusfabrikation und Fermentationstechnologie, Berlin.

Paris, D.F., Lewis, D.L. and Wolfe, N.L. (1975). Rates of degradation of malathion by bacteria isolated from aquatic systems, *Environmental Science and Technology* 9, 135-138.

Pawlowsky, U. and Howell, J.A. (1973). Mixed culture biooxidations of phenol. II. Steady state experiments in continuous culture, *Biotechnology and Bioengineering* 15, 897-903.

Pemberton, J.M. and Fisher, P.R. (1977). 2,4-D plasmids and persistence, *Nature, London* 268, 732-733.

Pfaender, F.K. and Alexander, M. (1972). Extensive microbial degradation of DDT *in vitro* and DDT metabolism by natural communities, *Agricultural and Food Chemistry* 20, 842-846.

Reich, M. and Bartha, R. (1977). Degradation and mineralisation of a polybutene film-mulch by the synergistic action of sunlight and soil microbes, *Soil Science* 124, 177-180.

Reineke, W. and Knackmuss, H.-J. (1979). Construction of haloaromatic utilising bacteria, *Nature, London* 277, 385-386.

Roper, M.M. and Marshall, K.C. (1974). Modification of the interaction between *Escherichia coli* and bacteriophage in saline sediment, *Microbial Ecology* 1, 1-13.

Sagoo, G.S. and Cain, R.B. (1979). Factors affecting the transfer of the catabolic plasmid specifying the utilisation of alklybenzenesulphonates between species of pseudomonas, *Society for General Microbiology Quarterly* 6, 17.

Salkinoja, M.S., Vaisanen, E. and Paterson, A. (1979).

Involvement of plasmids in the bacterial degradation of lignin-derived compounds. *In* "Plasmids of Medical, Environmental and Commercial Importance" (Ed. K.N. Timmis and A. Puhler), pp. 301-314. Elsevier/North-Holland Biomedical Press, Amsterdam.

Sayler, G.S. and Colwell, R.R. (1976). Partitioning of mercury and polychlorinated biphenyl by oil, water, and suspended sediment, *Environmental Science and Technology* 10, 1142-1145.

Senior, E., Bull, A.T. and Slater, J.H. (1976). Enzyme evolution in a microbial community growing on the herbicide Dalapon, *Nature, London* 263, 476-479.

Slater, J.H. and Bull, A.T. (1978). Interactions between microbial populations. *In* "Companion to Microbiology" (Ed. A.T. Bull and P.M. Meadow), pp. 181-206. Longman, London.

Slater, J.H., Lovatt, D., Weightman, A.J., Senior, E. and Bull, A.T. (1979) The growth of *Pseudomonas putida* on chlorinated aliphatic acids and its dehalogenase activity, *Journal of General Microbiology* 114, 125-135.

Slater, J.H. and Somerville, H.J. (1979). Microbial aspects of waste treatment with particular attention to the degradation of organic compounds. *In* "Microbial Technology: Current State, Future Prospects" (Ed. A.T. Bull, D.C. Ellwood and C. Ratledge), pp. 221-261. Cambridge University Press, Cambridge.

Stafford, D.A. and Callely, A.G. (1977). Microbiological and biochemical aspects. *In* "Treatment of Industrial Effluents" (Ed. A.G. Callely, C.F. Forster and D.A. Stafford), pp. 129-148. Hodder and Stoughton, London.

Stirling, L.A., Watkinson, R.J. and Higgins, I.J. (1976). The microbial utilisation of cyclohexane, *Proceedings of the Society for General Microbiology* 4, 28.

Sverdrup, H.W., Johnson, M.W. and Fleming, R.H. (1942). "The Oceans" Prentice Hall, Inc., Englewood Cliffs, N.J.

Tiedje, J.M. (1979). Report of Task Group VI: Persistence and extrapolation. *In* "Microbial Degradation of Pollutants in Marine Environments" (Ed. A.W. Bourquin and P.H. Pritchard) pp. 527-529. U.S. Environmental Protection Agency, Gulf Breeze, Florida.

Watson, J.R. (1977). Seasonal variation in the biodegradation of 2,4-D in river water, *Water Research* 11, 153-157.

Williams, P.P. (1977). Metabolism of synthetic organic pesticides by anaerobic microorganisms, *Residual Reviews* 66, 63-135.

Zoro, J.A., Hunter, J.M., Eglinton, G. and Ware, G.C. (1974). Degradation of p,p'-DDT in reducing environments, *Nature, London* 247, 235-237.

MICROBIAL ADAPTATION AND SELECTION

J. Howard Slater and Dianne Godwin

Department of Environmental Sciences
University of Warwick
Coventry CV4 7AL, England

INTRODUCTION

The recent past has seen the accumulation of a great body
of detailed information at the physiological, biochemical and
molecular levels which sheds light on the mechanisms of
adaptation and selection within the microbial world. An
understanding of the principles of microbial growth and the
significance of important growth parameters, such as the
maximum specific growth rate and the saturation constant, has
helped to explain why one organism (or mutant strain) may be
better adapted for growth in a particular environment than
another organism (or parent strain) (Powell 1958; Harder and
Veldkamp, 1971; Veldkamp and Jannasch, 1972; Harder, Kuenen
and Matin, 1977). However, many of these studies whilst
providing kinetic evidence for the selection of a particular
organism or strain, have failed to provide a detailed understanding of the mechanisms which give the selective growth
advantage. Studies of this kind, especially those based on
growth in continuous-flow culture systems, have demonstrated
important principles of considerable value to the microbial
ecologist seeking to explain the distribution and abundance
of particular microorganisms in their natural habitat. However,
it is important to recognise that in many instances the
conclusions drawn from laboratory-based studies may not have
been demonstrated, unequivocally, with microbial populations
growing in the natural environment.
    Another approach to microbial adaptation and selection
has come from studies at the biochemical and molecular level.
In a number of classic cases a great deal is known about the
molecular changes involved in the alteration of existing
metabolic capabilities resulting in the selection of mutant

strains with the ability to grow on novel growth substrates.
For example, much is known about the mutations involved in
uptake and regulatory mechanisms and the enzymes leading to
the metabolism of unnatural pentoses and pentitols, such as
D-arabinose and xylitol, by *Klebsiella aerogenes* (Mortlock
and Wood, 1964; Mortlock, 1976). Similarly, the evolution of
new metabolic activities by the wild-type amidase of *Pseudomonas
aeruginosa* has been extensively studied by Clarke and her
colleagues (Clarke, 1974; 1978). Again, however, these studies
have been undertaken almost exclusively under laboratory
conditions and so, whilst it is likely to be valid to assume
that events which occur in the laboratory will also occur
in growing populations in their natural environments, relatively
little is known about these processes outside the laboratory.

The present review is concerned with summarising the
evidence for the points mentioned above with reference, where
possible, to studies undertaken in the microorganism's natural
environment or in experimental systems designed to simulate
some of the important features of natural environments.
Furthermore we wish to discuss two additional factors, namely
extrachromosomal elements (plasmids) and microbial communities,
which more recently have been recognised as having significant
roles to fulfill in adaptation and selection. The discovery
of drug resistance plasmids and especially the recognition
that they are ubiquitous in bacterial populations and can be
readily transferred between the same and different species,
lead to the suggestion that they might have an important
function in microbial evolution (Anderson, 1966; 1968). More
recently these ideas have been extensively reviewed by Reanney
(1976, 1977 and 1978) with the conclusion that "extra-chromos-
omal elements provide the broad substratum of genetic plasticity
through which most adaptive processes work". Certainly
epidemiological studies of plasmid-mediated transfer of drug
resistance genes between populations of the same species or
different species has demonstrated the tremendous potential
for segregating and recombining different genes or groups of
genes in a way which could be extremely important in, for
example, the evolution of a novel metabolic sequence.

Furthermore the significance of plasmid-mediated gene
transfer might be even more important should it function
within a microbial community composed of two or more different
microorganisms which may have become associated in response
to unique environmental pressures. Certain basic interactions,
such as competition or prey-predator relationships, have for
a long time been recognised as important features of natural
populations since, for most habitats, it is almost inevitable
that different species will be in close proximity to one
another (Gause, 1934; Meers, 1973; Slater and Bull, 1978).

More recently studies with various types of continuous-flow
culture enrichment have indicated that beneficial associations
of microorganisms constituting stable entities, that is
microbial communities, can be isolated (Senior, Bull and
Slater, 1976; Daughton and Hsieh, 1977; Slater, 1978; Slater
and Somerville, 1979). Some microbial communities may
provide a permissive environment within which adaptation and
selection can occur. Others may themselves be the product
of adaptation and selection. From an ecological standpoint
it is important that the principles discovered on pure cultures
in the laboratory are examined with respect to natural
assemblages of interacting microbial populations.

SELECTION AND THE SIGNIFICANCE OF GROWTH PARAMETERS

Many growth-orientated studies have shown that "successful"
organisms, that is those which tend to dominate a particular
environment, are those which have the highest growth rate under
the prevailing conditions. In many cases the reasons for the
growth rate advantage enjoyed by one organism over others
have not been elucidated. So, for example, Jannasch (1967)
showed that a marine *Spirillum* sp. was well adapted for growth
in environments with low organic carbon concentrations whereas
at higher carbon levels, a marine pseudomonad grew considerably
faster than the *Spirillum* sp. One possible explanation
might have been that the *Spirillum* sp. possessed a better
uptake mechanism with a greater affinity for the carbon sources
at low concentrations compared with that of the *Pseudomonas*
sp. Determination of the saturation constant, K_s, for the
growth-limiting substrate, lactate, showed a three-fold lower
value for the *Spirillum* sp. ($K_s = 3.0 \times 10^{-5}$ M) compared with
that for the *Pseudomonas* sp. ($K_s = 9.0 \times 10^{-5}$ M). Similarly,
Megee, Drake, Fredrickson and Tsuchiya (1972) showed that
under conditions where glucose was a limiting resource (and
in the presence of other growth requirements, particularly
riboflavin), *Lactobacillus casei* was more competitive than
Saccharomyces cerevisiae, since the bacterium's K_s (glucose)
was over 200 times lower than that of the yeast. Similarly
Escherichia coli maintained a superior growth rate over
Azotobacter vinelandii under similar growth conditions
(Tsuchiya, Jost and Fredrickson, 1972).

More recently Kuenen, Boonstra, Schroder and Veldkamp
(1977) has shown that the growth rate of enrichment cultures
influenced the selection of the dominant population from
inocula taken from pond water under phosphate limited
conditions. At low growth rates ($0.03h^{-1}$) a *Spirillum* sp.
became the dominant population whereas at a ten-fold higher
growth rate ($0.30h^{-1}$) an unidentified rod-shaped organism
was selected. Subsequent studies showed that the *Spirillum*

sp. had a K_s (phosphate) approximately 2.5 times lower than
that for the unidentified bacterium. Moreover, the *Spirillum*
sp. appeared to have adapted for existence in low nutrient
conditions in general since it was found to be more competitive
than the rod organism at low dilution rates when potassium,
magnesium, ammonium, aspartate, succinate or lactate were
the growth-limiting substrates.

Similarly, without providing any detailed physiological
reasons, Harder and Veldkamp (1971) in an elegant series of
experiments showed that an obligately psychrophilic *Spirillum*
sp. was better adapted for growth at low temperatures compared
with a facultatively psychrophilic *Pseudomonas* sp. whereas
at higher growth temperatures the facultative organism was
more successful than the obligate psychrophile. These results
suggested that the low temperature specialist, in evolving for
growth in such an environment, carried a penalty rendering the
organism much less competitive against the facultative
psychrophile at higher temperatures. Conversely in retaining
a certain degree of growth temperature flexibility, the
facultative organism was disadvantaged for growth at low
temperatures.

More recently studies of competition between obligate and
facultative chemolithotrophic *Thiobacillus* spp. (Gottschal,
de Vries and Kuenen, 1979; Smith and Kelly, 1979) have shown
that the specialist organism, *Thiobacillus neapolitanus*,
competed successfully against the facultative strain,
Thiobacillus A2, under conditions which tended towards the
completely chemolithotrophic mode of nutrition, namely with
carbon dioxide as the sole carbon source and thiosulphate
as the energy source. However in two-membered mixed chemostat
cultures of *T. neapolitanus* and *Thiobacillus* A2 in medium
containing both acetate and thiosulphate, conditions were
readily obtained favouring the facultative organism and
resulting in the complete elimination of *T. neapolitanus*
(Gottschal *et al.*, 1979). Furthermore chemostat competition
studies between the facultative thiobacillus and a heterotrophic
organism, *Spirillum* G7, under conditions which were close to
complete heterotrophy, that is, low thiosulphate and high
acetate concentrations, showed that the broad spectrum
thiobacillus was at a disadvantage. These results have
clearly indicated that versatile organism are generally less
well adapted for extreme growth conditions but that they have
a strong selective advantage under fluctuating environmental
conditions which rarely give rise to completely favourable,
specialist conditions. This possibility was suggested by
Rittenberg (1972) and, at least in these cases, has been shown
to be broadly valid. However, Smith and Kelly (1979) noted
that under the chemostat conditions they studied, the less

competitive organisms were not removed entirely from the
growth environment, as predicted theoretically (Powell, 1958;
Slater, 1979). This phenomenon has been observed previously
in several different competition/selection experiments
(Melling, Ellwood and Robinson, 1977; Godwin and Slater,
1979; see later, p.149) and is worthy of further detailed
study since, from the viewpoint of maintaining species
diversity, it is important to understand why apparently
uncompetitive populations are retained within a particular
environment which might be expected to favour the selection
and complete domination of the most well adapted organisms.

*Selection and Physiological Reasons for Low Saturation
Constants*

   Although it may be difficult to show a direct, quantitative
relationship between an organism's growth-limiting substrate
affinity and a particular cellular function, it is clear
that there is frequently a relationship between K_s and the
affinity of a rate-limiting enzyme for the growth limiting
substrate. An excellent example comes from Hartley's group
(Hartley, Burleigh, Midwinter, Moore, Morris, Rigby, Smith
and Taylor, 1972) on competition between different strains
of *Klebsiella aerogenes* containing pentitol dehydrogenases
with different affinities for an unnatural substrate, xylitol.
The enzyme of the apparent strain had a K_m for xylitol of
approximately 1 M whilst that of a mutant had a K_m of 5 mM.
If to a culture of the parent organism growing on xylitol
as the limiting nutrient a few low affinity organisms were
added, then very rapidly the low affinity population
succeeded in dominating the culture and ousting the less
efficient parent strain.
   More recently, Matin and Veldkamp (1978) investigated the
physiological reasons for the domination of a small non-motile
Spirillum sp. at low growth rates in a lactate limited medium
over a motile *Pseudomonas* sp. which in its turn was selected
in preference to the *Spirillum* sp. at high growth rates.
The *Spirillum* sp. was found to have a K_s (lactate) of 2.3 x
10^{-5} M and a μ_{max} of 0.35 h^{-1} whilst the pseudomonad had a K_s
(lactate) of 9.1 x 10^{-5} M and a μ_{max} of 0.64 h^{-1}. Thus the
four-fold lower K_s value of the *Spirillum* sp. gave it a
selective advantage at low substrate concentrations ($\equiv$ low
growth rate) whilst the two-fold higher μ_{max} value ensured
the selective advantage of the pseudomonad at higher substrate
concentrations. It was also shown that the Michaelis-Menten
constant, K_m, for lactate transport for the *Spirillum* sp. was
three-fold lower than that for the pseudomonad (5.8 x 10^{-6} M
to 2.0 x 10^{-5} M). In addition the *Spirillum* sp. possessed a
higher V_{max} for lactate transport and had a greater surface

to volume ratio than the pseudomonad, two additional factors
which might possibly explain why the *Spirillum* sp. was
a better scavenger for lactate. The enzymes for lactate
metabolism were, however, more active in the pseudomonad
than in the spirillum further suggesting that the growth-
limiting step did involve lactate uptake rather than subsequent
substrate utilisation. Finally it was calculated that the
Spirillum sp. had a four-fold lower maintenance energy requirement than the pseudomonad (0.016 to 0.066 g lactate (g dry
weight)$^{-1}$ h^{-1}) which would also help to maximise substrate
use for biosynthetic purposes at low growth rates. These
results provide interesting evidence for the reasons behind
a selective growth advantage but there is a lack of such
detailed information in many other cases.

*Selection and Physiological Reasons for High Maximum Specific
Growth Rates*

Generally populations with the highest activity of an
enzyme involved in the rate-limiting step will be selected
in favour of lower activity organisms. So, for example,
Horiuchi, Tomizawa and Novick (1962) showed that under
lactose-limited growth conditions strains of *Escherichia
coli* containing highly elevated levels of the initial enzyme
involved in lactose metabolism, β-galactosidase – as much as
20% of the total protein was accounted for by this single
enzyme – were better adapted for growth under these conditions.
Conversely under high lactose concentrations, the hyper β-
galactosidase producing strains were much less successful
since there was no longer any advantage in synthesising high
levels of the primary, substrate-scavenging enzyme. The
precise mechanism resulting in the hyper producing strains
was not fully explained but it might have involved the
selection of mutants which were efficient at the transcription
and/or translation levels or which contained structural gene
duplication (or more). By either mechanism mutants would
result which would have a greater capacity to synthesise a
component, such as an initial catabolic enzyme present at
growth rate limiting levels, and hence effect an increase in
the organisms' growth rate.

The selection of a strain of *Pseudomonas putida* PP3 able
to use chlorinated alkanoic acids, particularly 2-monochloro-
propionic acid (2MCPA) and 2,2'-dichloropropionic acid (22DCPA),
further illustrates this point. The precise molecular mechanism
has not yet been elucidated but it could involve a mutation in
the regulatory mechanism resulting in a higher level of the
initial enzyme, dehalogenase, involved in halogenated alkanoic
acid metabolism. The parent organism, *P. putida* S3, growing
in the presence of 2MCPA contained low levels of dehalogenase

(0.3 - 0.8 µmol substrate converted (mg protein)$^{-1}$ h^{-1}) which permitted very slow (if any) growth on the chlorinated substrates. The mutant strain, *P. putida* PP3, had 2MCPA and 22DCPA dehalogenase activities which were 10 to 40 times greater than in strain S3, depending on the growth conditions (Slater, Lovatt, Weightman, Senior and Bull, 1979). These levels were adequate to account for the observed growth rates on these substrates. This example is important in an additional way since it illustrated the potential for adaptation by an organism which was a member of a naturally occurring mixed assemblage of microorganisms isolated with 22DCPA as the growth-limiting substrate (Senior, Slater and Bull, 1976). Furthermore the selection occurred despite the competitive presence of other 22DCPA utilisers in the original microbial community. It might have been anticipated that there were inadequate selection pressures to select for another 22DCPA utiliser and the exact reasons, either at the growth kinetic or dehalogenase kinetic levels, need to be examined more thoroughly. The selection of *P. putida* PP3 occurred under chemostat conditions which might have been expected to favour the selection of constitutive mutants, but this was not the case here since, in the absence of chlorinated alkanoic acids, *P. putida* PP3 did not synthesise the enzyme dehalogenase.

In subsequent experiments designed to select for mutants of *Pseudomonas putida* PP3 able to grow on novel chlorinated alkanoic acids, for example, 2-monochlorobutanoic acid (2MCBA), strains were selected with substantially elevated dehalogenase specific activities but these mutants also were inducible (A.J. Weightman and J.H. Slater, unpublished observations). For example, *P. putida* PP3 was grown in chemostat culture with a mixture of 2MCPA and 2MCBA, in the ratio of 2:5 respectively, as the combined growth-limiting substrate at a dilution rate of 0.075 h^{-1} (Fig. 1). After 2500 h of growth a new strain, *P. putida* PP310, was selected capable of utilising 2MCBA for biomass production and with a 2MCBA dehalogenase specific activity which was 10 fold higher than the maximum observed in the parent strain (1.8 compared with 18.6 µmol substrate converted (mg protein)$^{-1}$ h^{-1}). Previous work had shown that *P. putida* PP3 contained two distinct dehalogenase proteins (Weightman, Slater and Bull, 1979) and that over 80% of the 2MCBA dehalogenase activity was due to the Fraction I dehalogenase. It was not surprising to find that in *P. putida* PP310 the dehalogenase activity against the three major substrates for this dehalogenase were also increased by approximately 10 fold (Table 1). Moreover the ratio of MCA, 2MCPA, 22DCPA and 2MCBA activities remained constant

showing that the selection had not involved a change in the
dehalogenase in terms of altered substrate affinities.
Furthermore the selection had not involved an increase in the
DCA dehalogenase activity since this was associated with the
Fraction II dehalogenase.

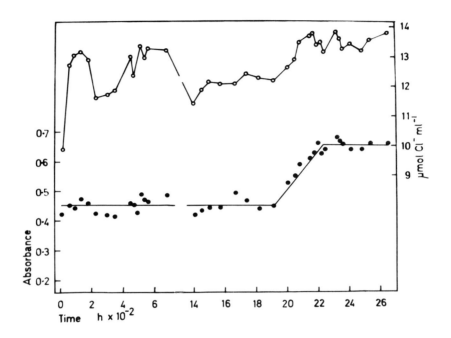

*Fig. 1.   The selection of Pseudomonas putida PP310 from
P. putida PP3 in a 2MCPA:2MCBA-limited chemostat culture
(see text for details)●, absorbance; ○, chloride release.*

Furthermore, the changed ratios with respect to DCA activity
were maintained when the new strain was grown in batch
culture (although the specific activities were characteristic-
ally lower) indicating that a permanent change had occurred
in the mutant (unlike the hyper β-galactosidase producing
strains described by Horiuchi *et al.*, 1962)).   It remains to
be seen whether these elevated Fraction I dehalogenase
activities were due to gene multiplication.

   Increased enzyme specific activities have been ascribed
to mutations resulting in the duplication of the structural
(enzyme) gene in other systems, for example, the ribitol
dehydrogenase of *Klebsiella aerogenes* (Mortlock and Wood, 1971;

Hartley *et al.*, 1972; Hartley, 1974; Rigby, Burleigh and
Hartley, 1974; Inderlied and Mortlock, 1977). One aspect of
these experiments was to determine whether or not the strong

TABLE 1

The Selection of Pseudomonas putida PP310
in 2MCPA:2MCBA Limited Chemostat Culture

Strain	Specific activity (μmol (mg protein)$^{-1}$ h^{-1}) (Relative activities with respect to 2MCPA)				
	MCA	DCA	2MCPA	22DCPA	2MCBA
PP3 (chemostat culture)	12.0 (2.26)	11.2 (2.11)	5.3 (1.0)	2.5 (0.47)	1.8 (0.34)
Chemostat culture after 2500 h	77.4 (1.65)	17.4 (0.37)	46.8 (1.0)	25.8 (0.55)	18.6 (0.39)
Batch culture inoculated from chemostat	22.2 (1.76)	7.2 (0.57)	12.6 (1.0)	8.4 (0.67)	6.0 (0.47)
PP310 (batch grown)	16.2 (1.93)	5.4 (0.64)	8.4 (1.0)	3.6 (0.43)	2.6 (0.31)

selective growth conditions obtainable in a chemostat culture
could result in the selection of mutant enzymes with altered
K_m or V_{max} characteristics towards the novel substrates.
Ribitol dehydrogenase has a K_m (ribitol) of approximately 1 mM
and a slight activity towards xylitol, K_m = 1 M (Hartley,
1974). Constitutive mutants were able to grow slowly on
xylitol, but in selection experiments designed to improve
the enzyme's K_m or V_{max} for xylitol, the only result was an
increase in the enzyme's specific activity with no change in
K_m or V_{max}. In one chemostat culture the parent strain with
a ribitol dehydrogenase activity of 1.0 unit which comprised
1.2% of the total cell protein, was initially replaced by a
mutant *K. aerogenes* strain which had a 4.5 fold increase in
ribitol dehydrogenase activity (6% of the total cell protein).
In turn this mutant was replaced by another with a level of
enzyme nearly 16 times that of the parent (18% of the total
cell protein). These results clearly demonstrated that the
favoured mutational response involved gene doubling rather
than a change in the structural protein.
  Other studies with this enzyme (Wu, Lin and Tanaka, 1968)

and other enzymes, especially the aliphatic amidases of
Pseudomonas aeruginosa (Clarke, 1974; 1979), have amply
demonstrated that mutants can be selected with enzymes
altered with respect to substrate activity. Hartley and
his colleagues (Hartley, 1974) have shown that continuous
culture populations subjected to increased mutagenesis (by the
presence of nitrosoguanidine) produced mutants with changed
enzyme specificity. This raises the important question of
what environmental pressures, other than substrate limitation
and the presence of a non-metabolisable carbon source,
influence the process of evolution. An important development
for the future will be to analyse systematically those factors
which influence the evolution of populations growing in their
natural environment or in laboratory systems which simulate
natural habitats. The experiments undertaken in continuous
culture to date have, however, illustrated one significant
point: namely that gene doubling can occur quite readily
and this is a process which, it has been argued, is a
prerequisite for the subsequent evolution of new enzymes with
novel catalytic functions (Hartley, 1974).

SELECTION DUE TO THE LOSS OF A METABOLIC CAPABILITY

In environments in which a particular metabolic capability
or activity becomes redundant, mutants which lack that
capability or activity may be at a selective growth advantage
and, under appropriate growth conditions, succeed in becoming
the dominant population.

Zamenhof and Eichhorn (1967) showed substantial growth
rate differences between a prototrophic strain of *Bacillus
subtilis* and several strains of amino acid auxotrophs. Under
carbon-limited conditions in chemostat culture in the presence
of the auxotroph's requirement, strains which had lost the
capacity to synthesise histidine or tryptophan separately
grew faster than the wild-type parent strain. Under these
conditions the auxotroph did not have to utilise a small,
but significant, fraction of the restricted growth resource
to synthesise the enzymes of a redundant pathway. Furthermore,
in the absence of the pathway none of the growth-limiting
substrate was unavailable in unnecessary intermediate metabolic
pools. Thus for the auxotroph a greater proportion of the
growth substrate was utilised for essential biosynthetic
functions, which resulted in an increased growth rate. It
might have been anticipated that, since the prototroph and
auxotroph strains were practically isogenic differing only
in a single biosynthetic function, the growth rate difference
would have been very small. Surprisingly this was not the
case and growth rate differences of up to 70% were reported.
More recently we have examined a similar type of inter-

action to determine the growth rate difference between a
prototroph and an auxotroph (Mason and Slater, 1979). In
glucose-limited chemostats in the presence of non-growth-
limiting concentrations of tyrosine, a tyrosine auxotrophic
strain of *Escherichia coli* displaced the isogenic prototrophic
parent. From the washout kinetics of the prototroph from
such a mixed culture it was possible to show that under
conditions which supported the growth of the *E. coli* tyrosine
auxotroph at a specific growth rate of 0.13 h^{-1}, the protrophic
strain was only capable of growth at 0.094 h^{-1}. Thus under
these conditions the auxotrophic strain had a 38% greater
specific growth rate, solely due to the loss of the tyrosine
biosynthetic capability. In these experiments, it is not
readily apparent why such a minor change in a single biosynthetic capability out of the total complex sequence of
reactions leading to new biomass production, should lead to
such large growth rate differences.

It has also been shown that organisms which wastefully
utilise growth resources, for example, strains of *Escherichia
coli* with the inability to regulate the synthesis of proline
and which excrete excess proline into the environment, have
a growth rate disadvantage compared with strains which maximise
the effective use of growth resources (Baich and Johnson,
1968). However the more complex the growth medium the less
of a disadvantage the wasteful metabolism is to the organism.
For example wild type and proline-excreting strains of
Escherichia coli have a growth rate difference of 1.8% (faster
for wild-type) in brain-heart infusion broth compared with
4.7% in mineral defined medium. The difference increased to
7.0% in defined minimal medium supplemented with proline,
presumably because the wild-type organism was able to repress
the operation of the proline pathway thereby making better
use of the available growth resources.

More recently it has been argued that the presence of
plasmids in populations growing in an environment in which
the functions coded for by the plasmid were not required,
constituted a major growth disadvantage (Melling *et al.*,
1977; Wouters, Rops and van Andel, 1978; Adams, Kinney, Thompson,
Rubin and Helling, 1979; Godwin and Slater, 1979; Jones and
Primrose, 1979). The presence of a drug resistance plasmid
in a bacterial population growing in a medium lacking any
drugs, represents an additional cellular component requiring
elemental and energy resources which otherwise might be
utilised for basic biomass production. Melling *et al.*, (1977)
examined the competitive growth of *Escherichia coli* strain
W3110 with and without plasmid RP1 which conferred resistance
to ampicillin (Ap), carbenicillin (Cb), tetracycline (Tc),
kanamycin (Kn) and neomycin (Nm). To steady state cultures of

E. coli W3110 (RP1) growing at D = 0.30 h^{-1} under magnesium-, carbon- or phosphorus-limited conditions a 1% inoculum of *E. coli* W3110 R^{-} population succeed in ousting the R^{+} parent population. Under carbon- or magnesium-limited conditions the R^{-} population was gradually eliminated. However, it has been shown that establishing competition experiments in this fashion markedly influences the outcome of competition (Godwin and Slater, 1979; Mason and Slater, 1979). Competition experiments have to be established either by inoculating the mixed culture simultaneously with chemostat grown organism or allow a period of non-competitive mixed culture growth, in order to eliminate difficulties due to different physiological states of the competing organisms.

A selective advantage can occur by the loss of a single plasmid-borne drug resistance marker (Godwin and Slater, 1979). Under both carbon- and phosphorus-limited growth conditions in a chemostat at any growth rate between D = 0.10 h^{-1} and 0.35 h^{-1}, *Escherichia coli* 1R713 (TP120A) which contained a plasmid coding for resistance to Ap, sulphonamide (Su) and streptomycin (Sm), was always more competitive than the parent, isogenic *E. coli* 1R713 (TP120) which contained a plasmid coding for resistance to Ap, Su, Sm and Tc. At the highest growth rate, 0.35 h^{-1}, conditions were established under phosphate-limited conditions which permitted the parent, four drug resistant strain to grow at $\mu = 0.22$ h^{-1} which therefore resulted in the washout of the tetracycline resistant strain. That is, the parent strain was only capable of growth at 67% of the rate of the tetracycline sensitive strain. At the lowest growth rate examined (0.10 h^{-1}) for the competitive, tetracycline sensitive population, conditions were established which permitted the parent population to grow at only 15% of the dominant population's growth rate. That is, the lower growth rate of the competitive population (and so the lower the concentration of the unused growth-limiting substrate), the greater the competitive disadvantage of the tetracycline resistance population. Indeed under carbon-limited conditions, the parent population was essentially eliminated as a non-growing population.

Under these conditions it seemed that the loss of the tetracycline resistance mechanism conferred a significant growth rate advantage. This may have been due to the loss of part of the plasmid DNA coding for tetracycline resistance and it has been shown that the TP120A plasmid (MW = 20.0 x 10^6 daltons) was nearly 40% smaller than the parent TP120 plasmid (MW = 31.6 x 10^6 daltons) (D. Godwin and J.H. Slater, unpublished observations).

Closed culture experiments showed that *E. coli* 1R713 (TP 120A) had a 20% greater μ_{max} than *E. coli* 1R713 (TP120). Furthermore, the plasmid-minus *E. coli* 1R713 had a

μ_{max} value which was greater than twice the rate of the four
drug resistant population, confirming the general conclusions
of other workers (Yokota, Kasuga, Kaieko and Kuwahara, 1972;
Inselburg, 1978; Nakazawa, 1978). Greater detailed experimentation, however, is required to explain the large growth
rate differences since these seem to be excessive simply in
terms of conserving limited growth resources, in replicating
a smaller plasmid.

The large growth rate differences, particularly at low
growth rates and low growth-limiting substrate concentrations,
suggested that it would be extremely likely that plasmid-
minus strains would be selected during chemostat culture.
For the system we have examined, however, this was not the
case since in many different selection experiments inititated
with the parent strain, *E. coli* 1R713 (TP120) (Ap, Sm, Su, Tc),
a plasmid-minus strain was selected on only one occasion after
over 700 h growth under phosphate-limited conditions. Plasmid
TP120 appeared to be even more stable under carbon-limited
conditions because Tc was the only resistance marker shown
to be lost. Under phosphate-limited conditions Ap was
regularly lost, as well as Tc, resulting in the selection of
a dominant population *E. coli* 1R713 (TP120D) (Su, Sm). It
is possible that the loss of Tc resistance genes (and perhaps
Ap resistance as well) was associated with the elimination
of a transposon from the plasmid (Cohen, 1976).

In all the selection experiments and in the two-membered
competition experiments, the uncompetitive populations were
always retained at very low levels ranging from 0.001% to 1%
of the total population. This should not occur in open growth
systems, the uncompetitive population should be washed out
completely. However, these observations have been reported
elsewhere (Melling *et al.*, 1977; Adams *et al.*, 1979) and
although growth system artifacts cannot be completely excluded,
the possibility does exist that some mechanism, as yet
undefined, operates to retain unwanted genes within the
population as a whole, despite any resulting growth-rate
disadvantage. For example, Adams *et al.*, (1979) have shown
that providing more than 50% of the population was composed
of *E. coli* RH202 (RSF 2124) which contained a plasmid coding
for Col-El bacteriocin and Ap resistance, then the plasmid-
minus *E. coli* RH202 was uncompetitive. In this case the
presence of the Col factor ensured that the plasmid-minus
strain did not have a growth rate advantage. Other workers
have suggested that drug resistance plasmids also code for
other functions which are important in ensuring the plasmids'
maintenance. Grabow, Middendorf and Prozesky (1973) showed
that under non-challenge conditions drug resistant coliforms
present in sewage maturation ponds increased from less than

1% of the total population of 2.5% as effluent flowed through
the ponds. It was suggested that, in addition to their drug
resistance, these strains also produced resistance mechanisms
against antibacterial materials, such as chlorellin produced
by unicellular algae. Such mechanisms could ensure the
continuing presence of a plasmid which may code for several
activities and functions not directly selected for.

MIXED MICROBIAL POPULATIONS AND GENE TRANSFER

Gene Transfer Mechanisms

It is well understood that a number of different mechanisms
exist which effect the transfer of DNA between microorganisms
of the same and of different species. These are transduction,
transformation and conjugation involving the transfer of
chromosome genes or simply plasmids. Relatively little is
known about the operation of gene transfer mechanisms between
populations under natural conditions. There is some
circumstantial evidence, particularly with respect to the
transfer of drug resistance mechanisms, suggesting that the
movement of genetic information occurs readily. For example,
the TEM-type β-lactamase which confers ampicillin resistance
has been shown to be identical from plasmids of 14 different
incompatibility groups found in 13 different bacterial genera
(Reanney, 1977). This has important evolutionary consequences
since it strongly suggests that once a successful protein
has evolved in one species, the original gene can be
distributed into many different species for their mutual
benefit. In principle, therefore, it is possible that a
successful gene could evolve only once instead of arising
independently in different species. The importance of gene
transfer seems to be well established for drug resistance
mechanisms but some of the more extravagant claims of its
importance ought to be tempered with some caution until it
has been conclusively demonstrated that gene transfer occurs
widely for other activities and functions. Furthermore,
whilst it may be true that "DNA transfer is a prime vehicle
of adaptive change in all prokaryotes" (Reanney, 1977), it
is important to recognise that a successful gene has to evolve
somewhere in the first place.

Morrison, Miller and Sayler (1978) have recently demonstrated
the operation of a transduction mechanism involving the transfer
of streptomycin resistance by the generalised transducing
phage F116 between populations of *Pseudomonas aeruginosa*
growing in a flow-through chamber located in a freshwater
reservoir. The frequencies of transduction ranged from 1.4 x
10^{-5} to 8.3 x 10^{-2} transductants per recipient and these
were similar to the rates of transduction under laboratory

conditions, providing clear evidence that transduction can
be an important mechanism under natural conditions. Transformation (Herdman, 1975) is not well characterised for
populations growing under natural conditions but here too the
evidence suggests that gene transfer by transformation for
species such as *Bacillus subtilis, Streptococcus pneumoniae*
and others may be a relatively frequent event (Saunders, 1979).
The importance of such transfer mechanisms ought not to be
minimised particularly with respect to populations growing
at low growth rates and so resulting in significant loss of
viability, cell lysis and the release of macromolecules,
including DNA, into the environment.

Much more is known about the potential of gene transfer
by conjugation. Anderson (1975) showed that wild-type enteric
bacteria growing in the human intestine acted as recipients
for the transfer of plasmid F-*lac*-Tc, although the frequency
of *in vivo* transfer was low in the absence of any antibiotic
selection pressures. Similarly Smith (1977a) showed that
substantial R factor transfer occurred *in vivo* in sheep
rumens in the absence of any antibiotic treatment provided
that the animals were starved for 24-28 h prior to being
inoculated with the drug resistant strains of *Escherichia
coli*. In another example, Novick and Morse (1967) showed
that the transfer of multiple drug resistance plasmids
between different strains of *Staphylococcus aerueus* occurred
in vivo in mice kidneys. Transfer of virulence from a
pathogenic to a non-pathogenic *Agrobacterium* sp. has been
demonstrated in mixed cultures inoculated onto tomato seedlings
(Kerr, Manigault and Tempé, 1977). The transfer of plasmids
possessing genes for enterotoxin production has been demonstrated
in vivo in pigs (Gyles, Falkow and Rollins, 1978). In
addition, many other studies have demonstrated the potential
of gene transfer between different species. Smith (1977b)
showed that R factor transfer could occur between *Escherichia
coli* and *Salmonella* species in the rumen of sheep. Lacy and
Leary (1975) showed the transfer of drug resistance plasmid
RP1 from *Pseudomonas aeruginosa* to *Escherichia coli* and then
into *Pseudomonas glycinea*. When this resistant strain was
inoculated onto *Phaseolus limensis* leaves and pods, RP1-
carrying strains of *Pseudomonas phaseolicola* were readily
isolated after a few hours of mixed culture growth. Drug
resistance factors have been shown to be transferred between
very different pairs of organisms, for example *Escherichia
coli* and *Caulobacter crescentus* (Ely, 1979) and catabolic
plasmids coding for toluene and xylene metabolism between
Escherichia coli and *Pseudomonas putida* (Jacoby, Rogers,
Jacob and Hedges, 1978).

In addition plasmids can mediate the transfer of genes

located in the bacterial chromosome (Holloway, 1979). For
example, Hedges and Jacob (1977) showed that up to 4% of the
chromosome of *Pseudomonas aeruginosa* PAC174 and carrying the
arg G gene (for arginine biosynthesis) was transferred to
an arginine auxotroph of *Escherichia coli* in a process
mediated by plasmid R68-44.

These selected examples serve to illustrate the vast
potential for various types of gene transfer and clearly
raise important evolutionary consequences. Whilst comparatively
little is known about the frequency of such events in mixed
natural populations, gene transfer does occur in natural
environments (Weinberg and Stotzky, 1972; Graham and Istock,
1979) and need to be systematically analysed to assess its
importance in producing adaptive changes in microbial
populations.

Evolution from Mixed Cultures or Microbial Communites

It is now quite clear that enrichment cultures frequently
result in the isolation of stable mixed cultures or microbial
communities which, as an assemblage, are better adapted for
growth under the enrichment conditions established than a
single microbial species (Slater, 1978; Slater and Somerville,
1979). In some cases it has been shown that the ability to
degrade a given carbon and energy source or simply to
transform a particular compound depends on a concerted
metabolic attack by the component species of the community.
It may be that alone none of the individual organisms can
grow on the carbon source, for example, linear alkylbenzenesulphonates (Johanides and Hršak, 1976), or that they do not
transform the compound in question to the maximum extent
compared with the complete mixture, for example the degradation of Diazinon (Gunner and Zuckerman, 1968) or the
metabolism of cyclohexanes (Beam and Perry, 1974).

If we consider a two-membered microbial community able to
utilise compound A as its sole carbon and energy source but
neither organism X or Y are capable of growing on compound
A, then it is possible to envisage a situation in which a
novel organism Z or Z' could evolve the capacity to grow on
A either by transferring the genes for enzymes a and b from
X into Y (organism Z) or by transferring the genes for
enzymes c and d from Y to X (organism Z') (Table 2). An
excellent example of the initial starting situation has been
shown by Daughton and Hseih (1977) with a microbial community
growing on the insecticide Parathion and which was isolated
by continuous-flow culture enrichment. The basic core of
the community involved two organisms, *Pseudomonas stutzeri*
and *Pseudomonas aeruginosa*, which together could use Parathion

TABLE 2

*The Selection of a Novel Metabolic Sequence
in a Single Organism from an Interacting
Microbial Community*

Pathway

$$A \xrightarrow{a} B \xrightarrow{b} C \xrightarrow{c} D \xrightarrow{d} E$$

Organism	Capability
X	Requires E for growth No growth on A Contains enzymes a and b Intermediate C accumulates; no growth on C
Y	Requires E for growth No growth on A Contains enzymes c and d Growth on C since E it can produce E
X + Y	Growth on A Complete pathways; enzymes a,b,c and d present
Z	Growth on A Complete pathway; enzymes a and b transferred from X to Y
Z'	Growth on A Compete pathway; enzymes c and d transferred from Y to X

as a carbon and energy source but separately neither could grow on the compound. It was shown that *P. stutzeri* possessed a Parathion hydrolysing capability yielding two products, diethylthiophosphate and ρ-nitrophenol, neither of which served as a carbon source for the organism. *P. aeruginosa* was unable to hydrolyse Parathion but could grow on one of the breakdown products, namely ρ-nitrophenol. In the complete community, excreted metabolites or cell lysis products served as the carbon source to maintain the growth of *P. stutzeri*. In the continuous-flow culture experiments reported by Daughton and Hseih (1977) there was no evidence to suggest that a single organism derived from the two pseudomonads and capable of growing individually on Parathion, did evolve.

Neverthless it is possible that such an organism could be
selected by transferring the Parathion hydrolysing capability
from *P. stutzeri* to *P. aeruginosa*. This could be achieved,
for example, by introducing a promiscuous plasmid into *P.
stutzeri* and selecting for strains of *P. aeruginosa*
containing the plasmid plus the relevant region of the
P. stutzeri chromosome containing the gene(s) for the
hydrolysis mechanism.

More recently Knackmus and his colleagues (Reineke and
Knackmus, 1979; Hartman, Reineke and Knackmus, 1979) have
demonstrated the central role plasmids may have in the
evolution of new strains of bacteria with novel catabolic
capabilities. A parent strain *Pseudomonas* sp B13 (WR1) was
shown to be able to grow on 3-chlorobenzoic acid or 4-chloro-
phenol but was unable to utilise 2-chlorobenzoic acid,
4-chlorobenzoic acid or 3,5-dichlorobenzoic acid. The
inability to utilise these chlorinated aromatics was due to
the specificity of the initial dioxygenase required to form
halogenated catechols. However, *Pseudomonas putida* mt-2 was
shown to contain a benzoate 1,2-dioxygenase, whose gene was
located on a TOL plasmid, which could metabolise chlorobenzoic
acids, including 4-chlorobenzoate and 3,5-dichlorobenzoate.
P. putida, however, could not grow on these compounds because
the strain was unable to metabolise further the chlorinated
catechols which were generated. In one enrichment experiment
the two organisms were grown in a chemostat initially with
3-chlorobenzoate and 4-methylbenzoate (Hartman *et al.*, 1979).
After 4 weeks operation, 4-chlorobenzoate was added as an
additional carbon source since alone it could not support
the growth of either organism. After another four week
period the culture was switched to 4-chlorobenzoate alone and
during the next twelve weeks 3,5-dichlorobenzoate was added
and eventually colonies able to grow on 3,5-dichlorobenzoate
alone were isolated. Eventually *Pseudomonas* sp (WR912) was
isolated and shown to be capable of growth on 3-chloro-,
4-chloro- and 3,5-dichlorobenzoate. This experiment strongly
suggested that there had been a selection of a strain
containing the initial dioxygenase enzyme from *P. putida* mt-2
and the chlorinated catechol degrading enzymes from
Pseudomonas sp B13 (WR1). Reineke and Knackmus (1979) showed
more directly that in Millipore filter mating experiments with
Pseudomonas sp B13 (WR1) and *P. putida* mt-2, exconjugants
such as *Pseudomonas* sp WR241 capable of growing on 4-chloro-
benzoate were isolated and analysis of this strain showed
that it was basically the same as strain WR1 and the ability
to utilise 4-chlorobenzoate (coded for by the *P. putida* TOL
plasmid) was the only new capability. It was conlcuded,
therefore, that *P. putida* mt-2 acted as a donor, transferring

its plasmid to the recipient strain.

Another possible example of such evolution is currently under investigation. *Pseudomonas putida* PP3 can dechlorinate monochloracetic acid but cannot grow on the breakdown product, glycollate. However, mixed cultures can be established in which *P. putida* PP3 growing on a mixture of 2MCPA and MCA cometabolises the second substrate producing glycollate which may be utilised by a glycollate utiliser which cannot grow on MCA. Currently we are examining the possibility of transferring dehalogenase gene(s) to the glycollate utilisers.

REFERENCES

Adams, J., Kinney, T., Thompson, S., Rubin, L. and Helling, R.B. (1979). Frequency-dependent selection for plasmid-containing cells of *Escherichia coli*, *Genetics* 91, 627-637.

Anderson, E.S. (1966). Possible importance of transfer factors in bacterial evolution, *Nature, London* 209, 637-638.

Anderson, E.S. (1968). The ecology of transferable drug resistance in the enterobacteria, *Annual Review of Microbiology* 22, 131-180.

Anderson, E.S. (1975). Viability of, and transfer of a plasmid from, *E. coli* K12 in the human intestine, *Nature, London* 255, 502-504.

Baich, A. and Johnson, M. (1968). Evolutionary advantage of control of a biosynthetic pathway, *Nature, London* 218, 464-465.

Beam, H.W. & Perry, J.J. (1974). Microbial degradation of cycloparaffinic hydrocarbons via cometabolism and commensalism, *Journal of General Microbiology* 82, 163-169.

Clarke, P.H. (1974). The evolution of enzymes for the utilisation of novel substrates. *In* "Evolution in the Microbial World", 24th Symposium of the Society for General Microbiology (Ed. M.J. Carlile and J.J. Skehel) pp. 183-217. Cambridge University Press, London.

Clarke, P.H. (1978). Experiments in microbial evolution. *In* "The Bacteria VI" (Ed. L.N. Orston and J.R. Sokatch), pp. 137-218. Academic Press, New York.

Cohen, S.N. (1976). Transposable genetic elements and plasmid evolution, *Nature, London* 263, 731-738.

Daughton, C.G. and Hsieh, D.P.H. (1977). Parathion utilisation by bacterial symbionts in a chemostat, *Applied and Environmental Microbiology* 34, 175-184.

Ely, B. (1979). Transfer of drug-resistance factors to the dimorphic bacterium *Caulobacter crescentus*, *Genetics* 91, 371-380.

Gause, G.F. (1934). "The Struggle for Existance", Dover Publications, New York.

Godwin, D. and Slater, J.H. (1979). The influence of the growth environment on the stability of a drug resistance plasmid in *Escherichia coli* K12, *Journal of General Microbiology* 111, 201-210.

Gottschal, J.C., de Vries, S. and Kuenen, J.G. (1979). Competition between the facultatively chemolithotrophic *Thiobacillus* A2, an obligately chemolithotrophic *Thiobacillus* and a heterotrophic spirillum for inorganic and organic substrates, *Archives for Microbiology* 121, 241-249.

Grabow, W.O.R., Middendorff, I.G. and Prozesky, O.W. (1973). Survival in maturation ponds of coliform bacteria with transmissible drug resistance, *Water Research* 7, 1589-1597.

Graham, J.B. and Istock, C.A. (1979). Gene exchange and natural selection cause *Bacillus subtilis* to evolve in soil culture, *Science* 204, 637-639.

Gunner, H.B. and Zuckerman, B.M. (1968). Degradation of 'Diazinon' by syngergistic microbial action, *Nature, London* 217, 1183-1184.

Gyles, C., Falkow, S. and Rollins, L. (1978). *In vivo* transfer of a plasmid possessing genes for *Escherichia coli* enterotoxin, *American Journal of Vetinary Research* 39, 1438-1441.

Harder, W. and Veldkamp, H. (1971). Competition of marine psychrophilic bacteria at low temperatures, *Antonie van Leewenhoek* 37, 51-63.

Harder, W., Kuenen, J.G. and Matin, A. (1977). A review: microbial selection in continuous culture, *Journal of Applied Bacteriology* 43, 1-24.

Hartley, B.S. (1974). Enzyme families. *In* "Evolution in the Microbial World", 24th Symposium of the Society for General Microbiology (Ed. M.J. Carlile and J.J. Skehel), pp. 151-182. Cambridge University Press, London.

Hartley, B.S., Burleigh, B.D., Midwinter, G.G., Moore, C.H., Morris, H.R., Rigby, P.W.J., Smith, M.J. and Taylor. S.S. (1972). Where do enzymes come from? *In* "Enzymes: Structure and Function", 8th FEBS Meeting (Ed. J. Drenth, R.A. Oosterbaan and C. Ch. Veeger), pp. 151-176. North Holland, Amsterdam.

Hartmann, J., Reineke, W. and Knackmus, H-J. (1979). Metabolism of 3-chloro-. 4-chloro- and 3,5-dichlorobenzoate by a pseudomonad, *Applied and Environmental Microbiology* 37, 421-428.

Hedges, R.W. and Jacob, A.E. (1977). *In vivo* translocation of genes of *Pseudomonas aeruginosa* onto a promiscuously transmissible plasmid, *FEMS Microbiology Letters* 2, 15-19.

Herdman, M. (1975). Transformation in the blue-green alga
    Anacystis nidulans and the associated phenomen of mutation.
    In "Bacterial Transformation" (Ed. L.J. Archer), pp. 369-
    386. Academic Press, New York.
Holloway, B.W. (1979). Plasmids that mobilise bacterial
    chromosome, *Plasmid* 2, 1-19.
Horiuchi, T., Tomizawa, J. and Novick, A. (1962). Isolation
    and properties of bacteria capable of high rates of
    β-galactosidase synthesis, *Biochemica et biophysica acta*
    55, 152-163.
Inderlied, C.B. and Mortlock, R.P. (1977). Growth of
    Klebsiella aerogenes on xylitol: implications for bacterial
    enzyme evolution, *Journal of Molecular Evolution* 9, 181-
    190.
Inselberg, J. (1978). Col E1 plasmid mutants affecting
    growth of an *Escherichia coli recB rec^C sbcC* mutant,
    Journal of Bacteriology 133, 433-436.
Jacoby, G.A., Rogers, J.E., Jacob, A.E. and Hedges, R.W.
    (1978). Transposition of *Pseudomonas* toluene-degrading
    genes and expression in *Escherichia coli, Nature, London*
    274, 179-180.
Jannasch, H.W. (1967). Enrichment of aquatic bacteria in
    continuous culture, *Archiv für Mikrobiologie* 59, 165-173.
Johnides, V. and Hršak, D. (1976). Changes in mixed bacterial
    cultures during linear alkybenzenesulphonate (LAS)
    biodegradation. *In* "Fifth International Symposium on
    Fermentation", (Ed. H. Dellweg), p. 426. Berlin,
    Westkreuz.
Jones, I. and Primrose, S.B. (1979). The survival of pBR
    322 in continuous culture. *Society for General
    Microbiology Quarterly* 6, 97-98.
Kerr, A., Manigault, P. and Tempe, J. (1977). Transfer of
    virulence *in vivo* and *in vitro* in *Agrobacterium, Nature,
    London* 265, 560-561.
Kuenen, J.G., Boonstra, J., Schroder, H.G.J. and Veldkamp,
    H. (1977). Competition for inorganic substrates among
    chemoarganotrophic and chemolithotrophic bacteria,
    Microbial Ecology 3, 119-130.
Lacy, G.H. and Leary, J.V. (1975). Transfer of antibiotic
    resistance plasmid RP1 into *Pseudomonas glycinea* and
    Pseudomonas phaseolicola in vitro and *in planta, Journal
    of General Microbiology* 88, 49-57.
Mason, T.G. and Slater, J.H. (1979). Competition between
    an *Escherichia coli* tyrosine auxotroph and a prototrophic
    revertant in glucose- and tyrosine-limited chemostats,
    Antonie van Leeuwenhoek 45, 253-263.
Matin, A. and Veldkamp, H. (1978). Physiological basis of

the selective advantage of a *Spirillum* sp. in a carbon-
limited environment, *Journal of Microbiology* 105, 187-197.
Meers, J.L. (1973). Growth of bacteria in mixed cultures,
CRC Critical Reviews in Microbiology 2, 139-184.
Megee III, R.D., Drake, J.F., Fredrickson, A.G. and Tsuchiya,
H.M. (1972). Studies in intermicrobial symbiosis.
Saccharomyces cerevisiae and *Lactobacillus casei, Canadian
Journal of Microbiology* 18, 1733-1742.
Melling, J., Ellwood, D.C. and Robinson, A. (1977). Survival
of R factor carrying *Escherichia coli* in mixed cultures
in the chemostat, *FEMS Microbiology Letters* 2, 87-89.
Morrison, W.D., Miller, R.V. and Sayler, G.S. (1978).
Frequency of F116-mediated transduction of *Pseudomonas
aeruginosa* in a freshwater environment, *Applied and
Environmental Microbiology* 36, 724-730.
Mortlock, R.P. (1976). Catabolism of unnatural carbohydrates
by microorganisms. *Advances in Microbial Physiology* 13,
1-53.
Mortlock, R.P. and Wood, W.A. (1964). Metabolism of pentoses
and pentitols by *Aerobacteria aerogenes*. Demonstration
of pentose isomerase, pentalokinase and pentitol dehydro-
genase enzyme families, *Journal of Bacteriology* 88,
838-844.
Mortlock, R.P. and Wood, W.A. (1971). Genetic and enzymatic
mechanisms for the adaptation to novel substrates by
Aerobacter aerogenes. *In* "Biochemical Responses to
Environmental Stress" (Ed. I.A. Berstein), pp. 1-14.
Plenum, New York.
Nakazawa, T. (1978). *TO1* plasmid in *Pseudomonas aeruginosa*
PAO: Thermosensitivity of self maintenance and inhibition
of host cell growth, *Journal of Bacteriology* 133, 527-535.
Novick, R.P. and Morse, S.I. (1967). *In vivo* transmission
of drug resistance factors between strains of
Staphylococcus aureus, Journal of Experimental Medicine
125, 45-58.
Powell, E.O. (1958). Criteria for growth of contaminants
and mutants in continuous culture, *Journal of General
Microbiology* 18, 259-268.
Reanney, D. (1976). Extrachromosomal elements as possible
agents of adaptation and development, *Bacteriological
Reviews* 40, 552-590.
Reanney, D. (1977). Gene transfer as a mechanism of microbial
evolution *Bioscience* 27, 340-344.
Reanney, D.C. (1978). Genetic engineering as an adaptive
strategy. *In* "Genetic interaction and gene transfer",
Brookhaven Symposia in Biology 39, 248-271.
Reineke, W. and Knackmus, H-J. (1979). Construction of

haloaromatics utilising bacteria, *Nature, London* <u>277</u>, 385-386.

Rigby, R.W.J., Burleigh, B.D. and Hartley, B.S. (1974). Gene duplication in experimental enzyme evolution, *Nature, London* <u>251</u>, 200-204.

Rittenberg, S.C. (1972). The obligate autotroph - the demise of a concept, *Antonie van Leewenhoek* <u>38</u>, 457-478.

Saunders, J.R. (1979). Specificity of DNA uptake in bacterial transformation, *Nature, London* <u>278</u>, 601-602.

Senior, E., Bull, A.T. and Slater, J.H. (1976). Enzyme evolution in a microbial community growing on the herbicide Dalapon, *Nature, London* <u>263</u>, 476-479.

Senior, E., Slater, J.H. and Bull, A.T. (1976). Growth of mixed microbial populations on the herbicide sodium 2,2'-dichloropropionate (Dalapon) in continuous culture, *Journal of Applied Chemistry and Biotechnology* <u>26</u>, 329-330.

Slater, J.H. (1978). The role of microbial communities in the natural environment. *In* "The Oil Industry and Microbial Ecosystems" (Ed. K.W.A. Chater and H.J. Somerville), pp. 137-154. Heyden and Sons, London.

Slater, J.H. (1979). Population and community dynamics. *In* "Microbial Ecology - a Conceptual Approach" (Ed. J.M. Lynch and N.J. Poole), pp. 45-63. Blackwell Scientific Publications, Oxford.

Slater, J.H. and Bull, A.T. (1978). Interactions between microbial populations. *In* "Companion to Microbiology" (Ed. A.T. Bull and P.M. Meadow), pp. 181-201. Longmans, London.

Slater, J.H., Lovatt, D., Weightman, A.J., Senior, E. and Bull, A.T. (1979). The growth of *Pseudomonas putida* on chlorinated aliphatic acids and its dehalogenase activity, *Journal of General Microbiology* <u>114</u>, 125-136.

Slater, J.H. and Somerville, H.J. (1979). Microbial aspects of waste treatment with particular attention to the degradation of organic carbon compounds. *In* "Microbial Technology: Current Status and Future Prospects, 29th Symposium of the Society for General Microbiology" (Ed. A.T. Bull, C.R. Ratledge and D.C. Ellwood), pp. 221-261. Cambridge University Press.

Smith, A.L. and Kelly, D.P. (1979). Competition in the chemostat between an obligately and a facultatively chemolithotrophic thiobacillus, *Journal of General Microbiology* <u>115</u>, 377-384.

Smith, M.G. (1977a). *In vivo* transfer of R factors between *E. coli* strains inoculated into the rumen of sheep, *Journal of Hygiene, Cambridge* <u>75</u>, 363-370.

Smith, M.G. (1977b). Transfer of R factors from *Escherichia coli* to Salmonella in the rumen of sheep, *Journal of Medical Microbiology* 10, 29-35.

Tsuchiya, H.M., Jost, J.L. and Fredrickson, A.G. (1972). Intermicrobial Symbiosis, Proceedings of the 4th International Fermentation Symposium: Fermentation Technology Today, pp. 43-49.

Veldkamp, H. and Jannasch, H.W. (1972). Mixed culture studies with the chemostat, *Journal of Applied Chemistry and Biotechnology* 22, 105-123.

Weightman, A.J., Slater, J.H. and Bull, A.T. (1979). The partial purification of the dehalogenases from *Pseudomonas putida* PP3, *FEMS Microbiology Letters* 6, 231-234.

Weinberg, S.R. and Stotzky, G. (1972). Conjugation and genetic recombination of *Escherichia coli* in soil, *Soil Biology and Biochemistry* 4, 171-180.

Wouters, J.T.M., Rops, C. and van Andel, J.G. (1978). R-plasmid persistance in *Escherichia coli* under various environmental conditions, *Proceedings of the Society for General Microbiology* 5, 61.

Wu, T.T., Lin, E.C.C. and Tanaka, S. (1968). Mutants of *Aerobacter aerogenes* capable of utilising xylitol as a novel carbon source, *Journal of Bacteriology* 96, 447-456.

Yokata, T., Kasuga, T., Kaieko, M. and Kuwahara, S. (1972). Genetic behaviour of R factors in *Vibrio cholerae*, *Journal of Bacteriology* 109, 440-442.

Zamenhof, S. and Eichhorn, H.H. (1967). Study of microbial evolution through loss of biosynthetic functions: establishment of 'defective' mutants, *Nature, London* 216, 456-458.

PHENOTYPIC RESPONSES TO ENVIRONMENTAL CHANGE

W.N. Konings and H. Veldkamp

Department of Microbiology, University of Groningen,
Kerklaan 30, 9751 NN Haren, The Netherlands

INTRODUCTION

The behaviour of a microbial cell is determined by the characteristics of its genome. This carries all available information needed to respond to changes in the physiochemical properties of its environment. No single microbial species expresses its entire genome under any set of environmental conditions. It has been estimated that in *Escherichia coli* growing in a mineral salts-glucose medium only 20-30% of the cell's DNA is actually utilised at least once per doubling (Kennedy, 1968; Bishop, 1969; cf. Koch, 1976). The genetic constitution of a microbial cell is not a given property which is not liable to change. In fact, most "pure" cultures are not really genetically homogeneous. Changes in environmental conditions therefore may give rise to selection of mutants. In this chapter, however, it is only the phenotypic responses which will be dealt with. Assuming a given genotype present in the cells of a bacterial population, they will all respond in a similar way to environmental change by expressing that part of the genotype that fits a given set of conditions, representing one of its phenotypes. A response to change always finds its origin on the metabolic level. Its overall result generally is reflected in a change of the specific growth rate which can be studied at the population level. Thus, if microbial ecology is defined as the study of the interrelations between microbes and their biotic and abiotic environment, there are always two components involved: microbial metabolism and its regulation, and concomitantly growth rate and changes in population density. In this chapter both aspects are dealt with. In the first part attention has been given to microbial strategies with respect to utilisation of various energy sources and the

concentrations thereof. The second part deals with the
properties of the cytoplasmic membrane which forms the barrier
between environmental chaos and internal order. As most
ecologists were supposed to have little knowledge of the
developments in this field of research, a rather extensive
introduction to this section has been given which is followed
by a number of hypotheses with respect to membrane reactions
in response to environmental change.

MICROBIAL STRATEGIES – THE UTILISATION OF ENERGY SOURCES

Every bacterial species is reactive in the sense that its
metabolic machinery is able to respond to chemical changes in
the environment. However, some species are more reactive
than others. Part of this reactivity is related to the
number of substrates that can be used as energy source. Some
species appear to be highly specialised, such as *Bacillis
fastidiosus*, discovered by den Dooren de Jong (1929), which
can only use uric acid and its degradation products allantoin
and allantoate as energy sources (Bongaerts, 1978; Vogels and
van der Drift, 1976). Other examples are encountered among
the strict anaerobes, for example, the purine-fermenting
clostridia (Barker, 1956; Vogels and van der Drift, 1976).
Among the chemoorganotrophic bacteria we find, at the
other extreme, species which are able to utilise a vast array
of organic compounds as carbon and energy sources. The first
to explore these abilities was den Dooren de Jong (1926). He
tested species of seven genera for their ability to dissimilate
organic compounds, exposing them to no less than 258 compounds.
Pseudomonas fluorescens and *Mycobacterium phlei* were the most
versatile, being able to utilise approximately 75 different
compounds. Some fifty substrates which could not be used by
any of the species tested were subsequently applied as a carbon
and energy source for a series of enrichment cultures which
resulted in the isolation of some other highly versatile
pseudomonads, *P. putida* and *P. acidovorans*. The work of den
Dooren de Jong formed the basis of an almost equally exhaustive
study on the aerobic pseudomonads by Stanier, Palleroni and
Doudoroff (1966), who described a still more versatile
organism, *P. multivorans* which can utilise more than a hundred
different organic compounds as carbon and energy sources.
When considering a particular habitat, the question arises
how diverse the bacterial strains present are with respect
to the numbers and kinds of organic compounds that can be
utilised as carbon and energy source. This question was
examined by Sepers (1979) in a study on ammonifying bacteria
occurring in the Hollands Diep, a eutrophic body of water in
the Delta area of the Netherlands. He established a number of
batch enrichments, each with one particular amino acid provided

as the energy, carbon and nitrogen source in a basal mineral medium and subsequently isolated the organisms which had become dominant. Similarly, bacterial strains were isolated from enrichments in continuous culture with glycine, aspartic acid, leucine or histidine as only organic compound presented growth-limiting concentrations. Finally a number of strains were isolated after directly inoculating agar plates containing a mineral base and either one particular amino acid (0.01-0.1% w/v) or casamino acids. Altogether 169 strains were selected on the basis of differences in colony type and 68 of these were subsequently tested for their ability to grow on 19 amino acids and an additional 23 organic compounds. It was found that of the 68 strains tested only two couples of strains were identical with respect to the spectrum of organic compounds that could be used as energy, carbon and nitrogen source. All strains were to a greater or lesser extent versatile with respect to the number of different organic compounds that could be utilised. No single strain was restricted to growth on only the amino acid in the medium from which it had been isolated. Neither could any one strain attack all 42 organic compounds tested. Table 1 shows the substrates tested and the percentage of strains able to grow on each substrate. No less than 83% of the strains could use 21-35 different organic compounds of the 42 tested.

TABLE 1

Substrate utilisation by ammonifying bacteria. Percentage of total number of strains tested (66) able to use a particular substrate. In the values given for the amino acids the strains isolated on each particular amino acid were not included (Sepers, 1979)

substrate	%	substrate	%	substrate	%
glycine	10	glutamine	72	fumarate	80
alanine	81	methionine	3	glutarate	60
valine	43	tryptophan	45	malate	98
isoleucine	56	phenylalanine	41	lactate	89
leacine	59	ornithine	53	glycollate	10
lysine	48	glucose	92	glycerate	95
arginine	70	sucrose	34	citrate	92
histidine	90	starch	10	α-ketoglutarate	90
proline	89	acetate	78	pyruvate	96
serine	48	propionate	54	glycerol	80
threonine	10	butyrate	77	benzoate	40
aspartic acid	97	oxalate	4	uric acid + NH_4^+	77
asparagine	89	malonate	27	uric acid - NH_4^+	79
glutamic acid	91	succinate	66	taurine	8

Plate counts of ammonifying bacteria were made on an agar medium containing mineral salts and either histidine (0.05% w/v) or casamino acids (0.05% w/v). Counts of "total" heterotrophs were made on an agar medium consisting of mineral salts, sodium lactate (0.04% w/v), yeast extract (0.02% w/v), beef extract (0.01% w/v) and casein hydrolysate (0.02% w/v). The number of colonies found with all plate counts were similar, indicating that most of the heterotrophs present could utilise one or more amino acids for growth.

One of the questions arising from this study is, of course, how so many organisms showing relatively little difference with respect to the ability to attack a variety of organic compounds can coexist. One of the answers must be that due to the heterogeneity of the environment containing countless tiny particles of varying chemical composition and due to diurnal and seasonal changes in temperature, every single strain or species is able to grow for short periods faster than its competitors. The variety should be due to differences in phenotypic responses to the fluctuating chemical and physical conditions in the microenvironments of the bacterial cells. A recent mathematical description of the effect of spatial inhomogeneities on the coexsitence of competing microbial populations has been given by Stephanopoulos and Frederickson (1979).

The above discussions might have given the impression that natural environments are mainly inhabited by microorganisms which are quite versatile with respect to the number of substrates which serve as energy source. This, however, is not at all true for all environments. In anaerobic sediments the greater part of the electrons arising from breakdown of organic compounds ends up either in sulphide or methane. And both *Desulfovibrio* species and all methanogenic bacteria can use only a very limited number of substrates for their energy metabolism. In this respect it is of interest that the genome size of *Methanobacterium autotrophicum* is two to three times smaller than that of *Escherichia coli* (Mitchell, Loeblich, Klotz and Loeblich, 1978).

Another group of specialised bacteria is formed by the obligate chemolithotrophs. Their metabolism, as encountered for instance in *Thiobacillus neapolitanus*, is quite rigid. A few reduced sulphur compounds can be used as the energy source and CO_2 is always the main carbon source. This type of organism is confronted with competition for reduced sulphur compounds with facultative chemolithotrophs such as *Thiobacillus* A2, isolated by Taylor and Hoare (1969). This *Thiobacillus* species can grow chemolithotrophically as well as heterotrophically under aerobic conditions with a large and chemically very diverse range of organic compounds. In addition it can

grow anaerobically on formate, acetate, gluconate or cyclohexane carboxylate with nitrate as the terminal electron
acceptor which is reduced to nitrogen. And finally, it can
grow mixotrophically under aerobic conditions, using both a
reduced sulphur compound and an organic substrate, such as
acetate, as the energy source. An example of the latter
behaviour is given in Figure 1 (Gottschal and Kuenen, 1980),
which shows the results of a batch culture experiment in
which *Thiobacillus* A2, pregrown on acetate, was exposed to a
mixture of acetate and thiosulphate. The phenotypic response
to this change shown by *Thiobacillus* A2 is the development of
thiosulphate-oxidising activity in the presence of acetate.
Throughout the experiment no CO_2 fixation occurred.

Organisms such as *Thiobacillus* A2 supposedly occur on the
surface of anaerobic sediments, where they have to compete for
sulphide and thiosulphate with obligate chemolithotrophs and
for organic substrates with obligate chemoorganotrophs.
Gottschal de Vries and Kuenen (1979) showed that under extreme
conditions, either "heterotrophic" or "autotrophic", *Thiobacillus*
A2 cannot successfully compete with the specialists which
cannot be beaten on their own grounds. The selective advantage
of *Thiobacillus* A2 became obvious, however, when both acetate
and thiosulphate were present. In the mixture of *Thiobacillus*
A2 and *Thiobacillus neapolitanus*, *Thiobacillus* A2 grew on
acetate and simultaneously consumed some of the thiosulphate.
The ratio between the population density of the coexisting
populations in a chemostat limited both by acetate and thiosulphate depended on the ratio in which both energy sources
were supplied. Above a certain acetate concentration
Thiobacillus A2 consumed virtually all thiosulphate and the
specialist *Thiobacillus neapolitanus* was washed out.

Similar observations were made with mixed chemostat cultures
of *Thiobacillus* A2 and a heterotrophic *Spirillum* species.
In both cases *Thiobacillus* A2 consumed two substrates simultaneously and the specialist only one. In a three-membered culture
Thiobacillus A2 could maintain itself over a wide range of
mixed substrate concentrations.

Observations of a similar kind were made by Laanbroek,
Smit, Klein Nulend and Veldkamp (1979) in a study of competition
between the specialist *Clostridium cochlearium* which can only
ferment glutamate and the more versatile *Clostridium tetano-
morphum* which can also ferment glucose. These examples show
that organisms with limited abilities to respond to changes
in their environment, as a consequence of a restricted genetic
potential, can only survive in places where the conditions
they are adapted to occur rather frequently. Or in other words,
a specialised metabolism has survival value in places where the
turnover rate of its energy substrate is high relative to that

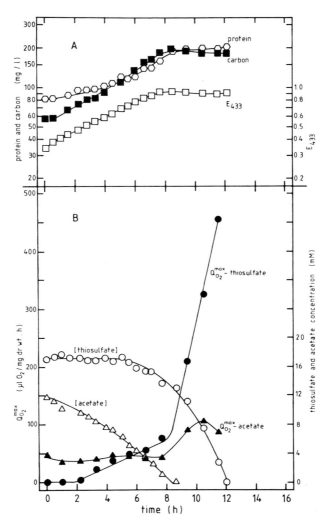

Fig. 1. Cell density, maximum substrate oxidation capacities
(Q_{O2}^{max}) *and substrate concentrations during growth of Thiobacillus*
A2 on a mixture of acetate and thiosulphate in batch culture.
The inoculum consisted of cells from an acetate limited
continuous culture at a dilution rate of 0.10 h^{-1}. (A) Protein
(○); Organic cell carbon (■); Absorbance (□); (B) Q_{O2}^{max} -
thiosulphate (●); Q_{O2}^{max}-acetate (▲); Thiosulphate concentra-
tion (○); Acetate concentration (△). (Data from Gottschal
and Kuenen, 1980).

MICROBIAL STRATEGIES - RESPONSES TO NUTRIENT CONCENTRATIONS

In a study on soil bacteria Winogradski (Winogradski, 1949) recognised two main groups with respect to preferred substrate concentrations, the zymogenous and the autochthonous bacteria. The former becoming dominant when readily degradable organic material (for example, peptone) had been added to soil, and the latter when the concentrations of readily available substrates were much lower. More recently the terms K- and r-strategists have been used in a similar way (Koch, 1979). K-strategists showing steady growth at very low substrate concentrations and having a very low threshold with respect to utilisation of nutrients. r-strategists exhibit a higher threshold and grow relatively fast at higher nutrient concentrations.

Each bacterial species can only respond to a relatively small number of conditions in such a way that it can grow faster than its competitors. In the same way as there do not exist organisms which are at the same time excellent thermophiles and psychrophiles, there do not seem to exist organisms that are both competitively successful at high as well as extremely low concentrations of nutrients. Even when a particular organism is able to grow both at relatively high and low substrate concentrations, this does not mean that it is competitively successful under both conditions. Being successful in surviving in a particular environment depends on the frequency with which a given set of conditions occurs which can be "translated" into relatively high growth rates. Most bacterial species seem to be able to grow under many sets of conditions, but only a limited number of these allows faster growth than that of competing species. One example is shown in Figure 2 which resulted from competition in the chemostat between an obligate psychrophile and a facultative psychrophile for limiting concentrations of lactate, which was the only carbon and energy source, at different temperatures. From these studies it can be predicted what the phenotypic responses of the two organisms will be if only lactate concentration and temperature change in their environment.

However, such data form only a fragment of the information needed to explain their behaviour in a more complex environment with many variables. At present we are hardly able to analyse more complex situations on the population level. And although a few attempts have been made to study the physiological background of relatively fast growth at extremely low nutrient concentrations (Matin and Veldkamp, 1978; Matin, 1979) and responses to fluctuating substrate concentrations (Koch, 1971; Koch, 1979), our knowledge of these aspects of phenotypic response to environmental change is still very limited. A

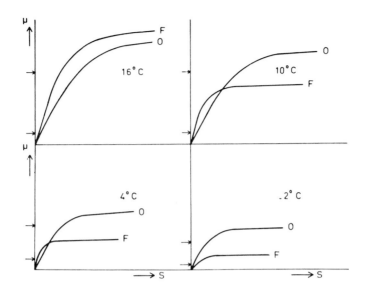

Fig. 2. *Specific growth rate of an obligately psychrophilic Pseudomonas sp. (O) and a facultatively psychrophilic Spirillum sp. (F) as a function of lactate concentration at different temperatures. The curves are schematic and based on two measurements each at the growth rates indicated by the arrows (Data from Harder and Veldkamp, 1971).*

profile of the expected properties which organisms adapted to life in nutrient-poor environments has been provided by Hirsch (1979).

The considerations given above also hold for eukaryotic microorganisms. Turpin and Harrison (1979) studied the effect of the degree of fluctuation ("temporal patchiness") of growth-limiting concentrations of ammonia on the competitive advantage of two diatom species. It was found that a continuous supply of ammonia favoured *Chaetoceros* sp., whereas *Skeletonema* sp., came to the fore when the same amount of ammonia was supplied at intervals. The longer the interval time, the greater was the relative abundance of the *Skeletonema* sp.

When the concentration of a growth rate-limiting substrate(s) decreases, the specific growth rate (μ) decreases concomitantly. However, the situation at the lower concentration range is still rather nebulous. Several possible situations are given in Figure 3. One could think of the possibility that μ decreases to zero at a finite concentration of the substrate

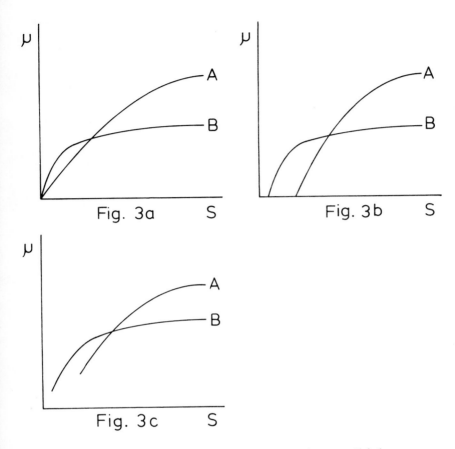

Fig. 3a S

Fig. 3b S

Fig. 3c S

Fig. 3. Hypothetical μ-s curves for organisms A (high μ_{max}, high K_s) and B (low μ_{max}, low K_s); substrate s is the energy source.

below which all available substrate is needed for maintenance purposes (Fig. 3b). Chesbro, Evans and Eifert (1979) grew *Escherichia coli* in a chemostat operating with glucose as the limiting nutrient under anaerobic conditions and with a 100% biomass feed back system. They observed that growth continued at any positive value of the glucose provision rate (Fig. 3a). The ratio between total count and viable count remained remarkably constant (0.93) at very low growth rates.

Law and Button (1977) reported a threshold value of glucose in a glucose-limited culture of a *Corynebacterium* sp. of 210 μg l^{-1} at a specific growth rate of approximately 0.02 h^{-1}

(aerobic; 25°C; pH 7.0). The μ-s relation presented corresponds
with Figure 3b. Tempest, Herbert and Phipps (1967) (cf.
Tempest and Neijssel, 1978) reported a minimum specific growth
rate in *Klebsiella aerogenes* (aerobic; 35°C; pH 6.8) of about
0.01 h^{-1}. The minimum growth rate was not greatly different
in ammonia-cultures as compared to glycerol-limited cultures.
It was suggested that changes in cellular RNA and ribosome
content may impose a lower limit to which growth can be
achieved (Fig. 3c).

Koch (1979) proposed that threshold values of nutrients
should be considered as an important ecological characterstic
of a bacterial species, which should be of decisive taxonomic
importance. However, in view of the rather unclear situation
with respect to threshold values and minimum growth rates this
does not seem to be very practicable at present. In addition,
a threshold value determined with a pure culture growing in
the presence of one carbon and energy source does not seem to
be of any ecological importance. The point is that the steady
state concentration of a particular growth-limiting nutrient
is highly dependent on the presence of additional utilisable
organic compounds. This was shown by Law and Button (1977)
in a study with a marine *Corynebacterium* sp. organism. When
this was grown in continuous culture with glucose as the
only carbon and energy source at growth rate-limiting
concentrations the minimum glucose concentration enabling
growth was 210 μg l^{-1}, which was surprisingly high for an
organism thriving in a nutrient-poor environment. However,
the steady state glucose concentration appeared to be very
much lower in the presence of additional growth-limiting
organic compounds which could be used as carbon and energy
sources.

When considering a parameter characteristic of bacteria
that grow at extremely low rates or dormant cells still able
to respond to sudden supplies of nutrients that enable growth,
the proton-motive force across the cytoplasmic membrane seems
to be of prime importance. Its value may also be applied to
distinguish cells which are no longer able to give a growth
response from those which are. This aspect of bacterial life
will be discussed in the next section.

THE CYTOPLASMIC MEMBRANE - BORDERLINE BETWEEN CELLULAR ORDER
AND ENVIRONMENTAL CHAOS

A brief outline will be given of the chemiosmotic theory
as developed by Mitchell (1966; 1968; 1970a,b; 1976), and of
some of its consequences. For further details and for
experimental evidence supporting this theory the reader is
directed to reviews which have appeared recently (Kaback,
1976; Hamilton, 1977; Konings, 1977; Harold, 1978; Konings

and Michels, 1980; Jones, this volume pp 193.

The chemiosmotic concept

A very elegant explanation for the mechanism of coupling between energy-generating and energy-consuming processes is the cytoplasmic membrane has been offered by the chemiosmotic theory. It was postulated that the energy-transducing systems, such as, electron transfer systems, ATP-ase complexes and solute transport systems, which are incorporated anisotropically in the membrane, act as electrogenic proton pumps. In the cytoplasmic membrane of bacteria these proton pumps translocate protons from the cytoplasmic side of the membrane to the external medium. The cytoplasmic membrane is practically imperm- eable to ions and in particular to protons and hydroxyl ions. Proton translocation by these electrogenic pumps will therefore result in the generation of two gradients. A pH-gradient (chemical gradient to protons; ΔpH) is formed as a result of alkalinisation of the cytoplasm due to removal of protons and acidification of the external medium due to accumulation of protons. An electrical potential ($\Delta\psi$) is formed because positive charges are removed from the cytoplasm and accumulate in the external medium.

Both gradients exert an inwardly-directed force on the protons, the proton-motive force ($\Delta\tilde{\mu}_H+$):

$$\Delta\psi \quad = \quad \Delta\tilde{\mu}_H+ \quad - \quad Z.\Delta pH$$

in which $Z = 2.303\ RT/F$; R is the gas constant, T, the absolute temperature and F, the Faraday constant. Z has a numerical value of about 60 mV at 25°C. The chemiosmotic hypothesis thus visualises the energy-transducing systems in the membrane as primary transport systems which convert chemical or light energy into electrochemical energy.

A second aspect of the chemiosmotic coupling theory is that it postulates that the proton-motive force drives energy- consuming processes in the membrane by a reversed flow of protons (Fig. 4). The energy of $\Delta\tilde{\mu}_H+$ is thus either converted into ATP by a reversed action of the ATP-ase complex or drives osmotic work such as the formation of solute gradients by secondary transport or drives mechanical work such as flagellar movements.

According to the chemiosmotic concept therefore the central intermediate between the energy-transducing processes in the cytoplasmic membrane is the electrochemical gradient of protons. Proton-motive force (pmf)-producing and pmf-consuming processes found in bacteria are listed;

PRIMARY TRANSPORT   SECONDARY TRANSPORT

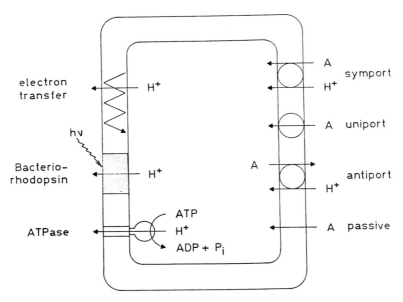

Fig. 4. Schematic presentation of primary and secondary transport systems in bacteria.

pmf-generating processes	pmf-driven processes
cytochrome-linked electron transfer	transhydrogenase reaction
ATP-hydrolysis	ATP-synthesis
bacteriorhodopsin-light interaction	secondary solute transport
efflux of metabolic end-products	flagellar movement

Energy-transducing systems

Aspects of three energy transducing systems which play an important role in bacterial energy metabolism will be discussed: electron transfer systems, Ca^{2+}, Mg^{2+}-activated ATP-ase complex and secondary solute transport systems.

Cytochrome-linked electron transfer systems A wide variety of different cytochrome-linked electron transfer systems is encountered in bacteria: respiratory chains with oxygen, nitrate or sulphate as electron acceptors, fumarate reductase systems and light-driven cyclic electron transfer systems.

All these systems are composed of several electron transfer
carriers, the nature of which varies considerably in different
organisms. One common feature of all electron transfer
systems is that they are tightly incorporated in the cytoplasmic
membrane. Another important general property of these systems
is that electron transfer results in the translocation of
protons from the cytoplasm into the external medium. Electron
transfer therefore leads to the generation of a proton-motive
force, internally alkaline and negative.

The mechanism of proton translocation by the electron
transfer system is at this moment hardly understood. Mitchell
(1966) postulated the redox loop model. The electron carriers
are located alternately at the inner and other surfaces of the
cytoplasmic membrane thus forming loops in the electron transfer
chain (Fig. 5). Each loop corresponds to a coupling site as
postulated by the chemical hypothesis of energy coupling
(Slater, 1953). In each loop two hydrogen atoms move from
the inner side of the membrane outward through a hydrogen
carrier (for instance, a quinone). Subsequently two
electrons pass back to the inside through an electron carrier
(for instance, a cytochrome) and two protons are liberated
into the external medium. For each redox loop, therefore
two protons are translocated from the cytoplasm into the
external medium per two electrons transferred. The total
number of protons that will be translocated per two electrons
transferred by the electron transfer chain will therefore
depend on the number of redox loops. The proton/2 e
stoichiometry per redox loop is currently subject of
turbulent discussion and evidence for stoicheimetries of 3
or 4 have been presented (see for instance Lehninger, 1979).

The maximum value that the proton-motive force can reach
is determined by the redox potential different ($\Delta E_o'$) between
the electron donor and acceptor and by the total number of
protons (n) translocated per two electrons transferred:

$$\Delta \tilde{\mu}_{H^+} = \frac{2}{n} \cdot \Delta E_o'$$

This equation allows us to calculate the maximum proton-
motive force that can be generated. In Table 2 this proton-
motive force has been calculated for three different electron
transfer systems and for three different n-values.

The available information about the proton-motive force
in aerobically grown cells indicates that this value is usually
between -200 and -250 mV. Proton-motive force values of
around -200 mV have also been found during nitrate respiration.
In cells carrying out fumarate reduction the values of the
proton-motive force appear to be lower (approximately -100 mV).
The number of protons which are translocated during electron
transfer are not known with certainty. In all three systems

TABLE 2

Electron transfer system	$\Delta E'_O$	maximum value of proton-motive force for n =		
		2	4	6
NADH to oxygen	1138	1138	569	379
NADH to nitrate	753	753	376	251
NADH to fumarate	353	353	176	118

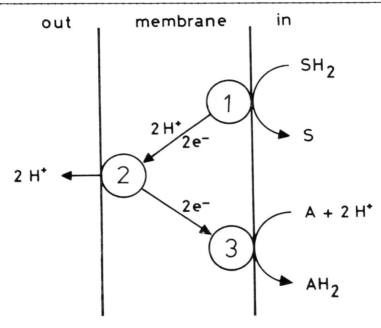

Fig. 5. Scheme of proton translocation during electron transfer via a loop mechanism. SH_2 and A represent the electron donor and acceptor, respectively. 1, 2 and 3 are components of the electron transfer system.

translocation of 6 protons per two electrons during NADH-oxidation would be sufficient to generate the observed proton-motive force. The proton-motive force that can be maintained in a cell will depend on the activities of proton-motive force-generating and proton-motive force-dissipating processes. Of the proton-motive force-dissipating processes it is especially the passive secondary transport processes (Fig. 4)

which are beyond control of the organisms. The uptake of
cations (H^+, K^+, Na^+) and the efflux of anions (such as anionic
pool constituents) will increase with increasing proton-motive
force. In order to maintain the internal environment as
constant as possible, bacteria have developed transport systems
which extrude cations such as Na^+ and Ca^{2+} by countertransport
(antiport) with protons. In a scheme as shown in Figure 6 the
entrance of one sodium ion into the cell requires the uptake
of one proton for removal. Bacteria have also developed
secondary transport systems which transport metabolic intermediates (such as glucose-6-phosphate, succinate, etc.) from the
medium back to the cytoplasm. These and other processes
result in the uptake of protons and/or charge. As a consequence

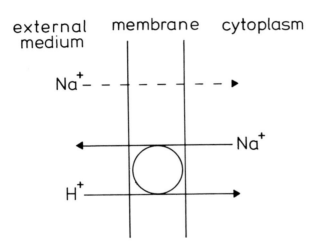

*Fig. 6. Entrance of Na^+ by passive secondary transport, uniport
or symport systems and extrusion by Na^+-H^+ antiport system.*

the steady state proton-motive force will seldom be in thermodynamic equilibrium with the redox potential different or with
the phosphate potential (see below).
    This also explains why respiratory control, that is, control
of electron flow by the proton-motive force, is rarely
observed in bacteria. In contrast, respiratory control is
found in organelles (such as mitochondria, chloroplasts)
which are surrounded by an almost constant environment.

The ATP-ase complex Hydrolysis of ATP by the membrane bound
ATP-ase complex results in the extrusion of protons from the
cytoplasm to the external medium. Two protons are probably

extruded per ATP hydrolysed but H^+/ATP stoicheiometries of four have also been reported.

The ATP-ase complex catalyses the following reaction:

$$ATP^{4-} + H_2O + n\ H^+_{in} \rightleftharpoons ADP^{3-} + P^-_i + n\ H^+_{out}$$

The maximal $\Delta\tilde{\mu}_{H+}$ that thermodynamically can be generated by this reaction is determined by the phosphate potential ($\Delta G'_{ATP}$) and the value of n:

$$\Delta\tilde{\mu}_{H+} = 1/n\ \Delta G'_{ATP}$$

in which $\Delta G'_{ATP} = \Delta G'_o - 2.3\ \dfrac{RT}{F}\ \log\ \dfrac{(ATP)}{(ADP).(P_i)}\ mV$

($\Delta G'_o$ = free energy under standard conditions and P_i = free inorganic phosphate concentration in cell). The free inorganic phosphate concentration is usually around 0.01 M At 25°C and an internal pH of 8, the $\Delta G'_o$ = 340 mV. Under these conditions $\Delta G'_{ATP}$ will be:

$$\Delta G'_{ATP} = -460 - 60\ \log\ (ATP)/(ADP)\ mV$$

When the H^+/ATP stoicheiometry = 2 the equilibrium value of $\Delta\tilde{\mu}_{H+}$ that can be reached will be: $\Delta\tilde{\mu}_{H+}$ = 230 + 30 log (ATP/ADP). This relationship is shown in Figure 7. This figure shows that under conditions of low $\Delta\tilde{\mu}_{H+}$ and high (ATP)/(ADP) a $\Delta\tilde{\mu}_{H+}$ can be generated by ATP-hydrolysis, while under conditions of low (ATP)/(ADP) and high $\Delta\tilde{\mu}_{H+}$ ATP-synthesis may occur. Such an equilibrium between the phosphate potential and the proton-motive force will only be reached when no other factors regulate the ATP-ase activity. Such a regulating factor (ATP-ase-inhibitor protein) has been reported for *E. coli* but it is not clear at this moment whether such regulation mechanisms are present in other organisms.

Solute transport   Solute transport across the cytoplasmic membrane of bacteria occurs by two major mechanisms:

   i)   Secondary transport systems: transport by these systems is driven by electrochemical gradients and transport will lead to the translocation of the solute in an unmodified form;

   ii)   Group translocation: a solute is substrate for a specific enzyme system in the membrane. The enzyme reaction results in a chemical modification of the solute and release of the products at the cytoplasmic side.

The only well established group translocation system is the phosphoenolpyruvate phospho-transferase system (PTS). The PTS is involved in the translocation and concomitant phosphorylation of several oligosaccharides in a number of facultative anaerobes but not in obligate aerobes. In most organisms which possess PTS activity only a few sugars are

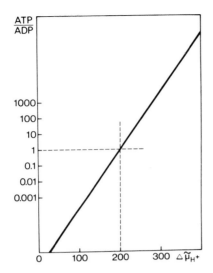

Fig. 7. ATP/ADP ratio under conditions that the $\Delta\tilde{\mu}_{H^+}$ is in equilibrium with the phosphate potential is given by line.

translocated by this system.

Most solutes are translocated across the cytoplasmic membrane by secondary transport either passively, without the involvement of specific membrane proteins, or facilitated by specific carrier proteins. This latter mechanism is often termed "active transport" (Fig. 4).

Mitchell (1968) visualised three different systems for facilitated secondary transport.

i)   "Uniport": only one solute is translocated by the carrier protein.

ii) "Symport": two or more different solutes are translocated in the same direction by the carrier protein.

iii) "Antiport": two or more different solutes are translocated by one carrier in opposite directions.

The driving forces for solute transport will depend on the overall charge and protons which are translocated, as well as on the solute gradient. The driving force is therefore composed of components of the proton-motive force and of the solute gradient and translocation of solute will proceed until the total driving force is zero. At that stage a steady state level of accumulation is reached. Figure 8 shows schematically four different translocation processes.

In example 1 transport of a neutral solute via passive secondary transport is shown. The driving force of this

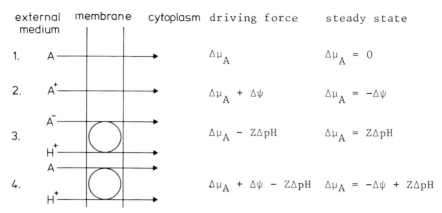

Fig. 8. Schematic presentation of four secondary transport processes. 1. Passive transport of a neutral solute; 2. passive transport of a cation; 3. facilitated transport of an anion in symport with one proton; 4. facilitated transport of a neutral solute in symport with one proton. For each process the driving force and the steady state level of accumulation are given.

transport will be supplied only by the solute gradient and will be equal to $\Delta\mu_A = Z\log A_{in}/A_{out}$. A steady state will be reached when the solute gradient is dissipated ($\Delta\mu_A = 0$) and the internal solute concentration equals the external concentration. Example 2 shows transport of a monovalent positively charged solute by passive secondary transport. This transport will not only depend on the chemical concentration gradient of solute but also on the electrical potential. The driving force therefore will be $\Delta\mu_A = Z\log A_{in}/A_{out} + \Delta\psi$ and a steady state will be reached when $Z\log A_{in}/A_{out} = -\Delta\psi$. Unless the electrical potential is dissipated by movement of other ions across the membrane in steady state the internal concentration of solute will not be equal to the external concentration. In the absence of active primary transport systems net transport will already stop when the internal solute concentration is lower than the external concentration. However, when a $\Delta\psi$ (interior negative) is generated by other systems accumulation of solute can occur.

In a similar way, as has been shown above, the driving forces and steady state levels of transport can be derived for other transport processes. In example 3 this is done for a symport system by which a negatively-charged solute is translocated together with one proton. The overall charge during this translocation process is zero and, due to the

translocation of one proton only, the ΔpH will contribute to
the driving force of this translocation process. In example
4 the driving force is determined for a neutral solute symp-
orted with one proton. In this process charge and protons are
translocated and consequently the total proton-motive force
($\Delta\psi$ and ΔpH) contributes to the driving force. In the
transport processes 2, 3 and 4 the level of accumulation of
solute will vary with the driving force supplied by the proton-
motive force. In order to maintain the proton-motive force
during transport of solute (interior negative and alkaline),
continuous proton extrusion by the primary transport systems
has to occur. In the absence of such activity the translocation
of solute will lead to a dissipation of the proton-motive
force.

When the proton-motive force is maintained at a certain
value by primary transport systems the level of accumulation
of a solute will depend on the mechanism involved in the
translocation. For instance, when the total proton-motive
force is −180 mV, composed of a $\Delta\psi$ of −120 mV and a ΔpH of
−60 mV the steady state level of accumulation of solute A in
example 2 will be 100-fold (−60 log A_{in}/A_{out} = −120); in 3
10-fold (−60 log A_{in}/A_{out} = −60) and in 4 1000-fold (−60 log
A_{in}/A_{out} = −180). Variations of the $\Delta\tilde{\mu}_{H}+$ therefore effect
the accumulation of solutes differently.

In 2, 3 and 4 a decrease of the driving force supplied by
the proton-motive force will lead to a decreased steady state
internal concentration and a decreased rate of accumulation.
In the absence of a proton-motive force solute will in all
cases equilibrate across the membrane. Because essentially
all solutes are to a certain extent membrane permeable, in the
absence of a proton-motive force leakage of internal metabolites will occur until the internal concentration equals
the external concentration. Ultimately, this leakage of
internal metabolites will lead to the death of the organism.

Carriers for facilitated secondary transport

Facilitated secondary transport of solutes across the
cytoplasmic membranes of bacteria is mediated by specific
proteins, the carriers (previously termed permeases). Cytoplasmic membranes usually contain many of these carrier
proteins, each having affinity for only one solute or a group
of structurally related solutes.

Carriers usually have a high affinity for their solutes
with affinity constants, K_s, values ranging between 10^{-5} and
10^{-6} M. However, affinity constants as low as 10^{-8} M have
been reported.

Secondary facilitated systems have been demonstrated for
a wide variety of solutes (for review see Konings, 1977):
    i)   amino acids: in a number of bacteria transport of
amino acids is mediated by at least 9 distinct carrier
proteins. Transport systems for amino acids have also been
demonstrated in bacteria which hardly utilise amino acids,
like *Thiobacillus neapolitanus* (Matin, Konings, Kuenen and
Emmens, 1974). This suggests that the presence of amino acid
transport systems is a common feature of bacteria.
    ii) Carboxylic acids: specific transport systems have
been demonstrated in a number of bacteria for monocarboxylic
acids (D-lactate, L-lactate, pyruvate), C_4-dicarboxylic acids
(L-malate, fumarate, succinate, oxalate), tricarboxylic acids
(citrate).
    iii) Sugars: transport systems for the following sugars
have been found: lactose, arabinose, glucuronate, gluconate,
D-galactose, D-fructose, L-rhamnose, D-glucose. A number of
bacteria possess, besides these secondary transport systems,
also group translocation for other sugars and in some cases
for the same sugars.
    iv)   Phosphorylated intermediates such as glucose-6-phosphate
and glycerol-3-phosphate.
    v)   Inorganic cations: specific transport systems have
been found for Na^+, K^+, Mn^{2+}, Ca^{2+}, Mg^{2+}.
    vi)   Inorganic anions: specific transport systems for
phosphate and sulphate have been demonstrated.
The number of different carriers present in the cytoplasmic
membrane will differ from organism to organism and will
depend in one organism on the growth conditions. For *E. coli*
it has been estimated that about 60 different carriers are
present in the cytoplasmic membrane. It seems reasonable to
state that specific transport systems are present for all
solutes which have to be accumulated (or excreted) at a high
rate. Most likely this is also the case for membrane permeable
solutes like acetate (permeable in the undissociated form),
ethanol or glycerol.
    A surprisingly high number of carriers are present
constutively in the cytoplasmic membrane. One may speculate
about the advantages for bacteria to possess always carriers
for several solutes. Possibly the possession of constitutive
carriers would enable an organism to scavenge intermediates
leaked passively out of the cell and/or allow the organism to
react rapidly to changes in the external medium. Besides
constitutive transport systems also inducible transport systems
are found in bacteria such as for lactose transport in *E. coli*
or for citrate transport in *B. subtilis*.
    It is important to realise that the carriers are embedded
in the cytoplasmic membrane and that the available space is

limited. When this is occupied by carriers and other membrane
proteins (as is most likely the situation) an increased number
of a specific carrier has to result in a decreased number of
other membrane proteins, unless the cell is able to increase
the surface area of the cytoplasmic membrane. Such changes of
surface area have been observed in bacteria.

The surface to volume ratio is a function of specific
growth rate; with decreasing growth rates, the surface to
volume ratio increases and this holds for all bacteria. However,
apart from this general phenomenon, the surface to volume ratio
is relatively large in bacteria that grow relatively fast at
extremely low concentrations of growth-limiting substrates.
This is shown in Table 3 for a *Spirillum* sp., and a *Pseudomonas*
sp., grown in a lactate-limited chemostat and which show
crossing μ-s curves; the spirillum growing faster at the
lower concentration extreme.

TABLE 3

Effect of growth rate on surface to volume ratio of a Spirillum
sp. (Sp.) and a Pseudomonas sp. (Ps.) grown under L-lactate
limitation (Data from Matin, 1979 and Matin and Veldkamp, 1978)

Dilution rate (h^{-1})	Steady state L-lactate concentration (μM)		Surface/volume (μm^{-1})	
	Sp.	Ps.	Sp.	Ps.
0.06	4	10	8.05	6.24
0.16	20	30	7.82	5.66
0.35 (batch)	<200	110	6.27	--
0.64 (batch)	–	<200	–	4.59

Regulation of transport can occur at the level of carrier
synthesis or at the level of carrier activity. Evidence has
been presented that regulation of activity is exerted by the
catalytic activity of protein components of the PTS (possibly
a regulatory protein) (Saier and Moczydlowski, 1978). Also
regulation by intracellular sugar phosphates has been suggested.

The action of the carriers is completely reversible and
symmetrical. During solute transport the energy of the proton-
motive force is used to generate a solute gradient and to
concentrate solute internally. The reversed process can occur
under conditions of an outwardly directed solute gradient and
a low proton-motive force. The solute gradient will then drive
solute efflux and a proton-motive force is generated because
this efflux is accompanied in many systems (Fig. 4) with
protons and/or charge. In other words, the energy of the
solute gradient is then used to generate a proton-motive
force.

In bacteria such situations are very common during fermentation. The end products of fermentation are continuously produced internally and excreted into the external medium. The energy yield by efflux of fermentation products can be quite considerable. Recently, Michels, Michels, Boonstra and Konings (1979) calculated, on theoretical grounds, that the additional energy yield during homolactic fermentation is in the order of 30% of the energy produced by substrate level phosphorylation. Otto, Sonnenberg, Veldkamp and Konings (1980) demonstrated that lactate efflux resulted in the generation of a $\Delta\tilde{\mu}_{H^+}$ in *Streptococcus cremoris*.

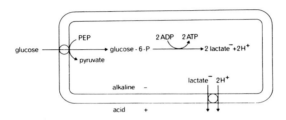

Fig. 9. Synthesis of ATP by substrate level phosphorylation and the generation of $\Delta\tilde{\mu}_{H^+}$ by lactate-proton efflux during homolactic fermentation.

The proton-motive force generated by efflux of fermentation products can be used to drive other energy consuming processes, such as solute transport and ATP-synthesis. The cell can therefore save its ATP produced by substrate level phosphorylation and use this predominantly for biosynthetic purposes. The proton-motive force generated by end-product efflux will reach its maximum value when it is in equilibrium with the end product gradient. The driving force for end-product efflux will then be zero, and consequently the internal concentration of end product will increase. The generation of a proton-motive force by other proton pumps, such as ATP-hydrolysis, will establish more rapidly this equilibrium. Such proton pump activity therefore would lead to high internal end product concentrations.

When the concentration of end products gradually increases in the environment, the end-product gradient decreases. In order to maintain a proton-motive force for a prolonged period of time, removal of end products from the environment is required. The presence of organisms which consume these will then be of benefit to the product-producing organism.

Thus, this relation between a product-producer and a product-consumer represents a case of mutualism; for both organisms the interrelation is profitable.

Interaction between energy-transducing processes

The information about energy-transducing processes presented above shows that these systems do not operate independently, but that they are closely coupled. This interaction between primary and secondary transport systems, substrate level phosphorylation and biosynthetic processes is given in the following scheme:

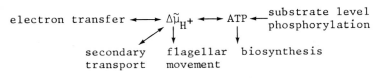

electron transfer $\longleftrightarrow$ $\Delta\tilde{\mu}_{H^+}$ $\longleftrightarrow$ ATP $\longleftarrow$ substrate level phosphorylation

secondary   flagellar   biosynthesis
transport   movement

This scheme shows that the proton-motive force plays a central role in the coupling between energy-transducing processes. In a cell these processes continuously attempt to reach thermo-dynamic equilibrium. The systems behave therefore like connected vessels. The scheme given above allows us to draw some general conclusions:

i)   Changes in any one of the processes will lead to changes in the other processes. The following section will discuss how a number of phenotypic responses of bacteria to changes in the environment can be explained by this inter-dependence.

ii) A simple relationship between electron transfer and ATP-synthesis does not exist and it is not justified to translate electron transfer activity into ATP-synthesis activity.

iii) In cells which are not energy-limited the flow of energy will be in the direction of ATP-synthesis and the rate of ATP-consumption by biosynthesis will determine to a large extent the rate of ATP-synthesis.

iv) Cells which cannot generate a sufficiently high proton-motive force by electron transfer or secondary transport have to use ATP for the generation of a $\Delta\tilde{\mu}_H^+$. Consequently less ATP becomes available for biosynthetic purposes.

Environmental effects on the proton-motive force

Many phenotypic responses of bacteria to changes in the environment are initiated by effects on the proton-motive force and consequently on the energy-flow. In this section we will consider the changes of the proton-motive force upon changes in the environment and on that basis predict what the phenotypic responses of the bacteria might be.

Shift from aerobic to anaerobic conditions   Under conditions
of sufficient energy supply a facultative aerobe will generate
a proton-motive force mainly by respiration. The proton-motive
force which can be maintained will be relatively high (-200 to
-250 mV). When the bacterium is shifted to anaerobic conditions
other primary transport systems have to be used in order to
maintain a proton-motive force. In some case a functional
nitrate respiration system will be preferentially used when
nitrate is present in the medium. The organism can thus
maintain a proton-motive force which is very close to the
level that is found under aerobic conditions (around -200 mV).
In the absence of nitrate respiration, the fumarate reductase
system can be used but the proton-motive force which can be
maintained will be significantly lower (around -100 to -120 mV).
This decrease of the proton-motive force will be accompanied
by a decrease in the internal concentrations of metabolic
intermediates, a decrease of ATP/ADP ratio (Fig. 7) and a
decrease of the growth yield and growth rate.

When the bacterium cannot mobilize either one of these
electron transfer systems, a proton-motive force can be
generated by efflux of metabolic end products. When this
mechanism also fails (for instance when the external concentration of end products is high), the organism has to use the
least attractive mechanism, which is ATP-hydrolysis. The
growth yield under these latter conditions will be lowest
because the ATP synthesized by substrate level phosphorylation,
becomes only partially available for biosynthetic purposes.

Shift from growth under energy limitation via dormancy to
death   When the energy supply is gradually decreased the
specific growth rate (μ) decreases. Initially, the organism
can maintain the proton-motive force at a high level but below
certain levels of energy supply the proton pump activity
decreases and consequently the proton-motive force also
decreases (growth phase, Fig. 10).

When the energy supply has stopped completely, μ will
become zero. The bacterium attempts to maintain the proton-
motive force as high as possible by degrading its reserve
material, if present, and some of its structural cell material
(transient phase). Gradually the proton-motive force decreases
and the cell enters a dormant phase in which its major concern
is to keep the proton-motive force sufficiently high in order
to prevent leakage of pool constituents into the environment.

A new steady state proton-motive force is established, the
value of which is determined by the equilibrium between proton-
motive force-generating processes (further breakdown of
structural cell material) and the unavoidable proton-motive
force-dissipating processes (leakage). In this stage the cell

has stopped energy-consuming processes, such as biosynthesis, ATP-synthesis, nitrogen fixation, motility, among others. Ultimately the cell will not be able to mobilize enough cellular constituents for this purpose and the proton-motive force decreases. Essential cellular constituents are lost and the cell loses its viability. Figure 10 schematically shows the expected variations of the proton-motive force under different energy-limiting conditions.

Growth-limitation by essential cations   The steady state level of the proton-motive force will be a balance between proton-motive force-generating activities and proton-motive force-consuming activities.

When a bacterium is confronted with decreasing external concentrations of an essential cation the organism has to increase the proton-motive force in order to maintain the internal concentration constant. This means that the organism has to increase the activities of the primary proton pumps. An increase of the proton-motive force, however, will also lead to an increase of the leakage processes (increase of cation permeation from outside to inside, and of anion permeation from inside to outside). A relatively small increase of the proton-motive force, therefore, requires a considerable increase of the primary pump activities. The primary pumps which in general the cell can effectively mobilize for this purpose, are the electron transfer systems. ATP-hydrolysis by the membrane-bound ATP-ase complex is not to be expected because only hydrolysis of ATP produced by substrate level phosphorylation will lead to a net extrusion of protons. (Hydrolysis of ATP-synthesised by the ATP-ase complex will only lead to extrusion of protons which were already taken up during the synthesis). Furthermore, ATP-hydrolysis under conditions of a high proton-motive force is energetically unfavourable and would require very high ATP/ADP ratios (Fig. 7).

An increase of electron transfer will require an increased dissimilation rate of the energy source and consequently an increase in the concentration of internal metabolites. Both this increase and the increased proton-motive force will lead to an increased leakage.

The response of a cell to growth-limitation by essential cations, therefore, would be an increased rate of oxidation of energy source, followed by the appearance of metabolic products in the medium. This is exactly what Hueting and Tempest (1979) observed in potassium-limited cultures of *Klebsiella aerogenes*.

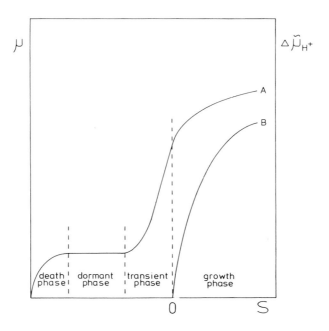

Fig. 10. Decrease of the specific growth rate µ of an energy-limited cell to zero (B) and the hypothetical change of proton-motive force in various phases (A). Substrate S is the energy source.

Shift from fresh water to sea water The major environmental change which confronts an organism upon transfer from fresh water to sea water, is an increase in the external NaCl concentration. The negatively charged chloride ion will be excluded from the cell by the proton-motive force (negative inside). The positively charged sodium ion will, however, be driven to the inside by the proton-motive force and by the sodium gradient. This results in sodium uptake which is undesirable for a bacterial cell. Bacterial cells possess secondary transport systems to extrude sodium ions in antiport with protons (see Fig. 6). The ultimate result of the increased external sodium concentration is thus an increased rate of proton influx and consequently an increased dissipation of metabolic energy.

In case the marine environment is energy-limited, this dissipation of metabolic energy will lead to a decrease of the proton-motive force, a decreased growth rate and ultimately in a cessation of growth and high death rate.

Marine microorganisms are also confronted with an inwardly
directed sodium gradient but these organisms have developed
answers to this problem (Lanyi, 1979). For instance: many
marine organisms possess Na^+ -solute symport systems by which
the energy of the sodium gradient is used for solute accumulation (for example, amino acids).

Similar effects as described for NaCl will be exerted by
increasing concentrations of external membrane permeable
weak acids (such as acetate, formate, propionate, etc). These
acids are membrane permeable in undissociated form and will
cross the cytoplasmic membrane in response to the pH-gradient.

Many phenotypic responses of bacteria can be predicted in
a similar way, as described above, on the basis of the effects
of changes in the environment on the proton-motive force or
its coupled energy-transducing processes. It should be stated,
however, that these predictions were made without taking into
account the numerous escape mechanisms which bacteria might
possess. Moreover, at this moment little information about
the level of the proton-motive force in bacteria is available.
However, there is no doubt that many environmental changes
primarily affect the proton-motive force and that changes in
this parameter often form the basis of the phenotypic responses
of a bacterial cell.

This field of research has hardly been explored. The
experimental tools to measure the proton-motive force in
bacteria are now available and there are no experimental
restrictions for studies on the effects of environmental
changes on the proton-motive force.

ACKNOWLEDGEMENTS

We are grateful to F.B. van Es, I. Friedberg, K.J. Helling-
werf, J.G. Kuenen and R. Otto for stimulating discussions and
critical comments. The help of R. Otto and Marry Pras in
preparing figures and manuscript, respectively, is greatly
appreciated. J. Gottschal and J.G. Kuenen as well as S.
Hueting and D.W. Tempest kindly allowed us to use some of
their unpublished data.

REFERENCES

Barker, H.A. (1956). "Bacterial Fermentations" John Wiley
    and Sons, Inc., New York.
Bishop, J.O. (1969). Interpretation of DNA-RNA hybridization
    data, *Nature* 224, 600-603.
Bongaerts, G.P.A. (1978). "Urid acid degradation by *Bacillus
    fastidiosus*" Ph.D. thesis, Nijmegen, the Netherlands.
Chesbro, W., Evans, T. and Eifert, R. (1979). Very slow growth
    of *Escherichia coli*, *Journal of Bacteriology* 139, 625-638.

Den Dooren de Jong, L.E. (1926). "Bijdrage tot de kennis van het mineralisatieproces" Ph.D. thesis, Delft, the Netherlands.

Den Dooren de Jong, L.E. (1929). Über *Bacillus fastidiosus*, *Zentralblatt für Bakteriologie, Parasitenkunde, Infektionskrankheiten und Hygiene (Abteilung II, Originale)* 79, 344-358.

Gottschal, J.C., de Vries, S. and Kuenen, J.G. (1979). Competition between the facultatively chemolithotrophic *Thiobacillus* A2, an obligately chemolithotrophic *Thiobacillus* and a heterotrophic *Spirillum* for inorganic and organic substrates, *Archives of Microbiology* 121, 241-249.

Gottschal, J.C., and Kuenen, J.G. (1980). Mixotrophic growth of *Thiobacillus* A2 on acetate and thiosulphate as growth limiting substrates in the chemostat, *Archives of Microbiology*, in press.

Hamilton, W.A. (1977). Energy coupling in substrate and group translocation, *Symposium of the Society for General Microbiology* 27, 185-216.

Harder, W. and Veldkamp, H. (1971). Competition of marine psychrophilic bacteria at low temperature, *Antonie van Leeuwenhoek, Journal of Microbiology and Serology* 37, 51-63.

Harold, F.M. (1978). Vectorial metabolism. *In* "The Bacteria" (Ed. I.C. Gunsalus), Vol. 6, pp. 463-521. Academic Press, New York.

Hirsch, P. (1979). Life under conditions of low nutrient concentrations, *In* "Strategies of Microbial Life in Extreme Environments" (Ed. M. Shilo), pp. 357-372. Verlag Chemie, D-6940 Weinheim, Germany.

Hueting, S. and Tempest, D.W. (1980). Influence of the glucose input concentration on the kinetics of metabolite production by *Klebsiella aerogenes* NCTC 418 growing in chemostat culture in potassium- or ammonia-limited environments, *Archives of Microbiology*, in press.

Kaback, H.R. (1976). Molecular biology and energetics of membrane transport, *Journal of Cellular Physiology* 89, 575-593.

Kennell, D. (1968). Titration of the gene sites on DNA by DNA-RNA hybridization, *Journal of Molecular Biology* 34, 85-103.

Koch, A.L. (1971). The adaptive responses of *Escherichia coli* to a feast and famine existence, *Advances in Microbial Physiology* 6, 147-219.

Koch, A.L. (1976). How bacteria face depression, recession and derepression, *Perspectives in Biology and Medicine* 20, 44-63.

Koch, A.L. (1979). Microbial growth in low concentrations of nutrients. *In* "Strategies of Microbial Life in Extreme Environments" (Ed. M. Shilo), pp. 261-279. Verlag Chemie, D-6940 Weinheim, Germany.

Konings, W.N. (1977). Active transport of solutes in bacterial membrane vesicles, *Advances in Microbial Physiology* 15, 175-251.

Konings, W.N. and Michels, P.A.M. (1980). Electron transfer driven solute translocation across bacterial membranes. *In* "Diversity of Bacterial Respiratory Systems" (Ed. C.J. Knowles). In press. CRC Press Inc., West Palm Beach.

Laanbroek, H.J., Smit, A.J., Klein Nulend, G. and Veldkamp, H. (1979). Competition for L-glutamate between specialised and versatile *Clostridium* species, *Archives of Microbiology* 120, 61-66.

Lanyi, J.K. (1979). Physicochemical aspects of salt-dependence in halobacteria. *In* "Strategies of Microbial Life in Extreme Environments" (Ed. M. Shilo), pp. 93-108. Verlag Chemie, D-6940 Weinheim, Germany.

Law, A.T. and Button, D.K. (1977). Multiple-carbon-source-limited growth kinetics of a marine coryneform bacterium, *Journal of Bacteriology* 129, 115-123.

Lechinger, A.L. (1979). Proton translocation sites in the respiratory chain and their stoichiometry, *Abstracts of the XIth International Congress of Biochemistry*, Toronto. pp. 414.

Matin, A. (1979). Microbial regulatory mechanisms at low nutrient concentrations as studied in chemostat. *In* "Strategies of Microbial Life in Extreme Environments" (Ed. M. Shilo), pp. 323-339. Verlag Chemie, D-6940 Weinheim, Germany.

Matin, A., Konings, W.N., Kuenen, J.G. and Emmens, M. (1974). Active transport of amino acids by membrane vesicles of *Thiobacillus neapolitanus*, *Journal of General Microbiology* 83, 311-318.

Matin, A. and Veldkamp, H. (1978). Physiological basis of the selective advantage of a *Spirillum* species in a carbon-limited environment, *Journal of General Microbiology* 105, 187-197.

Michels, P.A.M., Michels, J.P.J., Boonstra, J. and Konings, W.N. (1979). Generation of an electrochemical proton gradient in bacteria by the excretion of metabolic end-products, *FEMS Microbiology Letters* 5, 357-364.

Mitchell, P. (1966). Chemiosmotic coupling in oxidative and photosynthetic phosphorylation, *Biological Reviews* 41, 445-502.

Mitchell, P. (1968). "Chemiosmotic Coupling and Energy

Transduction" Glynn Research Ltd., Bodmin.

Mitchell, P. (1970a). Reversible coupling between transport and chemical reactions. *In* "Membranes and Ion Transport" (Ed. E.E. Bittar), Vol. 1, pp. 192-196. Wiley Interscience, New York.

Mitchell, P. (1970b). Membranes of cells and organelles: morphology, transport and metabolism, *Symposium of the Society for General Microbiology* 20, 121-166.

Mitchell, P. (1976). Possible molecular mechanisms of the proton-motive function of cytochrome systems, *Journal of Theoretical Biology* 62, 327-368.

Mitchell, R.M., Loeblich, L.A., Klotz, L.C. and Loeblich III, A.R. (1978). DNA organisation of *Methanobacterium thermoautotrophicum, Science* 204, 1082-1084

Otto, R., Sonnenberg, A., Veldkamp, H. and Konings, W.N. (1980). Generation of an electrochemical proton gradient in *Streptococcus cremoris* by lactate efflux, submitted for publication.

Saier, M.H. and Moczyslowski, E.G. (1978). The regulation of carbohydrate transport in *Escherichia coli* and *Salmonella typhimurium*. *In* "Bacterial Transport" (Ed. B. Rosen), Vol. 4, pp. 103-127. Marcel Dekker Inc., New York.

Sepers, A.B.J. (1979). "De aerobe mineralisatie van aminozuren in natuurlijke aquatische milieus" Ph.D. thesis, Groningen, the Netherlands.

Slater, E.C. (1953). Mechanisms of phosphorylation in the respiratory chain, *Nature* 172, 975-978.

Stanier, R.Y., Palleroni, N.J., and Doudoroff, M. (1966). The aerobic pseudomonads: a taxonomic study, *Journal of General Microbiology* 43, 159-271.

Stephanopoulos, G. and Fredrickson, A.G. (1979). Effect of spatial inhomogeneities on the coexistence of competing microbial populations, *Biotechnology and Bioengineering* 21, 1491-1498.

Taylor, B.F. and Hoare, D.S. (1969). New facultative *Thiobacillus* and a reevaluation of the heterotrophic potential of *Thiobacillus novellus, Journal of Bacteriology* 100, 487-497.

Tempest, D.W., Herbert, D. and Phipps, P.J. (1967). Studies on the growth of *Aerobacter aerogenes* at low dilution rates in a chemostat. *In* "Microbial physiology and continuous culture" (Ed. E.O. Powell, C.G.T. Evans, R.E. Strange and D.W. Tempest), pp. 240-254. Her Majesty's Stationary Office, London.

Tempest, D.W. and Neijssel, O.M. (1978). Eco-physiological aspects of microbial growth in aerobic nutrient-limited environments, *Advances in Microbial Ecology* 2, 105–154.

Turpin, D.H. and Harrison, P.J. (1979). Limiting nutrient patchiness and its role in phytoplankton ecology, *Journal of Experimental Marine Biology and Ecology* 39, 151–166.

Vogels, G.D. and van der Drift, C. (1976). Degradation of purines and pyrimidines by microorganisms, *Bacteriological Reviews* 40, 403–468.

Winogradski, S. (1949). "Microbiologie du Sol. Oeuvres Completes" Masson, Paris.

UNITY AND DIVERSITY IN BACTERIAL
ENERGY CONSERVATION

Colin W. Jones

Department of Biochemistry, School of Biological
Sciences, University of Leicester, Leicester LE 1 7RH
England.

INTRODUCTION

All cells need to conserve energy in the form of ATP which can subsequently be used to drive a variety of energy-requiring cellular processes. Three methods of energy conservation are available to bacteria, namely;

 i)   fermentation (that is, substrate level phosphorylation),

 ii) respiration (that is, oxidative phosphorylation) and

 iii) photosynthesis (that is, photophosphorylation).

Fermentation occurs during the anaerobic growth of many species of chemoheterotrophic bacteria in the absence of exogenous electron acceptors. Specific catabolic reactions, often linked to oxidation-reduction reactions, yield a variety of unstable acyl phosphates with high free energies of hydrolysis, each of which subsequently transfers its phosphoryl group to ADP with the formation of ATP. Any reducing power which is generated (for example, NADH, reduced ferredoxin) is transferred to a suitable endogenous oxidant (for example, pyruvate, acetaldehyde, butyraldehyde, acetone) or is released directly as hydrogen gas. Substrate-level phosphorylation is characterised by its soluble nature, its propensity to occur via covalent intermediates which are amenable to identification and isolation, and its essentially scalar properties. The myriad fermentation pathways and substrate level phosphorylation reactions which are exhibited by obligately anaerobic and facultatively anaerobic bacteria are well documented in a copious literature (Morris, 1975; Thauer, Jungermann and Decker, 1977).

Many chemoheterotrophs and some facultative phototrophs can conserve energy via respiration. Depending on the

species, these organisms oxidise a variety of organic
reductants of relative low redox potential which either
arise as intermediates of catabolism or are primary growth
substrates (for example, NADPH, NADH, succinate, lactate,
α-glycerophosphate). Oxidation occurs via a predominantly
membrane-bound respiratory chain and leads to the reduction
of an exogenous oxidant of higher redox potential (for
example, molecular oxygen, fumarate, carbon dioxide and a
variety of inorganic oxidants including nitrate, sulphate
and sulphite). Similarly, chemolithotrophs oxidise a variety
of inorganic substrates (for example, hydrogen, ammonia,
nitrite, sulphide, thiosulphate, sulphite and ferrous iron)
with the concomitant reduction of molecular oxygen or, less
frequently, nitrate or sulphate. The amount of energy which
is released by these various reactions, and which is therefore available to drive oxidative phosphorylation, is largely determined by the redox span over which electron transfers occurs. Thus, for example, the oxidation of hydrogen
by oxygen liberates considerably more free energy ($\Delta G^{O}{}' = -$
238 kJ mol^{-1}) than does the oxidation of NADH by fumarate
($\Delta G^{O}{}' = 68$ kJ mol^{-1}) and this is reflected in the different
ATP/2e$^-$ quotients and molar growth yields obtained. In
addition, it should be noted that many chemolithotrophs use
a considerable proportion of their conserved energy to drive
reversed electron transfer from their oxidisable substrate
to NAD$^+$, generating NADH which is used for the reductive
assimilation of carbon dioxide. However, in spite of these
variations and imperfections, the energy yield from respiration is generally significantly higher than from fermentation.
Three types of bacterial photosynthesis are known, namely;
     i)   bacteriochlorophyll-dependent anoxygenic photosyn-
thesis,
     ii) chlorophyll-dependent oxygenic photosynthesis and
     iii) bacteriorhodopsin-dependent anoxygenic photosynth-
esis.
     In the first two types, the absorption of solar energy
by a variety of photopigments causes a separation of electric
change with the resultant generation of a low redox potential
reductant and a high redox potential oxidant. Electron
transfer between these two entities allows free energy to
be released and subsequently conserved via photophosphorylation in a manner which is analogous to oxidative phosphorylation. In the third type of photosynthesis, conventional
redox carriers are absent and energy is conserved via a
light-dependent, bacteriorhodopsin-mediated H$^+$ pump linked
to a conventional ATP synthesising apparatus.
     Unlike substrate-level phosphorylation, oxidative phos-
phorylation and photophosphorylation are vectorial, membrane-

bound processes that are best described by chemiosmosis
(Mitchell, 1966; 1979; Garland, 1977; Konings and Veldkamp,
this volume pp. 170). In brief, this means that respiration,
photosynthetic electron transfer, the bacteriorhodopsin pump
and ATP synthesis are anisotropic processes which are linked
to a circulating H^+ current. Thus, both respiration and
photosynthesis cause the stoichiometric injection of H^+
across the osmotic barrier of the coupling membrane (expressed
where possible, as the $\rightarrow H^+/0$, $\rightarrow H^+2e^-$ or $\rightarrow H^+/e^-$ quotient),
so generating an electrochemical potential difference of
protons or promotive force ($\Delta\mu H^+$ or Δp; units: mV). The
latter is comprised of a chemical potential difference (ΔpH)
and an electrical potential difference or membrane potential
($\Delta\Psi$) which are related by the equation:
$$\Delta p = \Delta\Psi - Z.\Delta pH$$
Where Z = 2.303 RT/F, with R = gas constant (8.314 J mol^{-1}
K^{-1}), T = absolute temperature (K) and F - Farday constant
(96.6 kJ mol^{-1} equiv^{-1}), enables ΔpH to be expressed in
electrical units and has a value of approximately 59 mV at
25°. A multicomponent ATP phosphohydrolase complex spans
the coupling membrane and, during ATP synthesis, effects
the stoichiometric inward retranslocation of ejected protons
(expressed as the $\rightarrow H^+/ATP$ or $\rightarrow H^+/P$ quotient) under the
driving force of Δp (Fig. 1). Conversely, Δp can be
generated by H^+ ejection during ATP hydrolysis.

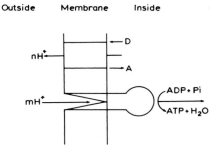

Fig. 1. Chemiosmotic aspects of energy transduction

The Δp which is formed at the expense of a suitable oxidant,
light or ATP, can also be used to drive other energy-
dependent membrane functions including reversed electron
transfer, cell motility, some forms of solute transport and
the synthesis of inorganic pyrophosphate in a few species
of phototroph.

RESPIRATION

Aerobic respiration in chemoheterotrophs and facultative phototrophs

The membrane-bound respiratory chains of these organisms contain the same basic types of redox carriers as those which are present in the mitochondria of higher organisms, namely: iron-sulphur proteins (Fe-S) and flavoproteins (Fp), known as the primary dehydrogenases, quinones, cytochromes and cytochrome oxidases. The primary dehydrogenases catalyse the oxidation both of endogenous and exogenous substrates with the concomitant reduction of either ubiquinone (Q) or menaquinone (MK). Subsequent electron transfer from quinol to molecular oxygen, with the concomitant formation of water, involves several non-autoxidisable cytochromes of the b and c types plus, finally, one or more cytochrome oxidases from a group of potential oxidases which includes cytochromes aa_3, o, d, a_1 and c_{co}. It should be noted, however, that the families *Streptococcus* and *Lactobacillus* do not normally synthesise cytochromes, so that during aerobic growth their respiratory systems are terminated by flavin oxidases which reduce oxygen to hydrogen peroxide. The potentially lethal effects of the latter are nullified by the action of catalase and peroxidase respectively. Many species of bacteria also contain a membrane-bound transhydrogenase which catalyses the reversible transfer of reducing equivalents between NADPH (NADP$^+$) and NADH (NAD$^+$) and, according to species, the transhydrogenase may or may not be linked to energy transduction.

One of the major characteristics of aerobic respiration in chemoheterotrophs is the immense interspecies variation in respiratory chain composition (Haddock and Jones, 1977; Jones, 1977). These variations are of two basic types, namely:

    i)   the replacement of one redox carrier by another with similar or different properties (for example, the replacement of Q by MK, cytochrome oxidase aa_3 by o, energy-dependent transhydrogenase by energy independent transhydrogenase) and

    ii)   the addition or deletion of certain redox carriers (for example, transhydrogenase and/or cytochrome c). In addition, variation in respiratory chain composition can also occur within a single species of bacterium as a result of changes in the growth environment, particularly the decreased availability of essential nutrients and alterations in the nature of the carbon and energy source. Thus, growth under oxygen-limited conditions is accompanied by the increased synthesis of alternative cytochrome oxidases (for example, cytochrome oxidase d relative to aa_3 or o or o relative to aa_3) which have higher affinities for molecular oxygen (Rice and Hempfling, 1978). Sulphate- or iron-limited growth results in depressed levels of iron-sulphur proteins and/or various cytochromes. Similarly, when the growth substrate is a powerful catabolite repressor, such as glucose, there is a general decrease in redox carrier concentrations which can be reversed by the addition of cAMP to the growth medium.

Aerobic respiratory chains exhibit considerable variation in their exact routes of electron transfer. All of them are branched at the level of the primary dehydrogenases, thus allowing reducing equivalents from different substrates to be channelled into a common quinone-cytochrome system which may be either linear or branched (Jones, 1977). Linear terminal pathways are rare and limited to the few systems which contain only one functional oxidase (for example, highly aerobic cultures of *Escherichia coli*). Branching is usually associated with the presence of two or more species of cytochrome oxidase and may (for example, *Azotobacter vinelandii*) or may not (for example, oxygen-limited cultures of *E. coli*) involve truly separate terminal pathways consisting of non-autoxidisable cytochromes and different cytochrome oxidases. There is compelling evidence that in the more complex branched systems, one branch usually contains cytochrome c and cyctochrome oxidase aa_3 or o, whereas in the other branch cytochrome b donates electrons directly to cytochrome oxidase d, aa_3 or o. It is likely that the major function of a branched respiratory system is to allow some flexibility in the exact pathway of electron transfer, thus enabling the organism to minimise the potentially deleterious effects of some growth environments (for example, the high respiratory activity of *A. vinelandii* during growth at high oxygen concentrations is required in order to protect an oxygen-sensitive nitrogenase) and to take maximum advantage of others (for example, electron transfer via cytochrome oxidase d rather than o in oxygen-limited cultures of *E. coli* enables the organism to complete successfully for the limited amount of acceptor).

A wide variety of experimental approaches have provided evidence that aerobic respiratory chains are organised assymetrically in the coupling membrane. The substrate-binding sites of transhydrogenase and most primary dehydrogenases and cytochrome oxidases are located on the cytoplasmic side of the membrane, whereas cytochromes o and c_{co} are located on the opposite surface. However, virtually nothing is known about the intermembrane location of the iron-sulphur proteins, quinones or b type cytochromes.

There is now ample evidence that respiration is accompanied by electrogenic H$^+$ translocation across the coupling
membrane (outwards in whole cells). → H$^+$/O quotients vary
from approximately 4 to approximately 8 g equiv H$^+$ (g atom
O)$^{-1}$ for the oxidation of endogenous substrates. Most experimental data for the oxidation of exogenous substrates
support the concept that aerobic respiration chains are
organised into two, three or four H$^+$-translocating segments
each of which translocates not less than 2H$^+$ per electron
pair transferred. Two of these segments are invarient,
namely, segments 1 (NADH dehydrogenase) and 2 (the central
quinone-b region). In contrast, segments 0 (nicotinamide
nucleotide transhydrogenase) and 3 (cytochrome c oxidase)
are predicated upon the ability of an organism to synthesise
an energy-dependent transhydrogenase and a high potential
cytochrome c linked to cytochrome oxidase aa_3 or o, respectively. It is likely that in some respiratory chains segments
2 and 3 are fused into a complex protonmotive quinone cycle,
thus circumventing repeated failures to detect a hydrogen
carrier for segment 3. This cycle translocates 4H$^+$ concomitant with the transfer of two electrons from quinol to
oxygen (Fig. 2a, b).

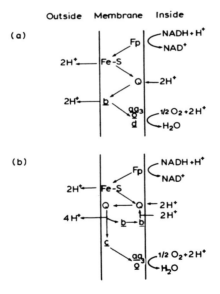

*Fig. 2. Aerobic respiration in chemoheterotrophs. (a)
cytochrome c absent, and (b) cytochrome c present.*

It should be noted that these H^+ -translocating segments are analogous to the classical energy coupling sites of the respiratory chain, in that :

  i) H^+ translocation at each segment can drive ATP synthesis under the appropriate conditions,

  ii) ATP hydrolysis can drive reversed electron transfer through each H^+ -translocating segment and

  iii) forward electron transfer through one H^+ -translocating segment can drive reversed electron transfer through another. This view is supported by measurements of molar growth yields which indicate that segments 1, 2 and 3 contribute fully to respiratory chain phosphorylation. On the other hand, there is no indication that segment 0 makes any significant contribution *in vivo*. Indeed the thermodynamic properties of the two redox couples involved, NADPH/NADP$^+$; $\Delta E'_o$ = +4 mV, suggest that the transhydrogenase may operate in reverse, that is, to generate NADPH at the expense of Δp generated by forward electron transfer through the other H^+ -translocating segments. It should be noted however, that segment 0 can drive ATP synthesis *in vitro* provided that the (NADPH) (NAD$^+$)/(NADP$^+$) (NADH) ratio is poised at a sufficiently high level.

Quantitation of ΔpH and $\Delta\Psi$ using methods based upon the transmembrane distribution of appropriate radioactive or fluorescent marker molecules following energisation under static head conditions (that is, in the absence of any competing, energy-dependent reactions) indicate Δp values in the range 175-230 mV for whole cells (outside acidic and electrically positive). ΔpH and $\Delta\Psi$ contribute variably to Δp according to the exact conditions of the assay, namely, the presence of ionophorous antibiotics or permeant anions, and the use of a significantly acid or alkaline pH.

Aerobic respiration in chemolithotrophs

Chemolithotrophs can use a wide variety of inorganic compounds as substrates for aerobic respiration, including ammonia and nitrite (*Nitrobacteraceae*), sulphide, thiosulphate and sulphite (*Thiobacillaceae*), ferrous iron (*Siderocapsaceae* and *Thiobacillus ferroxidans*)and hydrogen (representatives of the genera *Pseudomonas, Alcaligenes, Paracoccus* and *Nocardia*).

Since these substrates exhibit a very wide range of redox potential (for example, E'_o H$_2$/2H$^+$ =-420 mV: compared with Fe^{2+}/Fe^{3+} = +780 mV) their oxidation by molecular oxygen liberates differing amounts of energy and occurs via an assortment of respiratory chains. The latter contain flavoprotein dehydrogenases, quinones, cytochromes and cytochrome oxidases, which are occasionally supplemented

with novel electron carriers (for example, cytochromes P460
and a_1.Mo in *Nitrosomonas* and *Nitrobacter*, respectively, and
the copper protein rusticyanin in *T. ferroxidans*). It should
be noted, however, that oxidation of the higher redox potential substrates is independent of flavoproteins and quinones,
although these redox carriers function in energy-dependent
reversed electron transfer which generates the NAD(P)H which
is required for the assimilation of carbon dioxide by these
organisms (Aleem, 1977).

According to species, the oxidation of hydrogen occurs
initially via soluble and/or membrane-bound hydrogenases
which are linked to the reduction of NAD^+ or flavin respectively. Reoxidation of NADH or reduced flavin by oxygen
occurs via an H^+-translocating respiratory chain which
exhibits $\rightarrow H^+/O$ quotients of approximately 6 and 4 respectively. The resultant Δp is used principally for ATP synthesis and, in the case of those organisms with a flavin-
linked hydrogenase, to drive reduction of NAD^+ by reversed
electron transfer.

The oxidation of nitrite to ammonia in *Nitrosomonas*
with an overall reaction of $NH_3 + 1\frac{1}{2}O_2 \rightarrow NO_2^- + H_2O + H^+$,
occurs via hydroxylamine. However, at +899 mV the E_o' of
the ammonia/hydroxylamine couple is too high to allow
ammonia to be oxidised by molecular oxygen ($NH_3 + H_2O + \frac{1}{2}O_2^*$
$\rightarrow NH_2OH + H_2O^*$), and hydroxylamine to nitrite (E_o' NH_2OH
$NO_2 = +66mV$) involves the release of 2H, which are oxidised
via a flavoprotein dehydrogenase and a conventional quinone-
cytochrome system, and a second oxygenation reaction.
Corrected $\rightarrow H^+/O$ quotients of approximately 2 have been
observed for hydroxylamine oxidation which indicate that
the respiratory chain may be organised into only one H^+-
-translocating segment in spite of the large redox span
involved. The oxidation of nitrite to nitrate in *Nitrobacter*
with an overall reaction of $NO_2^- + H_2O + \frac{1}{2}O_2^* \rightarrow NO_3^- + H_2O^*$,
resembles the thermodynamically forbidden oxidation of
ammonia in *Nitrosomonas* in that it entails the initial
transfer of an oxygen atom from water into nitrite, followed
by the aerobic oxidation of the released 2H to form a
compensating molecule of water (Fig. 3a). The overall
reaction exhibits an $\rightarrow H^+/O$ quotient of approximately 2
which is commensurate with the redox potential of the nitrite/
nitrate couple, $E_o' = +420$ mV.

Much less is known about those respiratory chains which
catalyse the aerobic oxidation of reduced sulphur compounds
($S^{2-} \rightarrow S_2O^{2-} \rightarrow SO_3^{2-} \rightarrow SO_4^{2-}$) in the *Thiobacilli*. Reducing
equivalents from sulphide and thiosulphate enter the
respiratory chain at the level of flavin and cytochrome *c*
respectively, and are transferred to oxygen with the concom-

itant synthesis of ATP. In addition, some *Thiobacillus*
species are able to convert thiosulphate into two molecules
of sulphite from which they obtain a small amount of ATP via
substrate-level phosphorylation. The reducing power
released by the latter process is subsequently channelled
into respiration of the flavin level where it yields considerably more ATP via oxidative phosphorylation. Little is
known about the H^+-translocating abilities of these respiratory chains or about their spatial distribution in the
coupling membrane.

The simplest, and in many respects the most novel,
respiratory chain yet encountered in chemolithotrophs is
that which is present in the iron-oxidising bacterium *T.
ferroxidans*. This acidophile oxidises external Fe^{2+}, with
an overall reaction of $2Fe^{2+} + \frac{1}{2}O_2 + 2H^+ \rightarrow 2Fe^{3+} + H_2O$, via
a highly attenuated respiratory chain which has no H^+
-translocating capacity and which functions mainly to
catalyse inward electron transfer across the coupling membrane (Fig. 3b). The resultant consumption of $2H^+$ on the
inner surface concomitant with the reduction of molecular
oxygen to water maintains the cytoplasm at a more alkaline
pH than the growth environment ($\simeq$ pH 5.5 compared with
2.0). ATP synthesis occurs via inward H^+ movement through
the ATP phosphohydrolase complex under the driving force of
the large ΔpH.

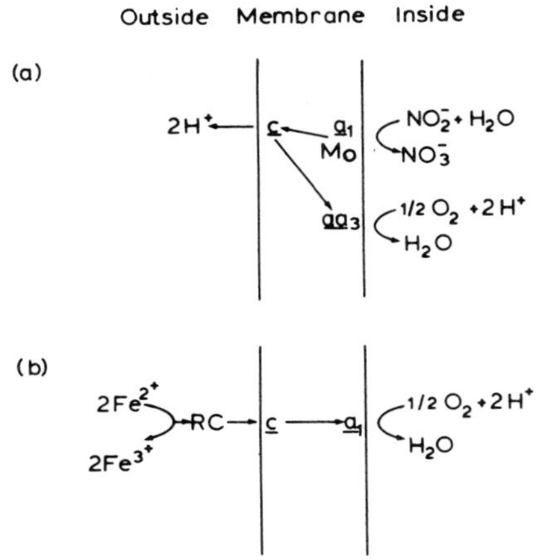

*Fig. 3. Aerobic respiration in chemolithotrophs. (a)
Nitrobacter and (b) T. ferroxidans (RC, rusticyanin).*

Anaerobic respiration

The ability to use a variety of organic or inorganic oxidants as terminal electron acceptors for respiration under anaerobic conditions is taxonomically fairly widespread. These oxidants include fumarate (many species of Gram-negative bacteria and some species of *Bacillus*), nitrate *(Enterobact-eriaceae*, some species of *Bacillus, Pseudomonas, Alcaligenes* and *Paracoccus)*, nitrite (as for nitrate except for the *Enterobacteriaceae*, that is, only the denitrifying bacteria), sulphate and sulphite *(Desolfotomaculum* and *Desulfovibrio)* and carbon dioxide *(Methanobacteriaceae)*. As with aerobic respiration in chemolithotrophs, both the amount of energy which is released during anaerobic respiration and the redox carrier composition of these respiratory chains is determined by the varied redox potentials of the electron acceptors (for example, E'_O SO_4^{2-}/SO_3^{2-} = -480mV compared with E'_O NO_3^-/NO_2^- = +420 mV). Anaerobic respiratory chains in general contain the same type of redox carriers as are present in aerobic systems, except that cytochrome oxidase(s) is replaced by the appropriate reductase and additional novel redox carriers are present in some systems (Krüger, 1977; Wood, 1978).

The reduction of fumarate by NADH or formate involves the respective dehydrogenase, a quinone (Q or MK), a b type cytochrome and fumarate reductase. Each overall reaction exhibits an $\rightarrow H^+/2e$ quotient of approximately 2. When NADH is the donor a complete H^+-translocating loop is probably involved and a similar loop operates for formate in some organisms but in others the respiratory chain only catalyses the inward transfer of electrons ($2H^+$ being released on the outside by the externally-located formate dehydrogenase).

Nitrate reduction utilises a similar respiratory chain to that which catalyses fumarate respiration, except that it is organised into two H^+-translocating segments and is terminated by nitrate reductase. The latter is known to have a transmembrane orientation which enables it to consume $2H^+$ on the cytoplasmic side of the membrane whilst reducing the nitrate on either side according to species, with an overall reaction of $NO_3^- + 2H^+ + 2e \rightarrow NO_2^- + H_2O$. This property maximises the efficiency of respiration-linked H^+ transloca-tion and eliminates from some species the need for a nitrate/nitrite transport system (Fig. 4). In contrast, nitrite respiration requires the additional presence of c type cyto-chrome, a copper protein (azurin) and nitrite reductase (cytochrome cd_1), all of which appear to be located in the periplasm or on the outer surface of the membrane. It is likely, therefore, that some of the protons which appear in

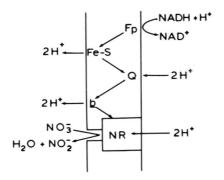

Outside   Membrane   Inside

Fig. 4. Nitrate respiration in E. coli

the periplasm during respiration are consumed in nitrite
reduction, with an overall reaction of $0.67NO_2^- + 2.67H^+$ +
$2e^- \rightarrow 0.33N_2 + 1.33H_2O$. The exact way in which the respiratory chain is organised to yield the observed $\rightarrow H^+/2e^-$
quotients of 3-4 has yet to be elucidated.

Sulphate respiration poses an initial thermodynamic
problem in that the E_o' of the sulphate/sulphite couple is
too low to allow reduction by known electron donors such as
hydrogen or lactate. This problem is overcome by activation
of the sulphate at the expense of two molecules of ATP to
form adenosine phosphosulphate (APS), which is of high
enough redox potential (E_o' APS/AMP + SO_3^{2-} = +60 mV) to allow
a substantial amount of energy to be released during its
reduction. The latter occurs via a variety of low potential,
soluble and membrane-bound redox carriers which include
ferredoxin, cytochrome c_3 and APS reductase. Thermodynamic
considerations predict a maximum of one molecule of ATP per
molecule of sulphite formed, that is, a net expenditure of
1 ATP for the overall reduction of sulphate to sulphite.
The further reduction of sulphite to sulphide E_o' SO_3^{2-}/S^{2-}
= -116 mV) is a six electron transfer reaction in which APS
reductase is replaced by bisulphite reductase, a novel enzyme
which contains a siderohaem prosthetic group. The complete
reduction of sulphite to sulphide at the expense of three
molecules of hydrogen liberates enough energy for the
synthesis of three molecules of ATP. The H^+-translocation
properties of these respiratory chains have not been
determined.

Anaerobic respiration from hydrogen to carbon dioxide is
also associated with ATP synthesis and leads to the eventual

liberation of methane, probably via the sequential formation
of aldehyde, alcohol and methyl compounds with coenzyme M.
Little is known about the redox carriers which are present
in these respiratory chains, except that a hydrogenase is
involved together with a novel deazaflavin-containing
electron carrier, coenzyme F_{420} (E'_o F_{420}/F_{420}^{2-} = -373 mV).
The possible organisation of these respiratory chains into
H^+-translocating segments has not yet been investigated in
detail.

PHOTOSYNTHESIS

The overall process of photosynthesis may be approximately
described by the equation:

$$2H_2A + CO_2 + \text{light energy} \rightarrow (CH_2O) + H_2O + 2A$$

where H_2A is a suitable organic or inorganic reducing agent,
for example, succinate, sulphide, thiosulphate or hydrogen
gas (purple and green bacteria), water (cyanobacteria and algae)
or amino acids (halobacteria). Oxygenic photosynthesis
occurs only when water is the reducing agent. In each group
or organisms, the major function of light is to drive cyclic
processes which lead to H^+ -translocation and hence photophosphorylation. Furthermore light is also responsible,
where appropriate, for the evolution of molecular oxygen
and for the generation of NAD(P)H via non-cyclic electron
transfer.

Bacteriochlorophyll-dependent anoxygenic photosynthesis

This type of photosynthesis is carried out by the purple
bacteria (*Rhodospirillaceae* and *Chromatiaceae*) and the
green bacteria (*Chlorobiaceae* and *Chloroflexaceae*). These
organisms contain similar types of redox carriers within
their light-dependent cyclic electron transfer systems,
namely; bacteriochlorophylls (BChl), bacteriopheophytins,
(BPh), carotenoids, quinones, iron-sulphur proteins and
cytochromes. The first three type of redox carriers are
organised within the coupling membrane into photosynthetic
units, each of which is comprised of multiple copies of one
or two species of light-harvesting (LH) complex plus a
single reaction centre (RC) complex. The light-harvesting
complexes contain bulk amounts of protein-bound bacteriochlorophyll and carotenoid which act as antennae for the absorption of radiant energy. The type of bacteriochlorophyll
(BChl a, b, c, d or e) is characteristic of particular
groups of purple and green bacteria and, in association with
the carotenoid content, determins both the colour of the
organism and the wavelength range of solar radiation which
can be used by that organism. The absorption spectrum of
each type of bacteriochlorophyll is modified by BChl-BChl,

BChl-protein and BChl-carotenoid interactions such that, for example, BChl a exhibits varying intensities of short wavelength (B800) medium wavelength (B820 or B850) and long wavelength (B870 or B890) bands in different organisms.

The absorption of a single photon by the light-harvesting pigments causes energy to be channelled initially into the singlet excitation of the longest wavelength constituent (for example, B870) and subsequently into the reaction centre complex. The latter conrains bacteriochlorophyll and bacteriopheophytin (in which the central Mg^{2+} atom of BChl is replaced by $2H^+$) in a specialised environment and each is bound to protein, often closely associated with other redox carriers. The reaction centre contains much less bacteriochlorophyll than the light-harvesting complexes since the LH BChl:RC BChl ratio ranges from 40-80 in the purple bacteria to 1000-1600 in the green bacteria. Each reaction centre complex in the purple bacteria contains four molecules of bacteriochlorophyll (for example, P870) and two molecules of the corresponding bacteriopheophytin. The energy received by the reaction centre is distributed between two of the four bacteriochlorophyll molecules, the so-called "special pair" $(P870)_2$, which enter the singlet excited state and subsequently transfer an electron to a single bacterio-pheophytin molecule. The electron is then passed to the primary acceptor (the Q or MK moiety of a quinone-iron complex; $E_m \leq -160$ mV), after which $(P870)_2^+$ is reduced at the expense of the primary donor (one of two identical c type cytochromes; $E_m = +300$ mV). Thus at the end of the overall photochemical process, the reaction centre components are in the form $c^+.(P870)_2.BPh.QFe^-$. Cyclic electron transfer is completed by the passage of an electron from QFe^- to c^+ via the quinone-cytochrome system (Dutton and Prince, 1978; Harold, 1978).

Much less is known about the early stages of photosynthesis in green bacteria. Light energy absorbed by the antennae photopigments is channelled to the bulk BChl c, d or e and is subsequently transferred via a BChl a-protein complex (B870) to the reaction centre bacteriochlorophyll (P840). The subsequent charge separation process in the reaction centre complex leads to the reduction of the primary acceptor (Fe-S; $E_m = -540$ mV) and to the oxidation of the primary donor (a c type cytochrome; $E_m = +145$ mV) and completed via the quinone-cytochrome system.

There is considerable evidence that cyclic electron transfer in purple bacteria leads to H^+-translocation across the coupling membrane and to the generation of a protonmotive force. Since cyclic electron transfer effects no net oxidation-reduction, it is virtually impossible to measure $\rightarrow$ H$_+^+$/e$_-^-$ quotients during continuous illumination. However, $\rightarrow$ H$^+$/e$^-$ quotients of approximately 2 have been determined following activation by flashes of light which are sufficiently brief (20 µs) to cause only one complete turnover of the photosynthetic apparatus. Since the c type cytochrome is loosely associated with the periplasmic surface of the coupling membrane, whereas the quinone-iron complex is quite close to the cytoplasmic surface, the initial photochemical reaction induces a separation of charge across the membrane (outside positive, inside negative). This is followed by net electron and H$^+$ transfer towards the periplasmic surface via the quinone-cytochrome system. The latter is probably organised into a complex, protonmotive cycle which is similar to that present in many respiratory chains (Fig. 5). The resultant Δp ranges up to 285 mV (inside alkaline and electrically negative in whole cells), to which ΔpH and $\Delta\Psi$ contribute variably according to the exact assay conditions.

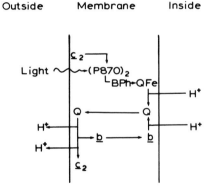

Fig. 5. Anoxygenic photosynthesis in Rhodospirillaceae

All species of photosynthetic bacteria require NADH for biosynthetic purposes. The reduction of NAD$^+$ by substrate of relatively high redox potential (for example, succinate, sulphide and thiosulphate) is light-dependent, the exact pathway of electron transfer being determined by the redox potential of the primary acceptor relative to that of the NADH/NAD$^+$ couple. In the purple bacteria, the E_m value of the quinone-iron complex is too high to allow direct reduction of NAD$^+$ and these organisms therefore use cyclic electron transfer to generate Δp which in turn drives reversed electron transfer from succinate to NAD$^+$. In contrast, the iron-sulphur protein which acts as the primary acceptor in green bacteria has an extremely low E_m value and is thus able to photoreduce NAD$^+$

directly, that is, without the mediation of Δp. During the photoautotrophic growth of either group of bacteria, NAD^+ is reduced by molecular hydrogen in a light-independent reaction.

Chlorophyll-dependent oxygenic photosynthesis

This type of photosynthesis is carried out by the *Cyanobacteriaceae*. The light-dependent energy conservation systems of these organisms differ from those of the purple and green bacteria in several important ways, including;

i)   the replacement of bacteriochlorophyll by chloropyll (which absorbs at lower wavelengths),

ii)   the supplementation of the antennae photopigments with the phycobiliproteins, namely, allophycocyanin, phycocyanin and phycoerythrin and

iii) the presence of an additional photosystem (photosystem II) which is reponsible for catalysing the oxidation of water with the concomitant release of molecular oxygen.

Absorption of light energy by the antennae photopigments of photosystem I leads eventually to the singlet excitation of a chlorophyll a molecule P700) in the reaction centre complex and hence to the reduction of the primary acceptor (probably a membrane-bound iron-sulphur protein; $E_m \leqslant 550$ mV) and to the oxidation of the primary donor (a c type cytochrome called cytochrome f; $E_m = +370$ mV). Subsequent electron transfer from the primary acceptor to ferredoxin and hence to $NADP^+$ generates the reducing power which is required for biosynthesis. The re-reduction of cytochrome f can occur either via cyclic electron transfer at the expense of reduced ferredoxin, or via non-cyclic electron transfer from a quinone (plastoquinone, PG; $E_m \simeq 0$ mV) which acts as the primary electron acceptor for photosystem II. The antennae apparatus of the latter transfers absorbed light energy to the reaction centre, and the electron which subsequently passes from chlorophyll b (P690) to plastoquinone is replenished via the oxidation of water to molecular oxygen. The non-cyclic transfer of two reducing equivalents from water to $NADP^+$ occurs therefore over a redox span of approximately -1 V, requires four quanta and is mediated by two separate photosystems. Both cyclic and non-cyclic electron transfer leads to the translocation of H^+ across the coupling membrane and hence to the generation of the proton-motive force which is responsible for ATP synthesis. Thus photosynthetic energy conservation in the cyanobacteria although showing some similarities to that which occurs in the green bacteria, is very closely related to green plant photosynthesis.

Bacteriorhodopsin-dependent photosynthesis

When the facultatively phototrophic halobacteria *Halobacteriaceae* are grown in the presence of light at low oxygen concentrations they synthesise a purple, membrane-bound photosystem which is composed principally of bacteriorhodopsin. No bacteriochlorophyll or conventional redox carriers are present and the carotenoids do not function as accessory photopigments (Oesterheldt, 1976). Bacteriorhodopsin (bR_{560}) consists of retinal covalently bound to the protein bacterioopsin via a Schiff base linkage. Low intensity light causes bR_{560} to undergo a 1,3-*cis* all-*trans* isomerisation to form the light-adapted purple complex (bR_{570}) which is subsequently converted via a high intensity photoreaction and a series of dark reactions, into a bleached complex (bM_{412}). The latter is finally reconverted to bR_{570} via further dark reactions. The Schiff base is protonated by bR_{570} ($-CH \cdot NH-$) but not in bM_{412} ($-CH=N-$) and so the cyclical light-induced interconversion of these two complexes is accompanied by the sequential release and uptake of a proton. Since the bacteriorhodopsin molecule spans the coupling membrane, the deprotonation-protonation cycle forms the basis of an outwardly-directed H^+ pump. Estimations of the resultant Δp indicate values of up to 230 mV (inside alkaline and electrically negative in whole cells) with contributions from both Δp and $\Delta \Psi$.

Assuming that $\not\!< 2H^+$ are required for the synthesis of one molecule of ATP (see Energy Transduction below), *Halobactereaceae* conserve incident light energy as ATP with an overall efficiency of approximately 5%, compared with approximately 21% in the *Rhodospirillaceae* and up to 18% during cyclic photophosphorylation in the *Cyanobacteriaceae*. However, it should be noted that this extremely low efficiency is partly alleviated by the savings in biosynthetic resources which accrue from having an H^+ translocation system which is comprised of only one protein instead of the multiprotein complexes which characterise the other two groups of phototrophs.

The halobacteria obtain NAD(P)H for biosynthesis in a manner which is analogous to that practised by most chemoheterotrophs, namely, by the light-dependent catabolism of the growth substrate.

ENERGY TRANSDUCTION

Although the protonmotive force can drive certain types of solute transport, reversed electron transfer and cell motility, its major function is to power the thermodynamically unfavourable synthesis of ATP from ADP and inorganic phosphate. This process of oxidative or photosynthetic phosphorylation transforms energy in the form of an electrochemical potential difference of H^+ across the coupling membrane (Δp) into the chemical energy of ATP in the cytoplasm, the phosphate potential (ΔGp), and is catalysed by the membrane-bound ATP phosphohyrolase complex. The latter consists of two multiprotein complexes (BF_0 and BF_1). BF_1 is located on the cytoplasmic side of the coupling membrane and characterised by being hydrophilic and is responsible for catalysing the synthetic and hydrolytic functions of the complex. In contrast, BF_0 is a hydrophobic, lipoprotein complex which spans the coupling membrane and facilitates the passage of H^+ to and from the active site of BF_1. The entire complex thus consists of a gated H^+ pump (BF_1) and an H^+ channel-filter (BF_0) which together are responsible for the catalysis, selectivity and regulation of H^+ translocation (Haddock and Jones, 1977; Downie, Gibson and Cox, 1979).

Purified BF_1 (MW=350-400x10^3) usually contains five types of non-identical subunits $\alpha, \beta, \gamma, \delta$ and ϵ probably in the ratio of 3:3:1:1:1 respectively. A wide range of biochemical and genetic evidence indicates that the α and β subunits are organised into a planar hexagon and that they form the active site of BF_1, possibly in association with the γ subunit which controls the passage of H^+ into and out of the $\alpha_3\beta_3$ complex. The δ and ϵ subunits facilitate the binding of $\alpha_3\beta_3\gamma$ to the BF_0 complex and thus probably form the H^+ translocating stalk which is clearly visible in electron micrographs of coupling membranes. It is possible that in some organisms the ϵ subunit also functions as in inhibitor of ATPase activity whilst leaving ATP synthesis unaffected. BF_0 usually contains two subunits ζ and η probably in the ratio of 3:6 respectively. The ζ subunits are responsible for binding BF_1 to BF_0 (via the δ ϵ stalk), whereas the η subunits probably comprise the H^+-translocating channel through the coupling membrane. The precise mechanism of H^+ translocation has not been determined.

There is now considerable evidence suggesting that the function of the $BF_0 \cdot BF_1$ complex varies with the mechanism of cellular energy conservation. Thus, during strictly fermentative growth ATP is synthesised by substrate-level phosphorylation and the $BF_0 \cdot BF_1$ complex acts as an H^+ rejecting ATPase in order to generate the Δp which is required for membrane energisation. In contrast, during growth in the presence of light or of an appropriate exogenous electron acceptor, energy is initially conserved as Δp and the $BF_0 \cdot BF_1$ complex acts as a conventional H^+ injecting ATP synthetase. These different physiological functions are reflected in the molecular properties of the $BF_0 \cdot BF_1$ complex. For example, the complex which is present in the fermentative anaerobe *Clostridium pasteurianum* contains fewer subunits, such as, the absence of the regulatory ϵ subunit, and is more simply organised than the corresponding complex from a respiratory organism. In addition, its kinetic properties are compatible with a hydrolitic rather than a synthetic function *in vivo*.

It is now clear that both respiration and photosynthesis can generate a trans- or intra-membrane Δp of up to 290 mV. Since the results of a variety of experiments have indicated that the net synthesis of ATP requires a minimum Δp in the range of 167–215 mV according to species, it would appear that the former value is quite sufficient to effect phosphorylation. Furthermore, phosphorylation appears to be capable of being driven by either ΔpH or $\Delta \Psi$ alone, provided that they are of the necessary magnitude and direction. As expected, the hydrolysis of ATP generates a Δp which is similar to that produced during respiration or photosynthesis.

Considerable effort is currently being expended on determining the $\rightarrow$ H^+/ATP quotient which is exhibited by the ATP phosphohydrolase complex. Direct measurements of pH changes associated with ATP hydrolysis by inside-out membrane vesicles (whole cells are impermeable to ATP) are technically extremely difficult to carry out and interpret and the results have generally indicated rather low values. More recently, emphasis has been placed on indirect determinations via parallel estimations of Δp and ΔGp since, at equilibrium:

$$H^+/ATP \quad = \quad \Delta Gp/F.\Delta p$$

where the Farday constant, F, allows ΔGp to be expressed in electrical units (mV). The most reliable results indicate $\rightarrow$ H^+/ATP quotients in the range of 2–3 g ion H^+ (mol ATP)$^{-1}$. Interestingly, comparison of the molar growth yields of bacteria growing aerobically and fermentatively point to $\rightarrow$ H^+/ATP quotients which are close to the $\rightarrow$ $H^+/2e^-$ quotients for each H^+-translocating segment in the respiratory chain (that is, not less than 2). Overall, therefore, these results suggest that the ATP/2e$^-$ quotients exhibited by respiration and bacteriochlorophyll-dependent photosynthesis are half, or slightly less than half, of the observed $\rightarrow$ $H^+/2e^-$ quotients.

There is now substantial evidence that the protonmotive force can also act as the energy source for the active transport of certain solutes, such as, glucose into aerobic bacteria, some other sugars, amino acids and inorganic ions. In each case the transported species is either neutral or positively charged, and moves under the force of the ΔpH and/or $\Delta \Psi$. Thus, anions and neutral molecules are cotransported with H^+ (for example, H^+.glutamate and H^+ glucose symports) in response to ΔpH and Δp respectively. Cations enter unaccompanied (for example, lysine$^+$ uniport) under the driving force of $\Delta \Psi$ and exit in exchange for protons (for example, $H^+ \cdot Na^+$ antiport) in response to Δp. As a result of such transport systems, (solute)$_{in}$/(solute)$_{out}$ ratios of up to 10^5 have been detected for various sugars and amino acids, these values being commensurate with the magnitude of the ambient Δp. Interestingly, the halophile *Halobacterium halobium* generates a substantial $\Delta \bar{\mu}_{Na^+}$ at the expense of its light-driven H^+ pump and an active Na$^+$ ·H$^+$ antiport which it subsequently uses in place of Δp to drive the accumulation of neutral and acidic amino acids via Na$^+$·symport.

In addition to exhibiting ATP phosphohydrolase activity, the facultatively phototrophic genera *Rhodospirillum* and *Rhodopseudomonas* catalyse the H$^+$-linked synthesis and hydrolysis of inorganic pyrophosphate. These reactions, which are catalysed by a membrane-bound pyrophosphate phosphohydrolase complex, are probably the last vestiges of an era, prior to the advent of ATP, when pyrophosphate was an important energy currency molecule in these organisms (Baltscheffsky and Battscheffsky, 1974).

Finally, it is becoming increasingly clear that bacterial motility also occurs at the expense of Δp, provided that the latter exceeds an ill-defined threshold value. However, the mechanism by which Δp drives flagellar rotation is currently unknown, as also is the precise method by which motility is controlled by external chemical stimuli (chemotaxis).

CONCLUSION

It is clear from the foregoing discussions that bacteria exhibit considerable unity in their basic mechanisms of oxidative or photosynthetic phosphorylation. In general, they transfer reducing equivalents from a low redox potential donor to a higher redox potential acceptor via a sequence of core redox components and in so doing generate a transmembrane or intermembrane protonmotive force which can drive ATP synthesis and other energy-dependent membrane functions. The halobacteria are of course rare exceptions to this generalisation in that they generate Δp via a mechanism which is independent of electron transfer.

In contrast, a more detailed examination of bacterial energy conservation systems reveals considerable diversity. Thus bacteria use a very wide variety of electron donors and acceptors according to species (the photosynthetic bacteria doing so far as to generate their own donors and acceptors for cyclic electron transfer as a result of initial photochemical events) and this property necessitates the synthesis of novel dehydrogenases and reductases which allow the substrates to interact with the thermodynamically appropriate portion of the core electron transfer system. Further diversity is also encountered in the spatial organisation of these electron transfer systems which enables the various donors and acceptors to be used with the maximum benefit to the organism in terms of energy conservation efficiency, biosynthetic economy or electron transfer activity.

REFERENCES

Aleem, M.I.H. (1977). Coupling of energy with electron
   transfer reactions in chemolithotrophic bacteria. *In*
   Microbial Energetics (Ed. B.A. Haddock and W.A. Hamilton),
   Society for General Microbiology, Vol, 27, pp.351-381.
   Cambridge University Press, Cambridge.
Baltscheffsky, H. and Baltscheffsky M. (1974). Electron
   transport phosphorylation, *Annual Review of Biochemistry*
   43, 871-897.
Downie, J.A., Gibson, F. and Cox, G.B. (1979). Membrane
   adenosine triphosphatases of prokaryotic cells, *Annual
   Review of Biochemistry* 48, 103-131.
Dutton, L.P. and Prince, R.C. (1978). Energy conservation
   processes in bacterial photosynthesis. *In* The Bacteria
   (Ed. L.N. Ornston and J.R. Sokatch), Vol. 6, pp.523-584.
   Academic Press, London and New York.
Garland, P.B. (1977). Energy transduction and transmission
   in microbial systems. *In* Microbial Energetics (Ed. B.A.
   Haddock and W.A. Hamilton), Society for General Microbiology
   Symposium Vol. 27, pp. 1-21. Cambridge University Press,
   Cambridge.
Haddock, B.A. and Jones, C.W. (1977). Bacterial respiration,
   Bacteriological Reviews 41, 47-99.
Harold, F.M. (1978). Vectorial metabolism. *In* The Bacteria
   (Ed. L.N. Ornston and J.R. Sokatch), Vol. 6, pp.463-521.
   Academic Press, London and New York.
Jones, C.W. (1977). Aerobic respiratory systems in bacteria.
   In Microbial Energetics (Ed. B.A. Haddock and W.A. Hamilton)
   Society for General Microbiology Symposium, Vol. 27,
   pp. 23-59. Cambridge University Press, Cambridge.
Krüger, A. (1977). Phosphorylative electron transport with
   fumarate and nitrate as terminal hydrogen acceptors. *In*
   Microbial Energetics (Ed. B.A. Haddock and W.A. Hamilton)
   Society for General Microbiology Symposium, Vol. 27, pp.
   61-93. Cambridge University Press, Cambridge.
Mitchell, P. (1966). Chemiosmotic coupling in oxidative and
   photosynthetic phosphorylation, *Biological Reviews* 41,
   445-502.
Mitchell, P. (1979). Compartmentation and communication in
   living systems. Ligand conduction : a general catalytic
   principle in chemical, osmotic and chemiosmotic reaction
   systems, *European Journal of Biochemistry* 95, 1-20.
Morris, J.G. (1975). The physiology of obligate anaerobiosis,
   Advances in Microbial Physiology 12, 169-246.
Oesterheldt, D. (1976). Bacteriorhodopsin as an example of a
   light-driven proton pump, *Angewandte Chemie (International
   edition in English)* 15, 17-24.

Rice, C.W. and Hempfling, W.P. (1978). Oxygen-limited
continuous culture and respiratory energy conservation in
Escherichia coli, Journal of Bacteriology 134, 115-124.
Thauer, R.K., Jungermann, K. and Decker, K. (1977) Energy
conservation in chemotrophic anaerobic bacteria,
Bacteriological Reviews 41, 100-180.
Wood, P.M. (1978). A chemiosmotic model for sulphate respiration,
Federation of European Biochemical Societies Letters 95,
12-18.

NUTRIENT AND ENERGY FLOWS THROUGH SOIL MICROBIAL BIOMASS

E.A. Paul and R.P. Voroney

Department of Soil Science, University of Saskatchewan
Saskatoon, Canada S7N OWO

INTRODUCTION

The recognition and recent studies of the role of living organisms (biomass) in carbon, nitrogen and phosphorus flows in nature have stressed the role of the microorganisms as sources and sinks for nutrients, in addition to their traditional role in transformation of the nutrient elements. Techniques for determining the size and activity of laboratory cultures have had to be altered for field measurements. Similarly, growth concepts developed using liquid culture studies should be applicable in the field after the effects of such factors as competition, the presence of inhibitors, physical adsorption and entrapment, and the broad variety of available energy sources have been considered.

Enough information is now available on microbial biomass and activity in terrestrial soils, sediments and aquatic systems that reasonably accurate mathematical descriptions of the pool sizes and processes involved can be made. Mathematical modelling is useful in that it allows one to describe the processes. It also provides for the testing of available data and concepts and thereby points out where further research is necessary.

MICROBIAL GROWTH AND CARBON TRANSFORMATION MODELS

One example of the role that microbial biomass and activity play in carbon cycling can be found in the description in the decomposition and stabilization of organic matter in grassland soils (Fig. 1). In compiling the information necessary to obtain a meaningful mathematical expression for the biological, chemical and physical processes involved in organic matter turnover under both native and cultivated conditions, it was found that reasonably good data are available for plant decomposition rates. Net decomposition rates in the laboratory, under optimum moisture and temperature conditions (van Veen and Paul, 1980), were found to be twice those in tropical soils of Nigeria (Jenkinson and Ayanaba, 1977). In turn the decomposition rates in the moist temperature climates of Europe were a quarter of those found in Nigeria (Jenkinson and Rayner, 1977). A Canadian grassland with a dry summer and cold winter had an average annual decomposition rate one-twentieth of that found in the laboratory under optimum conditions. The rates during the one or two warm moist months were higher than those found in Europe on soils.

Determination of plant decomposition rates requires a knowledge of not only net rates, that is, the amount of ^{14}C-carbon left behind but also involves correction for microbial growth and production of soil organic matter so that gross rates can be determined. Decomposition rates corrected for microbial utilisation of the substrate at reasonably high efficiencies (35 to 60%) show much faster transformation rates (Table 1) than usually described in the literature, and most field studies have not utilised short enough time intervals to obtain the adequate measurement of the kinetics involved during the early decomposition periods.

TABLE 1

*Organic matter levels, annual input of plant
residues and optimal decomposition rates
for a grassland soil (Sceptre)*

Material	Amount present (kg C ha^{-1} 80 cm^{-1})	Decomposition rate constants (day^{-1})
Litter	1,000	8×10^{-2}
Root	10,300	8×10^{-2}
Biomass	1,250	3×10^{-2}
Decomposable organic matter not protected	4,400	8×10^{-2}
Decomposable organic matter protected	44,100	8×10^{-4}
Recalcitrant organic matter not protected	17,000	8×10^{-6}
Recalcitrant organic matter protected	17,000	8×10^{-8}

MICROBIAL BIOMASS

The description of microbial growth and soil organic
matter formation (Fig. 1) has divided the plant residues into
two fractions; a decomposable fraction and one that is
recalcitrant. This agrees with present day concepts of humus
formation which indicate that recalcitrant materials, such as
lignins and polyphenols, enter the decomposable native soil
organic matter fraction. Physical-chemical condensation with
nitrogen rich products of microbial growth results in the
formation of resistant humic materials (Haider, Martin and
Filip, 1975).

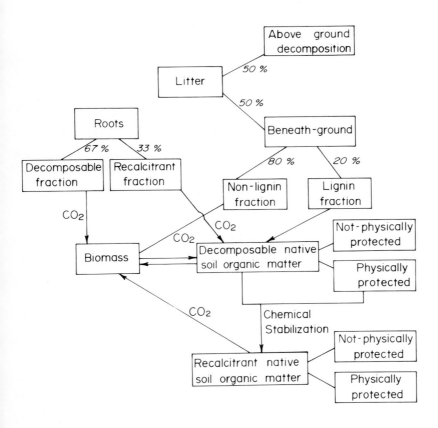

Fig. 1. Scheme of the long-term carbon turnover model.

In this study the modelling procedure was simplified by
using primarily first order kinetics rather than the more
complex hyperbolic equations involving microbial growth
(Michaelis-Menten or Monod equations). This proved feasible
because the soil biomass is so large (2% of the total carbon
and 4% of the soil nitrogen) and has such great growth potential
relative to the rather low amount of available substrate that
one long term basis, such as the 400 years with subdivisions of
0.1 year used in this model, its initial size seldom limited
microbial transformations.

Stabilized organic materials persist for long periods of
time in sediments as well as in terrestrial soils. Carbon
dating has indicated that the 40-60% of the carbon of terrestrial
soils which is non-acid hydrolyzable has persisted for at least
1000 years and thus has very low decomposition rates. In
natural systems, nearly all the reactions occur on or in solid
matrices. This includes colloids consisting of humic materials,
sesquioxides and fine clays. Amino acids added to microbial
cultures or to soil have high turnover rates (Schmidt, Putnam
and Paul, 1960; Marumoto, Kai, Yoshida and Harada, 1977).

The decay rate of amino acids in solution culture relative
to those produced and stabilized within the soil was used to
develop a soil physical protection factor (van Veen and Paul,
1980). Although the added acetate had disappeared from the
system within four days (data not shown), the total ^{14}C -carbon
remaining in the soil was stabilized at approximately 25% (Fig.
2). Microscopic examination indicated two populations of
microorganisms had developed sequentially during the one to
ten day incubation period before stabilization had taken place
(McGill, Paul, Shields and Lowe, 1973). There are many
conditions especially when glucose is added to moist field soils
where stabilization occurs at much higher levels (Ladd and Paul,
1973; Paul and McLaren, 1975; Wagner, 1975).

During computer simulation of the decay of the added ^{14}C-
acetate and of the transformations of the microbially produced
^{14}C-amino acids, the best fit of the data in Fig. 2 was obtained
by assuming 50-60% of the amino acids were protected. Non-
protected amino acids had a degradation rate of 0.3 day^{-1}.
Protection changed this by a factor ranging from 0.01 to 0.005
times the unprotected rate. It was found that the degree of
protection, not the initial decomposition rate, controlled the
amino acids present in the long term incubation (van Veen and
Paul, 1980). The 50-60% protection of organic materials agrees
with published data of the amount of organic materials closely
associated with soil solids (Legg, Chichester, Stanford and
De Mar, 1971; Anderson, 1979).

Physical protection controlled the decomposition rate of
both the decomposable and resistant compounds in the model

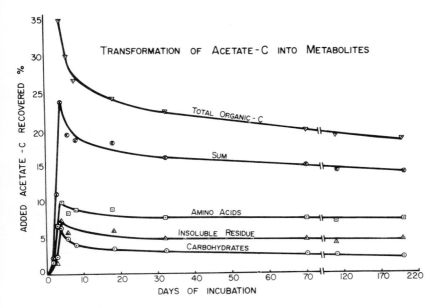

Fig. 2. Recovery of labelled carbon from soil during decomposition of ^{14}C-labelled acetate (date from Sørensen and Paul, 1961).

shown in Fig. 1. This differs from the model of Jenkinson and Rayner (1977) in which physical protection affected only the readily decomposable compounds. Materials such as clay-adsorbed amino compounds although degraded at much lower rates than free amino acids still persist in the soil for much shorter periods of time than humates which also may or may not be physically protected.

The alteration of physical protection by cultivation is one of the major factors controlling the availability of substrate to terrestrial microorganisms. More data on the interactions of organic materials with solid surfaces are required.

MICROBIAL GROWTH AND NITROGEN AND PHOSPHORUS CYCLING

The transformations of organic nitrogen are closely related to those of carbon. Processes such as dinitrogen fixation and denitrification are driven largely by the energy derived from the oxidation of carbonaceous materials. The C:N ratio of microorganisms can range widely but in most cases it ranges between 5 and 15. Mineralization-immobilization rates of nitrogen can be calculated once an understanding of microbial growth relative to carbon cycling is available (Paul and Juma, 1980).

Phosphorus transformations are not as closely related to carbon flows as are those involving nitrogen. However, phosphorus cycling studies indicate a high interrelationship between inorganic phosphorus availability and microbial activity (Hannapel Fuller and Fox, 1964). Simulation models of phosphorus cycling indicated that annual flow through the microbial biomass was greater than that through above and beneath ground plant phosphorus. The major controls in these models (Fig. 3) were inorganic phosphorus solubility, soil water content and rates of diffusion of phosphorus in soil (Stewart, Halm and Cole, 1973; Cole, Innis and Stewart, 1977). Microbial transformations of labile and stable organic phosphorus often produce the majority of the plant available solution phosphorus.

The principle pathways in the cycling of phosphorus within the detritus and sediment of aquatic systems (Parsons *et al.*, 1977, quoted by Fletcher, 1979) contained pools similar to those in soils. Zooplankton have long been considered to be a major factor in the mineralization of organic phosphorus in aquatic systems. Growth studies with mixtures of pseudomonads, amoeba and nematodes (Cole, Elliott, Hunt and Coleman, 1978; Coleman, Anderson, Cole, Elliott, Woods and Campion, 1978) have confirmed earlier suggestions (Cutler and Crump, 1935) that the soil fauna can play a major role in soil biomass dynamics and subsequently in the mineralization of nutrients such as organic phosphorus.

MICROBIAL GROWTH, BIOMASS AND PRODUCT FORMATION IN THE SOIL

The close interactions of microorganisms with soil solids and the low apparent growth rates of a very large soil population have meant that interpretation of the effects of microbial growth in terrestrial systems has lagged behind studies using chemostats and those in aquatic systems. Descriptions of microbial growth in sewage sludge, rumen and in single cell protein formation (Stewart, 1975; Harrison, 1978; Nagai, 1979) have added information that with proper interpretation should help in understanding the transformations in the solid-colloidal systems of soils and sediments.

The energy of organic substrates may be incorporated into biomass, evolved as heat, or incorporated into extracellular products as shown in the following equation (Erickson, 1979):

$$CH_mO_l + aNG_3 + bO_2 = y_cCH_pO_nN_g + zCH_rO_sN_t + cH_2O + dCO_2$$

CH_mO_l denotes the elemental composition of the organic substrate, $CH_pO_nN_g$ is the elemental composition of the biomass, and

$CH_rO_sN_t$ is the elemental composition of any extracellular products.

y_c is the fraction of utilized organic substrate carbon converted to biomass,

z is the fraction converted to extracellular products, and

d is the fraction converted to carbon dioxide.

PHOSPHORUS CYCLE

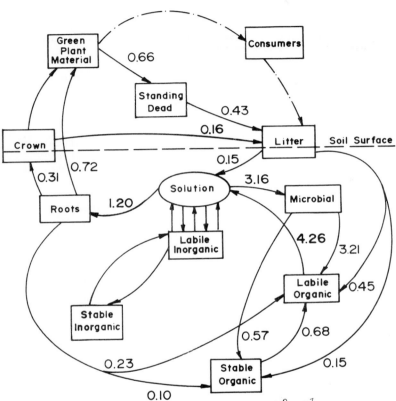

Fig. 3. *Predicted flow of phosphorus ($g\ m^{-2}\ y^{-1}$) between components of a native grassland ecosystem (drawn form Cole et al., 1977).*

The overall energy of consumption can also be divided into that utilized for growth and that consumed for maintenance of population (Pirt, 1965).

Substrate consumption = Rate of consumption + Rate of consumption
for growth   for maintenance

From this : $\dfrac{\mu x}{Y_e} = \dfrac{\mu x}{Y_{eg}} + mx$

where μ = specific growth rate,
         x = organism concentration,
         Y_e = yield of organisms,
         Y_{eg} = theoretical yield, and
         mx = maintenance energy.

Harrison (1978) in discussing the efficiency of microbial growth recognized that resting cells require little energy to maintain cell integrity. He suggested that maintenance energy may represent one or a number of reactions:
    i)   protection of cell integrity during active growth,
    ii) a constant leakage or wastage of energies from active cells due to inherent inefficiencies,
    iii) energy to maintain certain concentration gradients across the cell membrane, and
    iv) a constant death rate of the organisms.
The above equations describing microbial growth, maintenance energy and production formation were derived largely from chemostat studies. They have however been used by a number of authors to ask questions concerning the state and activity of soil biomass.
    The maintenance energy value of 0.001 h^{-1} suggested by Babiuk and Paul (1970) was not a miscalculation of the value of 0.04 h^{-1} obtained by Pirt (1965) as suggested by Lynch (1979). The use of the low value by Babiuk and Paul (1970) and Gray and Williams (1971) was based on the recognition that maintenance energies determined in liquid cultures during active growth could not apply to a total soil population.
    Parkinson, Domsch and Anderson (1978) also using the value of 0.001 h^{-1} concluded that with maintenance values of this magnitude approximately half of the carbon entering the soil would probably be utilized for maintenance, the rest would be available for growth of their measured population. Shields, Lowe, Paul and Parkinson (1973) calculated a maintenance value of 0.002 h^{-1} realizing that this value must also include cryptic growth and Behara and Wagner (1974) suggested a maintenance value of 0.003 h^{-1}. These estimates for the total soil population took into account the large proportion of live but relatively inactive cells which in the past could not be separated from the growing cells.
    Early authors considered the high values for biomass found by direct microscopy to include a large proportion of dead cells. Recent work with ATP and the chloroform incubation techniques indicates that even higher values for the soil biomass. Therefore, it does not take a great deal of calculation to show that the population of natural systems although alive

and resistant to decomposition is largely dormant with only small proportions of the population actively growing or even having maintenance energy rates comparable to those found in the laboratory. Where doubling times of soil organisms have been calculated, these have usually been estimated to range from 40 to 50 hours (Coleman *et al.*, 1978; Domsch, Beck, Anderson, Söderström, Parkinson and Trolldenier, 1979).

Van Veen and Paul (1980) in their modelling of microbial growth and soil organic matter formation stated that the concept of maintenance energy has been most useful to ecosystem microbiologists in asking pertinent questions concerning the size and activity of organisms in nature. However, not enough is yet known about these factors and cryptic growth in natural systems, therefore, in their mathematical analysis they incorporated the maintenance requirement of the population into the decay constant of soil organisms. This eliminated the need for separate maintenance energy constants for the various segments of the populations in various stages of activity.

THE SOIL BIOMASS

The biovolume in natural systems has classically been measured by direct microscopy. Visible light microscopy using cells stained with phenolic aniline blue (Jenkinson, Powlson and Wedderburn, 1976) or phase objectives (Parkinson, Gray and Williams, 1971) works well with the larger sized fungi. Fluorescence, particularly epifluorescence, makes it easier to see the smaller bacteria, and eliminates much of the interfacing background found in soils. Acridine orange has been utilized almost exclusively in aquatic studies. In soils, acridine orange, fluorescein isothiocyanate, magnesium sulfonic acid and europium chelate combined with a fluorescent darkeners have been utilized (Jenkinson and Ladd, 1980).

Attempts to measure that proportion of the population which is active hav included examination with the phase microscope. The use of nalidixic acid to prevent DNA formation, followed by staining with acridine orange (Kogure, Simidu and Tage, 1979) and the epifluorescent detection of INT formazan in respiring cells (Zimmermann, Hurriaga and Becker-Birck, 1978) have been some of the attempts to measure active segments of soil and sediment populations. The possibilities that fluorescein diacetate (FDA) which does not fluoresce can be taken up by bacterial cells where it is hydrolyzed to the fluorescent fluorescein was tested by Babiuk and Paul (1970). They found that the uptake of the ester was too small in other than very young cells to be an effective staining technique. More recently, Söderström (1979) suggested this as a useable technique for measurement of the viable portions of fungi. However, in a

comparison of methods for soil microbial populations (Domsch
et al., 1979) found FDA to operate with a very low recovery
of fungi.

The measurement of specific cell constituents such as
lipo-polysaccharides (Watson, Novitsky, Quinby and Valois,
1977), mumaric acid (Miller and Casida, 1970; King and White,
1977; Moriarty, 1977) and fungal glucosamines is limited by
the fact that these constituents occur in varying concentrations
in different portions of the population; they can also occur
as microbial products stabilized in soil organic matter.
However, these biochemical techniques could have specific uses
in microcosm studies and when using tracers so that radioactive
constituents produced during microbial growth can be isolated
from those generally occurring in soil organic matter.

The biovolume and not just the number of organisms must be
obtained in direct microscopy (Jenkinson and Ladd, 1980).
The conversion of biovolume to biomass has usually been
carried out using values that assume 80% moisture by weight
and 1.1 wet weight specific gravity. This resulted in a dry
weight specific gravity conversion factor of 0.22 or a dry
weight carbon content of approximately 0.11. Growth of soil
isolates at moisture stresses often found in a soil led van
Veen and Paul (1979) to suggest that the biovolume conversion
rates for fungi should be 1.44 times the usual quoted figures
of 0.22. They found very high densities for the bacterial
isolates studied, but their fungal data are in agreement with
values utilized by Parkinson *et al.*, (1978) and with density
gradient measurements of actual soil organisms (Faegri, Torsvik
and Goksθyr, 1977).

A number of authors have assumed both a high wet weight
specific gravity and a fairly high moisture content. According
to the calculations of van Veen and Paul, this is an anomoly
as a microorganism of 1.3 to 1.5 specific gravity must contain
less than the commonly assumed 80% moisture by weight. At
this moisture level, the volume of water within the cell would
be greater than the total volume of the cell originally
utilized for the specific gravity calculation.

The cell constituent, ATP, has recently been measured with
high recovery of internal standards and has shown high correla-
tions with other techniques (Paul and Johnson, 1978; Bullied,
1978; Jenkinson, Davidson and Powlson, 1979). Studies with
laboratory cultures (Lee, Harris, Williams, Armstrong and Syers,
(1971 a,b) and aquatic organisms (Holm-Hansen and Booth, 1966)
have indicated a general carbon to ATP ratio of 250:1 with the
content of ATP being much higher in actively growing cells than
in dormant cells. Jenkinson and Oades (1979) and Jenkinson
et al., (1979) compared the ATP content of the soil biomass
with the carbon measured by the $CHCl_3$-lysis incubation technique

and found much lower C-carbon:ATP ratios.

Jenkinson and his co-workers (Jenkinson, 1966, 1976; Jenkinson and Powlson, 1976) noted that ^{14}C-labelled microorganisms added to soil and lysed by agents such as $CHCl_3$ resulted in the evolution of fairly constant amounts of ^{14}C-carbon during subsequent incubation. Anderson and Domsch (1978a) found that labelled fungi lysed by $CHCl_3$ vapors were decomposed by a subsequent soil population and that 43% of the labelled CO_2 was evolved in 10 days; lysis of bacteria and subsequent incubation resulted in the evolution of 33% of the ^{14}C-carbon. Using a ratio of 1 to 3 for the distribution of bacterial and fungal carbon in the total soil biomass (Parkinson, 1973; Shields *et al.*, 1973), these workers calculated an average mineralization of the soil microbial population to be 41.1%. Thus the factor used to convert CO_2-C evolved to biomass was:

$$\text{Biomass C} = \frac{CO_2\text{-C evolved}}{K_c}$$

where K_c = 0.41.

The extent of decomposition of low levels of radioactive substrate has been effectively utilized in aquatic systems to indicate microbial activity (Wright and Hobbie, 1966). This has been found to be difficult to interpret for soils and sediments. However, Anderson and Domsch (1978b) utilizing saturating levels of glucose have developed a rapid respiration technique for total microorganisms and have suggested delineation of the fungi and bacteria by utilizing antibiotics.

Comparisons of biomass data from $CHCl_3$ fumigation, direct counts and ATP measurements have been made for a number of soils. The values for soil biomass summarized by Jenkinson and Ladd (1980) ranged from 200 to 3500 µg carbon g^{-1} soil (Table 2). There was a good correlation between all three techniques. The major exception being that the acid deciduous wood of England gave very low results for the $CHCl_3$ fumigation technique. The laboratory incubation studies differed from the field studies in that the microbial populations were measured at a time of flux after the addition of substrate to soils with large variations in available phosphorus. In the latter study, the direct counts of fungi and bacteria have been adjusted to take into account the higher specific gravities of soil organisms (van Veen and Paul, 1979) and to include the organisms other than fungi and bacteria said by Jenkinson *et al.*, (1976) to constitute 36% of the soil biomass. This technique, however, still tended to give lower results than the fumigation measurements. The ATP content was sensitive to phosphorus availability and thus showed greater variations.

TABLE 2

Estimates of soil microbial biomass by different methods
(µg biomass carbon g^{-1} soil)

Soil	Chloroform fumigation	Direct counts	ATP content
Field soil data (referenced in Jenkinson & Ladd, 1980)			
Continuous wheat + manure, England	560	500	430
Continuous wheat, no manure, England	220	170	170
Calcarears deciduous wood, England	1230	1400	1040
Old grassland, England	3710	2910	–
Acid deciduous, England	50	300	470
Secondary rain forest, Nigeria	540	390	–
Arable cropping, Nigeria	280	240	–
Laboratory incubation (referenced in Paul & van Veen, 1978)			
Parent material, low P	330	200	400
Parent material, high P	180	170	512
Parent material with clay, low P	250	190	175
Parent material with clay, high P	190	190	237

Oades and Jenkinson (1979) and Jenkinson *et al.*, (1979)
in studies of a heterogenous group of soils, found a linear
relationship (r = 0.98) between ATP content and biomass carbon
content as measured by CHCl$_3$-lysis-incubation. Domsch *et al.*,
(1979) compared 15 different techniques used in assessing soil
microbial biomass and activity. They also found that detailed
quantitative data can be obtained by direct microscopy, the
selective inhibition-incubation technique and ATP measurements.
Data from enzyme assays were, however, difficult to transform
to microbial biomass.

THE USE OF MICROBIAL BIOMASS STUDIES TO INTERPRET CARBON AND
NITROGEN TRANSFORMATIONS IN SOIL SYSTEMS

The recently developed techniques for measurement of the
soil biomass have made it possible to relate microbial growth
and energy flow in a number of systems. Earlier calculations
(Clark and Paul, 1970; Gray and Williams, 1971; Shields *et al.*,
1973) used the available data to obtain some indication of the

biomass size and turnover rate relative to energy supplies. More recent measurements have led to calculations of energy release and turnover rates within the rhizosphere (Barber and Lynch, 1977) in straw decomposition and during soil cultivation Harper and Lynch, 1979; Lynch and Panting, 1980), and in a number of forest soils (Parkinson *et al.*, 1978). When Jenkinson (1966) used the fumigation technique to trace the decomposition of labelled ryegrass residues, he found that 31 to 39% of the labelled carbon remaining in the soil after one year's incubation was present as microbial biomass; 19% was present as cells after 4 years incubation. Shields, Paul and Lowe (1974) added ^{14}C-glucose in the field. Their fumigation studies showed that 75% of the ^{14}C-carbon remaining in the soil after 90 days field incubation was present as soil biomass. The rest of the labelled carbon was assumed to be present as extracellular metabolites or lytic products stabilized in the soil system.

Recent incubation studies in our laboratory utilized the above techniques, tracers and mathematical analyses to determine the movement of added substrate carbon and mineral nitrogen to biomass and product carbon and nitrogen. Measurement of the biomass at various stages during the incubation of soil with ^{14}C-glucose requires reincubation of the samples for 10 days following fumigation. The period of incubation after fumigation and the decision as to what CO_2 evolution values to use as a control have been major problems in the interpretation of this technique. Jenkinson (1966) assumed that the CO_2 evolved from the non-fumigated sample should be subtracted from the CO_2 evolved from the chloroformed sample during a 10 day incubation after fumigation. Jenkinson and Powlson (1976) and Anderson and Domsch (1978b) recognizing that the control soil can evolve large amounts of CO_2 during the initial stages of incubation suggested that the substraction of CO_2 evolved during the 10 to 20 day incubation period.

The data in Fig. 4 show two methods of calculating both ^{14}C- and ^{12}C-carbon biomass during incubation with 1600 µg ^{14}C-glucose g^{-1} soil. Subtraction of the CO_2 evolved from the non-chloroformed soil proved difficult because the unfumigated soils had very high CO_2 levels. If the CO_2 content from the control was subtracted from the fumigated sample, the total soil biomass was found to rise from 425 µg carbon g^{-1} to 950 µg biomass carbon, with the non-labelled and labelled biomass present in approximately equal amounts (Fig. 4A).

Analysis of the glucose remaining in the soil (Fig. 5) showed that at the end of day 1, there was no glucose remaining in the soil although there is still a total of 950 µg ^{14}C-carbon g^{-1}. Figure 4A shows 500 µg of biomass ^{14}C-carbon at this time. One must conclude that the ^{14}C-carbon present was

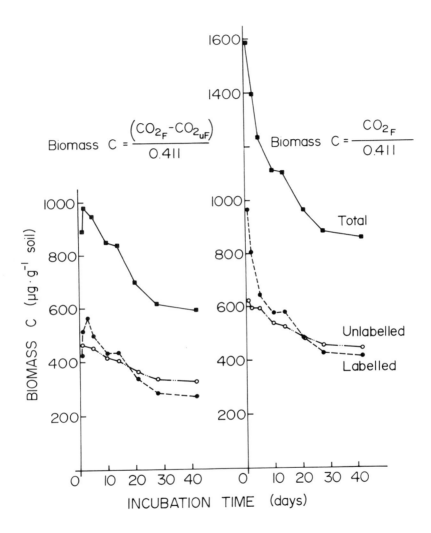

Fig. 4. *Soil biomass determined with the CHCl₃ lysis incubation technique with (A) and without (B) subtraction of unfumigated soil CO₂ levels.*

equally divided between the ^{14}C-biomass and ^{14}C-microbial products. It is difficult to envision such a large production of secondary metabolites or the occurrence of so many dead ^{14}C-cells immediately after the cessation of growth.

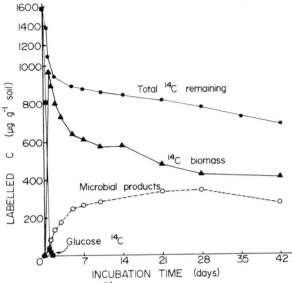

Fig. 5. Disposition of ^{14}C-carbon in soil.

Domsch and Anderson and Jenkinson and his co-workers in the development of the K factor for the CHCl$_3$-incubation technique added ^{14}C-labelled microorganisms. These were subsequently fumigated and the CO$_2$ evolution measured. Their calculations therefore did not involve a subtraction of control ^{14}C-CO$_2$ levels.

In our study, all the original ^{14}C-glucose had disappeared by day 1; the remaining ^{14}C-carbon (950 µg carbon g^{-1}) should be in microbial cells. Fumigation of these samples and incubation for 10 days should result in a K$_c$ value for *in situ* microorganisms. A total of 397 µg ^{14}C-CO$_2$ were evolved during the 10 day incubation after fumigation. The K$_c$ value when no control was subtracted would equal 397/950 = 0.41. This value is identical to the weighted average for bacteria and fungi in the study of Anderson and Domsch (1978a). The data in Fig. 4B was therefore assumed to represent a true measure of the ^{12}C-carbon and ^{14}C-biomass carbon remaining in the soil during continued incubation. It is of interest to note that the conversion of 1600 µg carbon g^{-1} soil to 950 µg biomass carbon results in a growth efficiency of 59%. Extracellular

polysaccharides such as capsular material if present were
considered to be a portion of the biomass.

The decay of the biomass carbon shown in a linear fashion
in Fig. 4 can best be described using first order reaction
kinetics. Using this procedure, day 1 of the incubation was
considered as day 0 for the decay of the biomass. The decay
of biomass in the control soil during incubation could be
represented by a single first order decay equation (Fig. 6).

$$\text{Biomass carbon} = 662\ e^{-4.5\ \times\ 10^{-3}\ t}$$

The decay of unlabelled biomass carbon, in the presence of
glucose yielded a curve when plotted on semilog paper;
exponential regression analysis and curve splitting techniques
were applied to differentiate the components. The slope of
the line describing the decay of each component represents
the first order net decay rate; the intercept represents the
amount present. The unlabelled biomass in the glucose
amended soil was separated into an active biomass accounting
for 12% of the total soil biomass with a decay rate of 1.27×10^{-1} day^{-1}. The decay rate constant for the resting biomass
was 5.5×10^{-3} day^{-1} which approximated the decay rate constant
of the biomass in the control soil.

The labelled biomass carbon could be separated into three
components (Fig. 6B). Component I at a level of 328 µg
carbon g^{-1} soil accounted for 40% of the glucose immobilized.

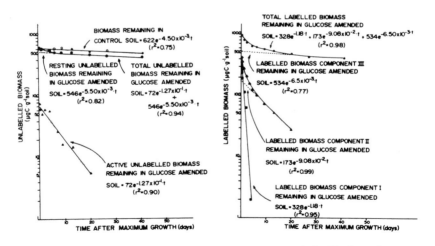

*Fig. 6. Decay rates and pool sizes of the labelled and
unlabelled biomass present in soil after growth on* [14] *C-
glucose.*

This decayed with a rate constant of 1.18 day^{-1}. Component I with a decay rate of 9.08×10^{-2} day^{-1} corresponded to the active unlabelled biomass. The decay rate constant of component III was 6.5×10^{-3} day^{-1} which is equivalent to that of the resting biomass.

The measurement of ^{14}C-biomass makes it possible to determine the production of microbial metabolites (Fig. 5). The decay of the ^{14}C-biomass during days 1 to 5 resulted in the rapid production of microbial products. They were probably largely the result of microbial death. After this period, both the total ^{14}C-carbon remaining and ^{14}C-biomass stabilized although the amount of microbial products remaining in the soil at any one time (net production) was found to decrease towards the end of the experiment.

CONCLUSIONS

The examples given for carbon and phosphorus flow through ecosystems indicate the significance of the microbial biomass as a source-sink as well as its importance in the various transformations involved. The concepts of microbial growth and activity developed from studies in pure microbiology must be applicable, perhaps with caution, to ecosystem studies. We have shown that techniques for the determination of biomass and activity exist. These together with tracer isotopes are making possible the characterization of nutrient flow in soils and sediments. However, an understanding of the behaviour of the microbial biomass within ecosystems is now limited as much by a lack of information concerning the inter-action of microorganisms with their solid environment as it is by the difficulty involved in their measurement and applying microbial growth concepts to complex systems.

We have found the most useful method for determining biomass to be the $CHCl_3$-fumigation incubation technique. This procedure is based on the somewhat difficult to believe observation that after lysis, bacteria and fungi are decomposed with the evolution of a relatively constant amount of CO_2. The measurement of the biomass-C produced during *in situ* growth on added ^{14}C-glucose indicated that previous suggestions that a control CO_2 should be subtracted from the total CO_2 evolved may be in error. Our data were most easily interpreted when control CO_2 levels were not subtracted. Also it is difficult to see how a non-fumigated sample can be a control for the fumigated sample where more than 99% of the organisms have been killed. Our K_c factor measured using a labelled *in situ* population agreed with the published value of 0.41. When this value was applied during extended incubation periods and the ^{14}C-biomass subtracted from ^{14}C-carbon remaining, it was found that the

production of microbial products nearly equalled the biomass
during extended incubation periods.

The separation of the unlabelled biomass into an active
and dormant fraction should make possible some estimates
concerning the percentage of the biomass carbon that was
activated by a single pulse of glucose. The unlabelled
biomass carbon, in the presence of glucose, was shown to be
comprised of two fractions. One decayed at the same rate as
the biomass in the unamended soil; the other decayed at the
same rate as the ^{14}C-biomass produced during growth on
glucose. Therefore, it should be possible to conclude that
the dormant biomass represented that proportion of the biomass
not activated by the glucose.

The above interpretations while indicating the potential
for biomass measurements must be used with caution. All the
data represented net rates. Simultaneous growth and decay
were not separated. Thus only the net outcome of the two
processes was measured. An estimate of the simultaneous
rates of growth and decay can be obtained by mathematical
simulation of the data presented in this paper in conjunction
with CO_2 evolution and specific activity measurements. If
the simulation accurately mimics the actual net rates for
^{14}C-carbon and ^{12}C-microbial biomass and ^{14}C-carbon and ^{12}C-
CO_2 evolved, the output from the computer can be used to
calculate the gross rates. The gross immbolization-minerali-
zation rates of ^{15}N in a similar experiment were calculated
by Paul and Juma (1980) to be 10 times as great as the net
rates.

The techniques are now becoming available for the development of an understanding of carbon, nitrogen and even sulphur
and phosphorus nutrient flow through the microbial biomass in
nature. As the understanding of this system evolves, there
must also be a feedback to microbial ecological studies in
the laboratory. More information is required for interpretation
of growth on and within soil and sediment solids. The present
information on resting cells and spores also cannot explain
how such a large population can persist for such extended
periods without the utilization of massive carbon supplies.

REFERENCES

Anderson, D.W. (1979). Processes of humus formation and
    transformation in soils of the Canadian Great Plains,
    Journal of Soil Science 30, 77-84.
Anderson, J.P.E. and Domsch, K.H. (1978a). Mineralization
    of bacteria and fungi in chloroform-fumigated soils,
    Soil Biology and Biochemistry 10, 207-213.
Anderson, J.P.E. and Domsch, K.H. (1978b). A physiological
    method for the quantitative measurement of microbial bio-

mass in soils, *Soil Biology and Biochemistry* 10, 215-221

Babiuk, L.A. and Paul, E.A. (1970). The use of fluorescein
isothiocyanate in the determination of the bacterial biomass of a grassland soil, *Canadian Journal of Microbiology*
16, 57-62.

Barber, D.A. and Lynch, J.M. (1977). Microbial growth in the
rhizosphere, *Soil Biology and Biochemistry* 9, 305-308.

Behara, B. and Wagner, G.H. (1974). Microbial growth rate in
glucose amended soil, *Soil Science Society of America
Proceedings* 38, 591-597.

Bullied, N.C. (1978). An improved method for the extraction
of adenosine triphosphate from marine segment and sea
water, *Limnology and Oceanography* 23, 174-178.

Clark, F.E. and Paul, E.A. (1970). The microflora of grassland, *Advances in Agronomy* 22, 375-435.

Cole, C.V., Elliott, E.T., Hunt, H.W. and Coleman, D.C. (1978).
Trophic interactions in soils as they affect energy and
nutrient dynamics, V. Phosphorus transformations, *Microbial
Ecology* 4, 381-387.

Cole, C.V., Innis, G.S. and Stewart, J.W.B. (1977). Simulation
of phosphorus cycling in a semiarid grassland, *Ecology*
58, 1-15.

Coleman, D.C., Anderson, R.V., Cole, C.V., Elliott, E.T.,
Woods, L. and Campion, M.K. (1978). Trophic interactions
as they affect energy and nutrient dynamics. IV. Flows of
metabolic and biomass carbon, *Microbial Ecology* 4, 373-380.

Cutler, D.W. and Crump, L.M. (1935). "Problems in soil
microbiology". The Rothamsted Monographs on Agricultural
Science, Longmans, Green and Co. Ltd., London.

Domsch, K.H., Beck, Th., Anderson, J.P.E., Söderström, B.,
Parkinson, D. and Trolldenier, G. (1979). A comparison
of methods for soil microbial population and biomass
studies, *Zeitschrift für Pflanzenernährung und Bodenkunde*
142, 520-533.

Erickson, L.E. (1979). Energetic efficiency of biomass and
product formation, *Biotechnology and Bioengineering* 21,
725-743.

Faegri, A., Torsvik, V. Lid and Goksøyr, J. (1977). Bacterial
and fungal activities in soil: separation of bacteria and
fungi by a rapid fractionated centrifugation technique,
Soil Biology and Biochemistry 9, 105-112.

Fletcher, M. (1979). Microorganisms in their natural environments - "The aquatic environment". *In* "Microbial Ecology-
A Conceptual Approach" (Ed. J.M. Lynch and N.J. Poole),
pp. 92-114. John Wiley & Sons, New York.

Gray, T.R.G. and Williams, S.T. (1971). Microbial productivity
in soil, *Symposium of Society for General Microbiology* 21,
255-286.

Haider, K., Martin, J.P. and Filip, Z. (1975). Humus bio-
chemistry. *In* "Soil Biochemistry" (Ed. E.A. Paul and
A.D. McLaren), Vol. 4, pp. 195-244. Marcel Dekker, New
York.

Hannapel, R.J., Fuller, W.H. and Fox, R.H. (1964). Phosphorus
movement in a calcareous soil. II. Soil microbial activity
and organic phosphorus movement, *Soil Science* 97, 421-427.

Harper, S.H.T. and Lynch, J.M. (1979). The kinetics of the
decomposition of straw in relation to the production of
phytotoxins. (In press).

Harrison, D.E.F. (1978). Efficiency of microbial growth. *In*
"Companion to Microbiology" (Ed. A.T. Bull and P.M. Meadow)
pp. 155-177. Longman, London.,

Holm-Hansen, O. and Booth, C.R. (1966). The measurement of
adenosine triphosphate in the ocean and its ecological
significance, *Limnology and Oceanography* 11, 510-519.

Jenkinson, D.S. (1966). Studies on the decomposition of
plant material in soil. II., *Journal of Soil Science* 17,
280-302.

Jenkinson, D.S. (1976). The effects of biocidal treatments
on metabolism in soil. IV., *Soil Biology and Biochemistry*
8, 203-208.

Jenkinson, D.S. and Ayanaba, A. (1977). Decomposition of
carbon-14 labelled plant material under tropical conditions,
Soil Science Society of American Journal 41, 912-915.

Jenkinson, D.S., Davidson, S.A. and Powlson, D.S. (1979).
Adenosine triphosphate and microbial biomass in soil,
Soil Biology and Biochemistry 8, 167-177.

Jenkinson, D.S. and Ladd, J.N. (1980). Microbial biomass in
soil - measurement and turnover. *In* "Soil Biochemistry"
(Ed. E.A. Paul and J.N. Ladd), Vol. 5, Marcel Dekker,
New York. (In press).

Jenkinson, D.S. and Oades, J.M. (1979). A method for measuring
adenosine triphosphate in soil, *Soil Biology and Biochemistry*
11, 193-199.

Jenkinson, D.S. and Powlson, D.S. (1976). The effects of
biocidal treatments on metabolism in soil. V., *Soil Biology
and Biochemistry* 8, 209-213.

Jenkinson, D.S., Powlson, D.S. and Wedderburn, R.W.M. (1976).
The effects of biocidal treatments on metabolism in soil.
III. The relationship between soil biovolume, measured by
optical microscopy, and the flush of decomposition caused
by fumigation, *Soil Biology and Biochemistry* 8, 298-305.

Jenkinson, D.S. and Rayner, J.H. (1977). The turnover of soil
organic matter in some of the Rothamsted classical experiments, *Soil Science* 123, 298-305.

King, J.O. and White, D.C. (1977). Muramic acid as a measure
of microbial biomass in estuarine and marine samples,

Applied and Environmental Microbiology 33, 777-783.

Kogure, K., Simidu, U. and Taga, N. (1979). A tentative direct microscopic method for counting living marine bacteria, *Canadian Journal of Microbiology* 25, 415-420.

Ladd, J.N. and Paul, E.A. (1973). Changes in enzymic activity and distribution of acid-soluble amino-nitrogen in soil during nitrogen immobilization and mineralization, *Soil Biology and Biochemistry* 5, 825-840.

Lee, C.C., Harris, R.F., Williams, J.D.H., Armstrong, D.E. and Syers, J.K. (1971a). Adenosine triphosphate in lake sediments. I., *Soil Science Society of America Proceedings* 35, 82-86.

Lee, C.C., Harris, R.F., Williams, J.D.H., Armstrong, D.E. and Syers, J.K. (1971b). Adenosine triphosphate in lake sediments. II., *Soil Science Society of America Proceedings* 35, 86-91.

Legg, J.O., Chichester, F.W., Stanford, G. and De Mar, W.H. (1971). Incorporation of ^{15}N-tagged mineral nitrogen into stable forms of soil organic nitrogen, *Soil Science Society of America Proceedings* 35, 273-276.

Lynch, J.M. (1979). Microorganisms in their natural environments; the terrestrial environment. *In* "Microbial Ecology - A Conceptual Approach" (Ed. J.M. Lynch and N.J. Poole), pp. 67-91. John Wiley & Sons, New York.

Lynch, J.M. and Panting, L.M. (1980). Cultivation and the soil biomass, *Soil Biology and Biochemistry*. (In press).

Marumoto, T., Kai, H., Yoshida, T.and Harada, T. (1977). Chemical fractions of organic nitrogen in acid hydrolysates given from microbial cells and their cell wall substances and characterization of decomposable soil organic nitrogen due to drying, *Communications in Soil Science and Plant Nutrition* 23, 125-134.

McGill, W.B., Paul, E.A., Shields, J.A. and Lowe, W.E. (1973). Turnover of microbial populations and their metabolites in soil, *Bulletin of Ecological Research and Communication (Stockholm)* 17, 293-301.

Miller, W.N. and Casida, L.E. (1970). Evidence for muramic acid in soil, *Canadian Journal of Microbiology* 16, 299-304.

Moriarty, D.J.W. (1977). Improved method using muramic acid to estimate biomass of bacteria in sediments, *Oecologia (Berlin)* 26, 317-323.

Nagai, S. (1979). Mass and energy balances for microbial growth kinetics. *In* "Advances in Biochemical Engineering II" (Ed. T.K. Ghose, A. Fiechter and N.B. Blakebrough), pp. 49-83. Springer-Verlag, Berlin.

Oades, J.M. and Jenkinson, D.S. (1979). The adenosine triphosphate content of the soil microbial biomass, *Soil Biology and Biochemistry* 11, 201-204.

Parkinson, D. (1973). Techniques for the study of soil fungi, *Bulletin of Ecological Research and Communication (Stockholm)* 17, 29-36.

Parkinson, D., Domsch, K.H. and Anderson, J.P.E. (1978). Die Entwicklung mikrobieller Biomassen im organischen Horizont-eines Fichtenstandortes, *Ecologia Plantarum* 13, 355-366.

Parkinson, D., Gray, T.R.G. and Williams, S.T. (1971). "Methods for Studying the Ecology of Soil Microorganisms". IBP Handbook No.19, Blackwell Scientific Publications, Oxford.

Paul, E.A. and Johnson, R.L. (1977). Microscopic counting and adenosine 5'triphosphate measurement in determining microbial growth in soils, *Applied and Environmental Microbiology* 34, 263-269.

Paul, E.A. and Juma, N.G. (1980). Mineralization and immobilization of soil nitrogen by microorganisms. *In* "Terrestrial Nitrogen Cycles - Processes, Ecosystem Strategies and Mangement Impacts" Workshop, Sweden.

Paul, E.A. and McLaren, A.D. (1975). Biochemistry of the soil subsystem. *In* "Soil Biochemistry" (Ed. E.A. Paul and A.D. McLaren), Vol. 3, pp. 1-36. Marcel Dekker, New York.

Paul, E.A. and van Veen, J.A. (1978). The use of tracers to determine the dynamic nature of organic matter. 11th International Society of Soil Science, Transactions (Edmonton), June, 1978, pp. 61-102.

Pirt, S.J. (1965). The maintenance energy of bacteria in growing cultures, *Proceedings of the Royal Society Series B* 163, 224-231.

Schmidt, E.L., Putnam, H.D. and Paul, E.A. (1960). Behaviour of free amino acids in soils, *Soil Science Society of America Proceedings* 24, 107-109.

Shields, J.A., Lowe, W.E., Paul, E.A. and Parkinson, D. (1973). Turnover of microbial tissue in soil under field conditions, *Soil Biology and Biochemistry* 5, 753-764

Shields, J.A. and Paul, E.A. (1973). Decomposition of ^{14}C-labelled plant material under field conditions, *Canadian Journal of Soil Science* 53, 297-306.

Shields, J.A., Paul, E.A. and Lowe, W.E. (1974). Factors influencing the stability of labelled microbial materials in soils, *Soil Biology and Biochemistry* 6, 31-37.

Söderström, B.E. (1979). Some problems in assessing the fluorescein diacetate-active fungal biomass in the soil, *Soil Biology and Biochemistry* 11, 147-148.

Sørensen, L.H. and Paul, E.A. (1971). Transformation of acetate carbon into carbohydrate and amino acid metabolites during decomposition in soil, *Soil Biology and Biochemistry* 3, 173-180.

Stewart, C.S. (1975). Some effects of phosphate and volatile
    fatty-acid salts on growth of rumen bacteria, *Journal of
    General Microbiology* 89, 319-326.
Stewart, J.W.B., Halm, B.J. and Cole, C.V. (1973). Nutrient
    cycling. I. Phosphorus. Matador Project Technical Report
    No.40. Canadian Committee on International Biological
    Programme, University of Saskatchewan, Saskatoon.
Veen, J.A. van and Paul, E.A. (1979). Conversion of biovolume
    measurements of soil organisms, grown under various moisture
    tensions, to biomass and their nutrient content, *Applied
    and Environmental Microbiology* 37, 686-692.
Veen, J.A. van and Paul, E.A. (1980). Organic carbon dynamics
    in grassland soils. I. Background information, *Canadian
    Journal of Soil Science* (Submitted).
Wagner, G.H. (1975). Microbial growth and carbon turnover.
    In "Soil Biochemistry" (Ed. E.A. Paul and A.D. McLaren),
    Vol. 3, pp. 269-305. Marcel Dekker, New York.
Watson, S.W., Novitsky, T.J., Quinby, H.L. and Valois, F.W.
    (1977). Determination of bacterial number and biomass in
    the marine environment, *Applied and Environmental Micro-
    biology* 33, 940-946.
Wright, R.T. and Hobbie, J.E. (1966). The use of glucose
    and acetate by bacteria and algae in aquatic ecosystems,
    Ecology 47, 447-464.
Zimmermann, R., Hurriaga, R. and Becker-Birck, J. (1978).
    Simultaneous determination of the total number of aquatic
    bacteria and the number thereof involved in respiration,
    Applied and Environmental Microbiology 36, 926-935.

MINERALIZATION AND THE BACTERIAL CYCLING OF CARBON, NITROGEN AND SULPHUR IN MARINE SEDIMENTS

B. B. Jørgensen

Institute of Ecology and Genetics
University of Aarhus, Ny Munkegade
DK-8000 Aarhus C, Denmark

INTRODUCTION

The cycling of elements such as carbon, nitrogen, phosphorus and sulphur in aquatic ecosystems is biologically regulated by two main processes: assimilation of inorganic nutrients by the photosynthetic organisms and the subsequent mineralization by the heterotrophs. The mineralization processes in the oxic waters have been described by different authors in terms of the overall stoichiometry (for example, Richards, 1965):

$$(CH_2O)_{106}(NH)_{16}H_3PO_4 + 138\ O_2 \rightarrow$$
$$106\ CO_2 + 122\ H_2O + 16\ HNO_3 + H_3PO_4$$

This stoichiometry reflects the average, elemental composition of living organic matter in the plankton community. The equation applies to the aquatic ecosystem as a whole, but a similar equation can be used to describe the mineralization taking place within the individual heterotrophic organism. In the aerobic environments, the organic matter produced is exploited through complex food webs. Each type of organism consumes and assimilates only a small selected fraction of the organic pool. But during their respiration, the aerobic organisms perform a mineralization which in principle follows the equation above. So, although the food webs are very complex in terms of ecological and morphological species diversity, they are also simple in the sense that all the organisms basically have the same respiratory metabolism. It is only for this reason that Richards' equation can be used to describe the mineralization which is taking place at many levels of organisation in natural oxic waters.

Such simple stoichiometric models do not apply to aquatic

sediments where both aerobic and anaerobic processes take
place. Among the additional complicating factors here is the
important role played by sulphur compounds, not as a constituent
of the living biomass, but as respiratory electron acceptors
and carriers of detrital energy. The organic carbon follows
more elaborate, chemical pathways during its heterotrophic
breakdown in the sediment. It may be oxidized to CO_2 in the
anaerobic sediment and then again reduced to CH_4 by methanogenic
bacteria. The methane may later be reoxidized to CO_2 at the
sediment surface and the CO_2 may be assimilated into living
organic matter by chemolithotrophic bacteria.

The high complexity of the mineralization processes in
sediments is largely due to the restricted metabolic potential
of the individual anaerobic microorganisms. A complete
mineralization does not take place within one organism, but
only within associations of physiologically different types
of microorganisms. Each cell exploits only a part of the
detrital energy and supplies the rest, in the form of excreted
metabolic end products, to the next step in the detrital food
chain. Thus, the anaerobic detrital energy flow is carried by
small extracellular molecules, which to a large extent are
inorganic.

It is the aim of the present paper to discuss some of the
interactions between the carbon, nitrogen and sulphur cycles
associated with mineralization of organic matter in sediments.
The quantitative significance of the processes will be
demonstrated from data obtained in Danish coastal sediments by
the microbial ecology group of the Institute of Ecology and
Genetics, Aarhus University.

AEROBIC AND ANAEROBIC RESPIRATION

The bacterial detritus food chains require an initial
breakdown of the organic matter to smaller molecules. A net
oxidation of the organic compounds involves a respiratory
metabolism. Oxygen, nitrate, sulphate and carbon dioxide
(CO_2) can function as electron acceptors for different types
of respiring organisms (Fig. 1). They form a vertical
sequence in their main activity distribution in the sediment
which is determined by a combination of factors, such as:
differences in requirements for specific redox levels of the
environment; differences in energy yield of the respiration
types; competition for electron donors; and differences in
concentration and distribution patterns of the electron
acceptors. These factors are partially interrelated and it
is difficult to assess the importance of any one of them.

The energy yield from respiration decreases from O_2, via
NO_3^- and SO_4^{2-} to CO_2 as the electron acceptors gradually
become less oxidizing. Parallel to the decrease in energy

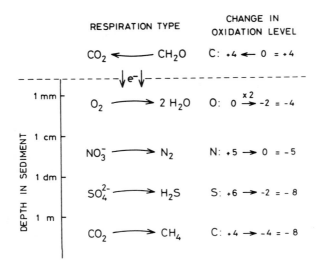

Fig. 1. Main distribution of different types of respiration in a marine sediment.

yield from their respiration, the organisms become increasingly restricted in the range of energy substrates which they can utilize. The aerobes as well as the denitrifiers, which are facultative anaerobes, are very versatile with respect to organic energy sources. The sulphur reducers (*Desulfovibrio* spp.) can utilize only a limited number of organic molecules such as lactate, ethanol, succinate, malate and, in addition, hydrogen, but they oxidize neither carbohydrates nor hydrocarbons. As they do not possess a complete enzyme system of the TCA cycle they cannot catalyze a complete substrate oxidation to CO_2. They produce acetate which can be oxidized by other sulphate reducers (*Desulfotomaculum* spp.). Only one substrate, H_2, is sufficiently reducing to be used as an electron donor for CO_2 reduction by the methanogenic bacteria. These, however, can also produce methane from acetate, methanol and formate.

The sulphate reducers and the methanogens are dependent on substrates which are products of fermentation processes in the reducing sediment. They even seem to compete for two of the most important products, namely acetate and hydrogen. The energy yield in fermentations is quite low. As this is also the case in sulphate reduction, and even more so in methanogenesis, a significant fraction of the detrital energy

is conserved in the end products, methane and sulphide. This
energy can be utilized by aerobic bacteria if the products
diffuse up to the sediment surface. Oxygen and nitrate
respiration are energetically too efficient to yield products
(H_2O and N_2) which can be biologically exploited.
    The transfer of detrital energy in H_2S, H_2 or NH_4^+ shows
that the energy flow in anaerobic environments is not as closely
linked to the carbon flow as it is in aerobic environments.

RESPIRATION IN MARINE SEDIMENTS

    Although the vertical sequence of respiration types shown
in Fig. 1 was suggested many years ago, direct measurements
to document such a zonation in sediments have been, and are
still, very scarce. Among the many recent developments which
have advanced our understanding of the respirations in
sediments are:
    i)   the use of microelectrodes to determine the
distribution of oxygen in surface layers with depth intervals
of less than one millimeter (Revsbech, Jørgensen and Blackburn,
1980; Revsbech, Sørensen, Blackburn and Lomholt, 1980);
    ii) the acetylene blockage technique to measure *in situ*
rates of denitrification (Sørensen, 1978);
    iii) studies in the field or in sediment suspensions of
the interrelations between sulphate reduction and methane
production (Abram and Nedwell, 1978a,b; Oremland and Taylor,
1978; Winfrey and Zeikus, 1977).
    Direct, simultaneous measurements of the depth distribution
of sulphate reduction and methanogenesis from CO_2 reduction
have not yet been published. Measurements have, however, been
made of the oxygen, nitrate and sulphate respiration rates
and distributions in Danish fjord sediments (Sørensen,
Jørgensen and Revsbech, 1979). Figure 2 shows examples of
the distribution of the four electron acceptors. Oxygen
generally disappeared already at 5 mm depth or less. In a
very irregular pattern it also penetrated deeper into the
sediment, but only along animal burrows and worm tubes.
Nitrate had a maximum concentration just below the sediment
surface and reached down to 7 cm depth. This is a rather
deep penetration, as nitrate was most often found to be
restricted to the upper 1-2 cm. Sulphate was the most abundant
electron acceptor and was still not exhausted at 2 m depth,
while bicarbonate (HCO_3^-) increased steadily with depth. The
latter shows, as expected, that more CO_2 is produced during
anaerobic mineralization than is used for methanogenesis.
Table 1 shows normal concentrations and penetration depths
of the electron acceptors in coastal environments.
    The corresponding rates of oxygen, nitrate and sulphate
respiration are shown in Figure 3. The depth distribution of

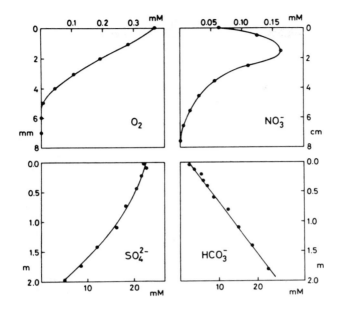

Fig. 2. Distribution of the four electron acceptors used for respiration in coastal sediments, Denmark. Note the difference in depth scales. (O_2 and NO_3^-: date from Randers Fjord, Sørensen et al., (1979); SO_4^{2-} and HCO_3^-; data from Great Belt, Jørgensen, unpublished observations).

TABLE 1

Electron acceptors in coastal waters and sediments

Electron acceptor	Concentration (mM)	Depth of penetration
O_2	0.3	a few mm
NO_3	0.0 - 0.1	a few cm
SO_4^{2-}	25	a few m
HCO_3^-	1	increases with depth

the rate of oxygen uptake within the 5 mm oxic zone is difficult to determine. If the rate was uniform within this

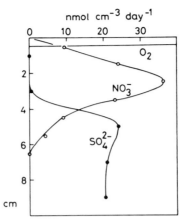

Fig. 3. Oxygen, nitrate and sulphate respiration rates in a
marine sediment of Randers Fjord, Denmark. Oxygen respiration
in animal burrows is not shown. (Data from Sørensen et al.,
1979).

zone it could be calculated from the uptake per unit area.
The resulting rate, 6,000 nmol O_2 cm^{-3} day^{-1}, is more than
100-fold higher than both denitrification and sulphate
reduction. Although the true rate must be somewhat lower due
to the extension of the oxic zone into worm tubes etc., the
calculation demonstrates the high metabolic intensity at the
aerobic sediment-water interface. The nitrate respiration
was very well developed in this sediment due to the large
pool of nitrate (cf. Fig. 2). Sulphate reduction started
below the denitrification maximum but there was clearly an
overlap between the two processes. The depth sequence of the
three respiration types shown in Figure 3 is thus in accordance
with the energy yields of the processes.

Methane generally accumulates only below the depth at
which sulphate is depleted (Fig. 4), although it may also be
produced at a low rate within the sulphate zone (Oremland
and Taylor, 1978). Depth profiles such as those in Figure 4
indicate that methane is oxidized in the lower sulphate zone
(Reeburgh and Heggie, 1977).

STOICHIOMETRY OF MINERALIZATION

In order to compare the quantitative importance of the
three respiration types for the oxidation of organic matter
in the sediment, the stoichiometry of mineralization must be
considered. From the changes in oxidation level, which are
shown in Figure 1, the respiration of substrates composed of
carbohydrate (for example, glucose) can be calculated:

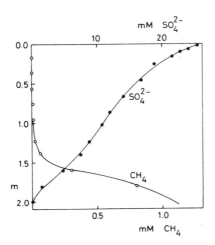

Fig. 4. Distribution of sulphate and methane in the porewater of a marine sediment, Great Belt, Denmark. (Data from B.B. Jørgensen, unpublished observations).

Aerobic respiration (4 C : 4 O);

$$C_6H_{12}O_6 + 6\ O_2 \rightarrow 6\ CO_2 + 6\ H_2O$$

Denitrification (5 C : 4 N);

$$5\ C_6H_{12}O_6 + 24\ NO_3^- + 24\ H^+ \rightarrow 30\ CO_2 + 12\ N_2 + 42\ H_2O$$

Sulphate reduction (8 C : 4 S);

$$C_6H_{12}O_6 + 3\ SO_4^{2-} + 6\ H^+ \rightarrow 6\ CO_2 + 3\ H_2S + 6\ H_2O$$

The sulphate reducers cannot carry through a complete glucose oxidation. The last process therefore proceeds in several steps. Combined, these may give the equation above, for example;

Lactate fermentation: $C_6H_{12}O_6 \rightarrow 2\ CH_3CHOHCOO^- + 2\ H^+$

Desulfovibrio: $2\ CH_3CHOHCOO^- + SO_4^{2-} \rightarrow 2\ CH_3COO^- + 2\ HCO_3^- + H_2S$

Desulfotomaculum: $2\ CH_3COO^- + 2\ SO_4^{2-} \rightarrow 4\ HCO_3^- + 2\ HS^-$

In the experimental measurements of respiration rates, the oxygen uptake is found per unit area of sediment while denitrification and sulphate reduction are calculated per unit volume at different depths. By summing the last two rates over the whole sediment column, a rate per area can also be obtained. The yearly average rates of oxygen uptake and sulphate reduction were found to be 34 and 9.5 mmol m^{-2} day^{-1},

respectively, in a Danish fjord, Limfjorden (Jørgensen, 1977).
The rate of denitrification in Limfjorden was about 0.5-1 mmol
m^{-2} day^{-1} as a yearly average (T.H. Blackburn, personal
communication). Similar rates were found also in two other
Danish fjords (Sørensen *et al.*, 1979).

In Table 2 these rates have been recalculated as carbon
oxidation equivalents according to the stoichiometry of the
respiration processes. The table shows that denitrification
is of little importance for the overall mineralization of
organic matter. Sulphate reduction on the other hand plays
a quantitatively important role in this respect as will be
discussed below.

The measurements of denitrification and sulphate reduction
are true measures of bacterial respiration as the processes
require biological catalysis. This is not the case with the
oxygen uptake. Oxygen reacts chemically with some of the
reducing compounds which are produced in the sediment. It
is therefore only a part of the oxygen uptake which is due
to respiration, the rest is chemical reduction.

TABLE 2

Mineralization rates in sediments
of Limfjorden (mmol m^{-2} day^{-1})

Process	Reduction rate	CH_2O oxidation equivalents
$O_2 \rightarrow H_2O$	34	34 (-17, -4)
$NO_3 \rightarrow N_2$	0.5 - 1	1
$SO_4^{2-} \rightarrow H_2S$	9.5	19

Hydrogen sulphide is the most important compound to
chemically reduce oxygen (Jørgensen, 1977). Its contribution
to the oxygen uptake can be calculated from the measured rate
of sulphate reduction after subtracting the amount of sulphide
which is bound within the sediment by iron in the form of
pyrite. In Limfjorden only 10% went into pyrite while the
rest was reoxidized at the sediment surface. Some of the
reoxidation may not be a direct reaction with oxygen as the
O_2 and H_2S zones often did not seem to overlap. (Fig. 5).

Iron may perhaps be an important electron carrier between
H_2S and O_2, being slowly transported by bioturbating animals.
Ferric iron is reduced in the sediment to ferrous iron by
iron at the sediment surface. Iron was present in the sediments
at a concentration of 200 μmol cm^{-3}, that is, almost a
thousand-fold higher than oxygen at the surface.

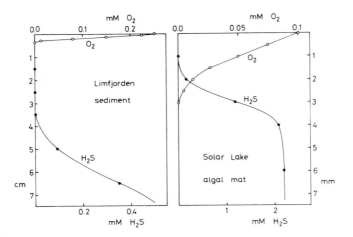

Fig. 5. *Distribution of oxygen and hydrogen sulphide in a marine sediment and in an algal mat of a saline lake. The algal mat profiles were recorded during a transitition from dark to light. (Limfjorden: data from Jørgensen, 1977; and N.P. Revsbech, unpublished observations; Solar Lake: data from Jørgensen et al., 1979).*

In sediments with higher bacterial activity, such as dense algal mat communities, a clear overlap between O_2 and H_2S was found (Fig. 5).

Irrespective of the mechanism of sulphide oxidation, 90% or 8.5 mmol H_2S m^{-2} was daily reoxidized at the surface, thereby consuming 2 x 8.5 = 17 mmol O_2 m^{-2} day^{-1}:

$$H_2S + 2 O_2 \rightarrow SO_4^{2-} + 2 H^+$$

This means that half of the total oxygen uptake of the sediment is used for sulphide oxidation and only 17 mmol are used for direct oxidation of organic carbon (cf. Table 2). It also means that the oxygen uptake measurement includes about 90% of the anaerobic respiration by the sulphate reducers. Oxygen uptake does not include denitrification since N_2 is not reoxidized, but denitrification was only equivalent to 3% of the oxygen uptake and is thus not significant in this context. The uptake of 34 mmol O_2 m^{-2} day^{-1} is therefore representative of the total mineralization of organic matter in the sediment. The sulphate reduction of 9.5 mmol m^{-2} day^{-1}, equivalent to 19 mmol O_2, shows that a little over 50% of the detritus oxidation is carried out by the sulphate reducing bacteria.

CHEMOLITHOTROPHIC BACTERIA

The energy yield during sulphate reduction is rather small and most of the potential chemical energy of the substrate is left in the produced H_2S. Since sulphate reduction accounts for 50% of the mineralization, this means that probably 30-40% of the total energy flow in the sediment ecosystem is carried by H_2S.

If the reoxidation of H_2S was only chemical, this energy would be lost for the benthic community. An unknown fraction of the energy is, however, conserved by bacterial oxidation which is known to take place in marine sediments. Chemolithotrophic bacteria, for example, thiobacilli, oxidize sulphide or intermediate products such as elemental sulphur or thiosulphate. They use the energy and reducing power to assimilate CO_2. From standard values of growth yield, a maximum CO_2 assimilation of 5 mmol C m^{-2} day^{-1} has been calculated for Limfjorden (cf. Fenchel and Blackburn, 1979). This is about 15% of the average daily input of organic matter to the sediment. In order to demonstrate such a potential contribution of H_2S to the carbon and energy flow of sediments, it is necessary to experimentally determine the relative contribution of biological and chemical oxidation. This has not yet been done.

Similar calculations as those for H_2S can be made for the oxidation of ammonia by the nitrifying bacteria (for example, *Nitrosomonas* sp., *Nitrobacter* sp.). Ammonia is produced during the breakdown of detrital protein. It is partly reincorporated by the decomposer organisms and partly it diffuses to the sediment surface where some is lost to the water column and the rest is reoxidized at an average rate of 2 mmol m^{-2} day^{-1} (T.H. Blackburn, personal communication). As ammonia does not react spontaneously with oxygen, this oxidation is biological and may lead to a bacterial CO_2 assimilation of about 0.4 mmol m^{-2} day^{-1} (cf. Fenchel and Blackburn, 1979). This is only 2% of the carbon budget of the sediment and is therefore less significant in the energy balance. The amount of oxygen consumed for the process is 4 mmol m^{-2} day^{-1} or a little more than 10% of the total oxygen uptake. When this is also subtracted from the oxygen uptake in Table 2 the remaining aerobic respiration is now only 13 mmol m^{-2} day^{-1}.

This calculation does not quite hold when the chemolithotrophic oxidation of sulphide or ammonia is taken into account. This will here be shown for the case of H_2S oxidation. Due to the chemosynthesis of the sulphide oxidizing bacteria, the CO_2, which they reduce to form organic carbon, partly takes over the function as an electron acceptor for H_2S oxidation. Thus, if the potentially maximum chemosynthesis of about 5 mmol C m^{-2} day^{-1} took place then 5 mmol less of O_2 would be used for

H2S oxidation.

This complex relation is illustrated in Figure 6, where all the numbers have been calculated as carbon oxidation equivalents. If no chemosynthesis took place, the budget is similar to that discussed for Table 2. Numbers in brackets show the situation where the maximum chemosynthesis took place. In this situation, a correspondingly larger fraction of the oxygen uptake was used for aerobic respiration than was expected in the first calculations. It would be important to know the actual CO_2 reduction in the investigated sediments, but methods to determine chemosynthesis still need to be developed.

The chemosynthetic sulphide oxidation with CO_2 does not change the applicability of the oxygen uptake as a measure of the overall mineralization in the sediment. The electrons and thus the energy, which were carried by the equivalent amount of H_2S, remain in the sediment bound in the newly formed bacterial biomass. The bacteria are later mineralized and their organic matter is oxidized, thus ultimately consuming O_2. This is included in the shift towards a 5 mmol increase in aerobic respiration. In the mineralization budget of the whole sediment, chemosynthesis functions as a delay in the oxidation of the total organic pool of CO_2 by oxygen.

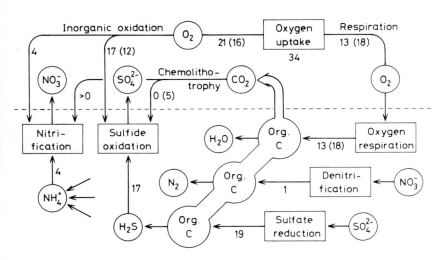

Fig. 6. Carbon, nitrogen and sulphur transformations of importance for the minerlization in a marine sediment. The numbers indicate process rates (mmol m^{-2} day^{-1}) calculated as carbon oxidation equivalents. Numbers in brackets depict the case of maximum chemolithotrophic CO_2 assimilation. Data from

Limfjorden, Denmark, (Fenchel and Blackburn, 1979; Jørgensen, 1977).

Chemosynthesis does not provide new energy or increase the total pool of oxidizable matter as all the electrons and the energy are originally derived from the organic detritus settling on the bottom.

Bacterial photosynthesis, which takes place in some protected, coastal sediments, can completely change the mineralization budget. The purple and green bacteria and the cyanobacteria can use H_2S to reduce CO_2 photosynthetically. As the energy is derived from sunlight, they can convert all the reducing power of H_2S into bacterial biomass. Thereby a closed carbon-sulphur cycle can theoretically exist within the sediment which does not require uptake of oxygen. The bacterial photosynthesis uncouples the close connection between oxygen and the mineralization in sediments. The oxygen exchange will underestimate the carbon flow in proportion to the photosynthetically oxizied H_2S. In natural sediments, however, there will inevitably be a flow of H_2S which is oxidized by O_2. This drains reducing power from the carbon-sulphur cycle which must be resupplied from external sources of organic matter, for example, from detritus or from algal photosynthesis.

There is no possibility of an internal carbon-nitrogen cycle in sediments similar to that of sulphur since there seem to be no bacteria which can use ammonia as electron donor for photosynthesis.

REFERENCES

Abram, J.W. and Nedwell, D.B. (1978a). Inhibition of methano-
    genesis by sulphate reducing bacteria competing for
    transferred hydrogen, *Archives of Microbiology* 117, 89-92.
Abram, J.W. and Nedwell, D.B. (1978b). Hydrogen as a substrate
    for methanogenesis and sulphate reduction in anaerobic
    saltmarsh sediment, *Archives of Microbiology* 117, 93-97.
Fenchel, T. and Blackburn, T.H. (1979). "Bacteria and mineral
    cycling". Academic Press, London.
Jørgensen, B.B. (1977). The sulphur cycle of a coastal marine
    sediment (Limfjorden, Denmark), *Limnology and Oceanography*
    22, 814-832.
Jørgensen, B.B., Revesbech, N.P., Blackburn, T.H. and Cohen,
    Y. (1979). Diurnal cycle of oxygen and sulphide micro-
    gradients and microbial photosynthesis in a cyanobacterial
    mat sediment, *Applied and Environmental Microbiology* 38,
    46-58.
Oremland, R.S. and Taylor, B.F. (1978). Sulphate reduction
    and methanogenesis in marine sediments, *Geochemica et
    Cosmochimica Acta* 42, 209-214.

Reeburgh, W.S. and Heggie, D.T. (1977). Microbial methane consumption reactions and their effects on methane distribution in freshwater and marine environments, *Limnology and Oceanography* 22, 1-9.

Revsbech, N.P., Jørgensen, B.B. and Blackburn, T.H. (1980). Oxygen in the seabottom measured with a microelectrode, *Science*, in press.

Revsbech, N.P., Sørensen, J., Blackburn, T.H. and Limholt, J.P. (1980). Distribution of oxygen in marine sediments measured with microelectrodes, *Limnology and Oceanography* in press.

Richards, F.A. (1965). Anoxic basins and fjords. *In* "Chemical Oceanography" (Ed. J.P. Riley and G.S. Skirrow), Vol. 1, pp. 611-645, Academic Press, London.

Sørensen, J. (1978). Denitrification rates in a marine sediment as measured by the acetylene inhibition technique, *Applied and Environmental Microbiology* 36, 139-143.

Sørensen, J., Jørgensen, B.B. and Revsbech, N.P. (1979). A comparison of oxygen, nitrate and sulphate respiration in coastal marine sediments, *Microbial Ecology* 5, 105-115.

Winfrey, M.R. and Zeikus, J.G. (1977). Effect of sulphate on carbon and electron flow during microbial methanogenesis in freshwater sediments, *Applied and Environmental Microbiology* 33, 275-281.

SOME ASPECTS OF THE INTERACTION
OF MICROBES WITH THE HUMAN BODY

G.W. Jones

*Department of Microbiology, University of Michigan,
Ann Arbor, Mi 48109, U.S.A.*

INTRODUCTION

The interaction of man and microbes is a subject almost
entirely devoted to disease and the agents of disease. In
consequence, human pathogens are the better understood microorganisms of the human body and the ways in which these organisms establish themselves in the body are known in the
greater detail.

Some microbes are undoubtably more pathogenic than others
but given the right circumstances and access to appropriate
sites many organisms, whether or not normal inhabitants of
the body, can cause disease (von Graevenitz, 1977). Pathogens, therefore, are unique only in the sense that they alter
their habitat to such an extent that their presence and activity become inimicable to the host and disease ensues.
Such may arise in any of several ways. For example, an organism may colonize a body site that normally accommodates
few if any bacteria; the colonization of the small intestine
by *Vibrio cholerae* (Gorbach, *et al.*, 1970) or colonic mucosal cells by *Shigella* species (Gemski and Formal, 1975) are
cases in point. Alternatively, a microbe may be translocated from its usual habitat to another site; such is the case
in the translocation of rectal *Escherichia coli* to the urinary tract (Roberts *et al.*, 1975). Even such alterations in
the habitat that allow certain species to proliferate beyond
acceptable limits may be the basis for disease; colitis
caused by toxigenic *Clostridium difficile* in antibiotic
treated persons (Bartlett, 1979) appears to arise in this
way. Essentially, therefore, in those instances when an organism acts as a pathogen, its behaviour conforms closely
with that expected of an allochthon.

Survival on the body surface rests upon the ability of the microbe to overcome all the environmental forces that impede colonization. This is equally true of pathogens and non-pathogens and, in principle at least, it may be supposed that solutions to these problems in general are no different in pathogens than they are in non-pathogens. Only in their specifics, dictated by the habitat, may solutions differ. Within the tissues, however, overtly pathogenic organisms do differ from non-pathogens in that they are more thoroughly equipped to resist and overcome the normal defence mechanisms of the body.

Studies on man are of a restricted nature and most on the human microflora are often confined to the examination of single microbial traits that may account for the behaviour of a particular organism in the body. Such studies, which are often based initially on clinical and pathological observations on man, take one of several of the following forms: Experiments on animals infected with microorganisms of human origin; investigation of the protective host response as an indicator of the traits necessary for the survival of a microbe in the body; comparison of presumed virulent and avirulent forms in man and other animal species; *in vitro* studies with body tissues and fluids the results of which may proffer explanations of microbial colonization (i.e. adhesion to host tissues or resistance to host immune systems); *in vitro* studies of microbial interactions. The interpretations of such studies are often difficult and the assumption that a single microbial trait can account for the pathogenic nature of the microbe or can determine whether or not an organism colonizes a particular habitat, is in general, without foundation. Under some conditions admittedly, a single ability of the microbial cell does seem to outweigh others in importance. Indeed, one may suppose that successful colonization of a body site is not a single event but depends on a sequence of several phases, each of which must be accomplished in turn before viable populations of the organism become established. Firstly, in order to maintain itself in the human population the microbe must be able to withstand transmission directly from one human host to another or to the human body via some other habitat; survival in the latter may be of equal importance to that in the human body. Secondly, microbes must be transferred with some efficiency either passively to a particular site of the body or must in some way be able to locate a particular site after transfer. Thirdly, most bacteria appear to need to localize on or in particular tissues of the body, a process that may involve adhesive and/or invasive properties. Fourthly, the bacteria must be able to proliferate in the host habitat in the face of many antimicrobial

influences of host origin and originating from other members
of the microbial flora. The following brief account of the
microbial flora of man is written with this sequence of
events in mind and with regard to the obvious questions
raised by such complex behaviour.

THE FLORA AND FAUNA OF THE HUMAN BODY

The human body plays host to a vast array of microbes from
mites to fungi, protozoans, bacteria and viruses (Marples,
1965, Gorbach *et al.*, 1973, Drasar and Hill, 1974, Moore and
Holdeman, 1974a, Noble and Somerville, 1974, Gibbons and van
Houte, 1975, Kligman *et al.*, 1976, Savage, 1977). The majority of microbial cells are intimately associated with the
surfaces of the body, often in complex communities characteristic of a given site.

It is apparent that some sites of the human body support
larger numbers of microbes and a greater variety of species
than do others. The large intestine, for example, harbors
more bacteria of more diverse species than does the skin
(Moore and Holdeman, 1974b, Kligman *et al.*, 1976). Even
smaller numbers of organisms are recovered from the proximal
small intestine of healthy persons (Gorbach *et al.*, 1967) and
these only appear when digesta is released from the stomach
(Drasar *et al.*, 1969); the population is transient and quite
different from that of the large intestine except under ab-
normal conditions when the flora of the large intestine
colonizes this site (Gorbach *et al.*, 1970, Mallory *et al.*,
1973). The diversity of the microbial flora and the unique-
ness of the flora of particular surfaces is a reflection of
the many available habitats of the body and the numerous
niches each habitat provides. Habitats include almost all
the surface areas of the human body and in pathological states,
the intracellular habitat of the host cells and the body
fluids. Besides the apparent differences between surfaces
such as the keratinized layers of the skin and the various
mucosal surfaces of the oral cavity, and of the respiratory,
gastrointestinal and urogenital tracts, no surface provides
a uniform habitat across its whole extent. The surfaces of
the living epithelial cells are complex (Jones, 1977) and
moreover, are modified by a covering of mucus gel. Even the
skin which appears uniform is not so, but differs subtly
from site to site to the extent that the microbial populations found on the forehead, axillar region and interdigital
webs of the feet are different (Marples, 1965, Noble and
Somerville, 1974, Kligman *et al.*, 1976, Tachibana, 1976).
Much of the pertinent information has been reviewed elsewhere,

The organisms of the distal small intestine and the large intestine constitute a permanent flora which inhabit the superficial mucus on the epithelial surface (Plaut, *et al.*, 1967, Nelson and Mata, 1970). The oral bacteria, in contrast, are firmly attached to the host cells and teeth and display marked affinities for particular surfaces (Gibbons and van Houte, 1975). Such microbial assemblages are often complex. The microbes of dental plaque, for example, are present in densely packed aggregates within which discrete alignments of particular cell types become apparent as a sequence of species accrue to existing deposits (Socransky *et al.*, 1977); this is sometimes achieved by the formation of intermicrobial sites of adhesion (Cisar *et al.*, 1979), and as the plaque ages by alterations in spatial arrangement of the microbial cells (Tinanoff *et al.*, 1976, reviewed by Gibbons and van Houte, 1980).

Presumably the normal flora of the body evolved with the human species and as each new niche became available so it was filled by a member of the existing flora or by microbes from sources outside the body. In general, this has led to an eventual amicable and stable association between host and flora. Only with the advent of a breakdown in the forces that constrain populations, with the translocation of a species to another site in the body or the introduction of a pathogenic allochthonous species do disturbances to the equilibrium become apparent. Indeed, under constant conditions and over the short term populations are maintained with periodic changes in types rather than species (Robinet, 1962, Cooke *et al.*, 1969). Changes in microbial population, however, may occur as the individual ages and the environment of the body surfaces change (Weinstein *et al.*, 1936, Smith, 1962, Smith, 1965, Kligman *et al.*, 1976). Oral streptococci which colonize the surface of the teeth, for example, appear as components of the oral flora only after tooth eruption and disappear again upon removal of dentures (Hardie and Bowden, 1974).

At most surface locations of the body the microorganisms are parasitic in that their nutrition ultimately depends on substrate produced by the host. Microbes that may be exceptions to this are those of the oral cavity and the alimentary tract. However, feeding by intubation does not modify the oral flora (Gibbons and van Houte, 1975) and although some intestinal organisms utilize complex polysaccharides of plants (Salyers, *et al.*, 1977) diet has little influence on the overall composition of the faecal flora of comparable groups of adults (Finegold *et al.*, 1977, Hentges *et al.*, 1977, reviewed by Savage, 1977). This, however, is not to imply that dietary changes do not influence the microbial type and the number of niches in the intestine. Indeed, the numbers of a species may respond to changes in diet (Crowther, 1971) and diet can influence profoundly microbial populations of the infant intestine (Mata *et al.*, 1972). *Lactobicillus bifidus* predominates in the faecal flora of infants fed human milk whereas Gram-negative anaerobes are the most common organism in the faeces of infants fed a diet based on cows milk. The suppression of the Gram-negative flora in breast-fed infants appears to be due to the end-products of the growth of *L.bifidus* in human milk which creates conditions that are not conducive with the growth requirements of Gram-negative bacteria (reviewed by Bullen *et al.*, 1973). The activities of the host may also influence microbial populations. The continual cleansing of surfaces by the movement of mucus and/or fluids (Gibbons and van Houte, 1975) is a major hindrance to colonization as may be the continual sloughing of cells from the surface of the body together with their attached population of microbes (Gibbons and van Houte, 1975). Unfavourable physiochemical states such as acid pH of the stomach (Gianella *et al.*, 1972) or the water limitations of the skin also constrain populations (Marples, 1965, Kligman *et al.*, 1976). Cyclic physiological changes in the host, moreover, influence bacterial numbers as appears to be the case in short term fluctuations in the numbers of trypsin-sensitive types of *Neisseria gonorrhoea* that occur during the menstrual cycle (James and Swanson, 1978). Moreover, *Bacteroides melanogenicus* increases in numbers during the second trimester of pregnancy, possibly due to its utilization of the increased amounts of available oestrogens (Kornman and Loesche, 1980). Climatic and geographical locations may also influence microbial populations of the skin and intestine respectively (Drasar and Hill, 1974, Noble and Somerville, 1974).

The exclusion of particular microorganisms from habitats of the body has been ascribed to many activities including inhibitory fatty acids of the gut and the skin, competition for substrate and bacteriocins for example (Aly *et al.*, 1972, reviewed by Savage, 1977). It seems unlikely, however, that single mechanisms such as these can account for antagonism in all habitats and under all conditions. Rather, a complex of mechanisms is a more likely situation (Savage, 1977). There is no succinct explanation of the competition between *Staphylococcus aureus* strains (Shinefield *et al.*, 1975), for example, other than that the first strain to become established is the one that survives. In the gnotobiotic mouse the numbers of *E.coli* in the large intestine are not suppressed to the levels found in the conventional mouse unless many species of indigenous bacteria also are present (Freter and Abrams, 1972). Moreover, it is of some significance that the

suppression of the numbers of introduced bacteria is much
increased by the presence of antibody directed against the
introduced bacteria and that such antibody is more stable in
the presence of the gut flora than it is in its absence
(Fubara and Freter, 1972).

In summary, therefore, the microbial flora of the human
body is composed of numerous species which often exist as
assemblages on the surface of the body. Communities may be
simple and consist of a few species or complex and composed
of many intimately associated species. Usually, however,
such communities are characteristic of particular sites on
the body. Microbial populations are constrained by the activity of other microbes and by the host but the exact interplay
and nature of the regulatory forces are far from understood.
The means and ways by which microbes initially establish
colonies on the surfaces of the body and within the host
cells, and the factors that influence such are of great importance and will be discussed in the following sections.

THE PROCESS OF COLONIZATION

The human body is neither an enduring nor an unchangeable
habitat for microbes. Their survival within the human population, therefore, very much depends on the effective transfer
to and from individuals and upon the ability of the microorganism to locate and then to establish itself at some site
in the body that serves as its habitat.

Transmission

The ways in which microorganisms are transferred to the
human body are not known in any great detail. In part the
mode of transfer depends on the origin of the microbe and the
site of the body to which it is transferred. What may be expected, however, is that successful organisms have evolved to
a stage where essential transmission is sufficiently efficient in maintaining the population and that the natural
history of the organism in the body is geared to its mode of
transmission. Three forms of transmission are readily envisaged. Firstly, direct transfer of organisms by the contamination of one individual by another. Secondly, transfer
from or via some source other than another person. Thirdly,
transmission through the agency of vectors particularly
arthropods that feed on blood.

Rapid transfer from body surface to body surface is evident
in the colonization of new-born infants by organisms of
maternal origin. Certain readily recognizable types of
faecal *Escherichia coli* are perhaps transferred in this way

(Ørskov and Sørensen, 1975) as are group B streptococci of the
vagina at parturition (Patterson and Batool Hafeez, 1976).
However, direct transfer to the identical body site does not
necessarily occur and it would seem that a change in habitat
must also involve an initial change in the behaviour of the
microbe involved. Less direct transmission than these may
subject the organism to the possible lethal effects of dessication for example. Of some interest, therefore, is the recognition of a plasmid than codes for properties that increase
the resistance of staphylococcal cells to such environmental
forces (Grinsted and Lacey, 1973).

The ecology of microorganisms that are transferred to the
body from or via sources other than the body is illustrated
by species of *Vibrio*. Certain types of *Vibrio cholerae*
(water vibrios, non-cholera vibrios or non-agglutinable
vibrios) are inhabitants of surface waters (Kaper *et al.*,
1979) whereas other types (cholera vibrios) colonize and
cause disease in the human small intestine. The latter are
occasionally isolated from water and this is presumed to be
due to faecal pollution. Taxonomically the two forms differ
only in respect of a single cell surface antigen (Jones, 1980).
It is possible that *V.cholerae* occupies an equivalent aquatic
niche to *Vibrio parahaemolyticus* because both species adsorb
to chitin (Kaneko and Colwell, 1978, Nalin *et al.*, 1979)
which, in the case of *V.parahaemolyticus*, is probably responsible for the ability of this organism to colonize the surfaces of copepods that inhabit shoreline waters (Kaneko and
Colwell, 1978). *V. parahaemolyticus* as *V.cholerae*, upon
transfer from aquatic sources to the human intestine grows
and causes disease. In both instances, however, pathogenic
types are uncommon members of aquatic vibrio populations.
The route of transmission of *V.cholerae* exemplifies the classical cycle of oral ingestion and faecal excretion; the latter possibly being much enhanced by the induction of diarrhoea
by the enterotoxin of *V.cholerae*. The species, however, is
not restricted to the intestine of man but appears to be
adapted to survive in two apparently distinct habitats i.e.
the human intestine and surface waters. Whether or not the
habitat of the human intestine is essential to the survival
of the pathogenic types is a question yet to be answered, as
are the questions of whether or not cholera vibrios are distinct forms of *V.cholerae* which are evolving from water
vibrios to eventually become a member of the autochthonous
flora of the body. However, such speculation may be based on
the erroneous assumption that *V.cholerae* constitutes a single
species because we are at present unable to measure in sufficient numbers those subtle differences that may account for
the different behaviour patterns of cholera and non-cholera

vibrios.

In the transmission of microbes, the animal vectors may function simply as a vehicle of transfer which disseminates organisms that contaminate the body of the vector; the involvement of domestic flies in the spread of enteric pathogens has been implicated in this way many times. Other vectors, however, may have a more essential role to play. Mosquito species, for example, act not only as vectors of many malarial parasites but also as hosts that provide a habitat in which these organisms complete their sexual cycle before asexual reproduction is accomplished in the secondary host (Garnham, 1966). The helminth *Loa loa* exhibits what appears to be a marked adaptation to a parasitic life and transmission by blood-feeding tabanid flies in that the appearance of the parasite in the blood follows a cycadian cycle that coincides with the feeding periods of the vector (reviewed by Krinsky, 1976). In contrast to these examples the transmission of *Borrelia* species to the human habitat by tick vectors is probably of little consequence to this bacterial species because these bacteria are able to maintain themselves by transovarian transmission in their tick hosts alone (Roberts and Robinson, 1977).

Adhesive events

The establishment of localized microbial populations on the surface of the body results from adhesive phenomena that bring about the attachment of the microbial cell to the surface of the host cell or the lodgement of the microbe in mucus secretions. Such, however, may not be the initial event of colonization. It is very possible that microbes initially adsorb temporarily to a surface while adhesive materials are either synthesized and/or recognise and bind to their surface receptors (*loc.cit.*). Moreover, recent studies have shown that *V.cholerae* displays a marked chemotaxis towards mucosal surfaces; in the absence of this response the success of colonization is much reduced (Freter *et al.*, 1979).

The act of adhesion does not solely depend upon the synthesis of adhesive substances (adhesins) by the microbe but also depends upon the existance of tissue surface receptors (Jones, 1977). However, the existance of both tissue receptors and microbial adhesins is no guarantee that colonization ensues. The yeast *Torulopsis pintolopesii*, for example, adheres readily to certain cells of the mouse stomach *in vitro* and colonizes these cells in germ-free mice; in the conventional mouse, however, only lactobacilli colonize the surface composed of these cells (Nobuo *et al.*, 1979). Conversely, the fact that a microbe *in vitro* does not produce

adhesive substances in no way means that the bacteria do not
synthesize adhesins *in vivo* nor that the microbe is incapable
of adhering to tissue surfaces of the host (Moon *et al.*, 1979).
What has been observed in most instances, however, is a close
correlation between the colonization of a particular body surface and the ability of that organism *in vitro* to adhere to
the tissue cells that compose that surface (Gibbons and van
Houte, 1975).

The adhesins   The adhesive properties of microbes which have
been reviewed thoroughly in recent years (Gibbons and van
Houte, 1975, Jones, 1977, Beachey, 1980), are inherent properties of many species. In some instances adhesive activities
are known only from *in vitro* investigations or deduced from
the locations of the microbes near to tissue surfaces.
Neither constitute proof of adhesive activity *in vivo*.   Indeed, interpretation of results from *in vitro* studies in some
cases is difficult, not because the existance of adhesive
activity is in doubt, but because firstly, their synthesis and
activity *in vivo* is unknown and secondly, more than one adhesin may be present in a culture (eg cultures of both *Neisseria
gonorrhoea* and *E.coli* can produce two distinct adhesins (reviewed by Jones, 1977).
     Microbial adhesins are diverse chemically and morphologi-
cally and in their modes of action.   Adhesins may be composed
of proteins, carbohydrates or lipids, and each may have a
characteristic form (Jones, 1977).   Examples of three distinctly different adhesive systems are given below; greater
details may be obtained from the literature cited.
     The adhesive system of *Streptococcus mutans,* although per-
haps not uniquely different from the adhesive systems of
other microbes of the oral cavity, is distinctly different
from those known adhesive systems of microbes that colonize
other sites of the human body (Gibbons and van Houte, 1975,
Montville *et al.*, 1978, Gibbons and van Houte, 1980, reviewed
by Hamada and Slade, 1980).   After initial adsorption to the
tooth surfaces (Gibbons and van Houte, 1975), adhesion involving the deposition of dextrans onto the surfaces of the
teeth results.   The dextrans, which are derived from dietary
sucrose are rich in $\alpha-(1\rightarrow3)$ linkages and are in consequence
water insoluble.   The adhesion of the streptococcus is then
promoted by the binding of the bacteria to the dextran deposits
on the teeth.   Prerequisites for this adhesive system, therefore, are the production of water insoluble dextrans, the
availability of surfaces to which dextrans may accumulate and
the means of binding the microbial cell to these dextrans.
     The adhesive system of *Streptococcus pyogenes* is quite dif-
ferent from that of *S.mutans.   S.pyogenes,* and most strepto-

cocci that have been studied, have a surface fimbrilar layer
(Gibbons and van Houte, 1975, Jones, 1977) in which is located
an adhesive system associated with the attachment of the bacterium to animal cell surfaces. Originally, the adhesive properties of *S. pyogenes* were tentatively ascribed to the M protein of the coccal surface (Ellen and Gibbons, 1974). More
recently, such properties are thought to reside in the lipid
moiety of the lipoteichoic acid (reviewed by Beachey, 1975).
The physicochemical problems of the latter concept (Jones,
1977) have been resolved recently and it is now proposed that
the lipid is the functional adhesin but M protein stabilizes
the lipid in the aqueous phase until such time that the latter
interacts with the host cell membrane (personal communication
by E.H. Beachey).

The proteinaceous adhesins of *E.coli* and related organisms
are again different from the two systems outlined above. The
majority of *E.coli* strains produce common or type 1 fimbriae
(Duguid and Old, 1980) which produce a highly characteristic
adhesive reaction, but which have as yet no clearly defined
function. Some pathogenic strains produce additional adhesive
substances which are now known to be involved in adhesive
events some of which probably contribute to the ability of
these organisms to colonize the mucosal surfaces of either the
intestine (Evans and Evans, 1978, Evans *et al.*, 1978, Satterwhite *et al.*, 1978) or the urinary tract (Svanborg Edén, 1978,
Svanborg Edén and Hansson, 1978).

Adhesin receptors   It is assumed that microbes are able to
adhere to a surface of the body because of the existance on
the surface of receptors that bind specifically to the microbial adhesins, and thereby bring about the fixation of the
organism. Receptors may be natural components of the animal
cell or components of the surfaces of other bacteria afixed
to the surface. Receptors for the most part, however, remain
unidentified and their existance can only be deduced from one
of the following types of observation. Firstly, bacteria *in
vitro* attach preferentially to cells of particular animal
species and/or tissues more so than to others. Secondly,
particular tissue cells produced by some individuals differ
from the normal for such cells of the species in that they
prevent microbial attachment. From this it may be supposed
that subtle differences in the surface composition of cells
exist that determine whether or not a cell can function as a
suitable substratum. Thirdly, adhesion may be inhibited
specifically in the presence of materials that are normally
found as components of the animal cell surface. Fourthly,
the direct attachment of bacteria to immobilized punative receptors of adhesion have been demonstrated. Such evidence

indicates strongly that the normal constituents of animal cell surfaces such as the oligosaccharides or even single monosaccharide residues of the cell glycocalyx, and the phospholipids and sterols of the cell membrane, act as receptors for microbial adhesins (reviewed by Jones, 1977).

Ecological considerations of adhesion

It is well documented that adhesion prevents the physical removal of microbes from their habitats by the shear forces of liquids that move across the tissue surface (Gibbons and van Houte, 1975, Jones, 1977). Adhesion may also provide nutritional advantages to the organism. *Mycoplasma pneumoniae,* for example, appears to gain such an advantage from its close association with its host cell (Gabridge and Dee Barden Stahl, 1978) as does *V.cholerae* from its intimate association with mucosal surfaces (Freter *et al.,* 1979). Not all bacteria in a population, however, need be adhesive. Theoretically, this may facilitate survival of a population on a surface because the surface cells of the host are continually shed and lost with their microbial communities. Subsequently, unattached bacteria of the population may reinitiate colonization by adhering to the newly exposed surface (Gibbons and van Houte, 1975). Likewise, the release of non-adhesive bacteria from populations on a surface may also facilitate the spread of microbial populations from host to host (Jones, 1980). Little is known about the regulation of adhesin synthesis but it is predictable that in environments outside the body host-specific adhesins serve little or no function and their synthesis accordingly may be expected to be repressed. Such environmental regulation may be reflected in the cessation of host-specific adhesion synthesis by *E.coli* strains at temperatures less than 37°C (Jones and Rutter, 1972, Evans *et al.,* 1978).

It is apparent that the adhesins and the mechanisms of adhesion differ markedly from species to species and even within a species such as *E.coli* (Jones, 1977, Duguid and Old, 1980). The origins and the evolution of these systems, therefore, are probably very distinct (Jones, 1977). Some adhesin genes, however, are readily transmissable and may in consequence spread to new populations of bacteria, giving rise to the colonization of new habitats by such populations. The appearance of the K99 adhesin in porcine strains of *E.coli* when previously this antigen appeared to be confined to bovine strains (Moon *et al.,* 1979) is perhaps a case in point. Not all bacteria have evolved their own adhesive systems but rather have evolved means of using the adhesive systems of other bacteria (McCabe and Donkersloot, 1977).

Presumably mixed populations of adhesive bacteria exist on

apparently homogeneous surfaces because each microbial species
is attached to its own distinct receptor (Moon *et al.*, 1979).
If each species does occupy a discrete 'spatial niche' it follows that colonization would be restricted if two microbes competed for the same receptor (Davidson and Hirsl, 1975) and by
distribution of available receptors. For example, in pigs
that do not synthesize the receptor for the adhesin of an
enteropathogenic *E.coli*, colonization by this organism, and
hence disease, is much reduced (Rutter *et al.*, 1975). Unlike
some plant-microbe associations (Lippincott and Lippincott,
1980), in man and some other animals, microbial adhesion
appears to be entirely the function of the microbe. From this
one is tempted to conclude that the bacterium is the principal
beneficiary. Certainly there is no evidence at this time that
man benefits from such associations. Indeed, when the microbe
involved is pathogenic the consequences for the individual may
be disastrous. Under pressure from pathogenic microorganisms,
therefore, it is predictable that given the right circumstances,
populations of the host species will arise that are resistant
to colonization by the pathogen; the sickle cell aneamia trait
is probably the result of such a change in a human population
(Trager, 1979). Likewise, the adhesion of a pathogen may be
prevented if the adhesin receptors are absent. This is clearly the case in some populations of disease resistant pigs
which lack receptors for the adhesins of particular pathogenic
E.coli (loc. cit.). Similar differences within human populations are also apparent. For example, the adhesion of pathogenic *E.coli* to cells of the urogenital tract and the colonization of such surfaces by these organisms is less pronounced
in disease resistant individuals than in others susceptible
to urinary tract infections (Stamey and Sexton, 1975, Fowler
and Stamey, 1977). The extent of bacterial communities attached to the mucosal surface of the body may be regulated by the
host in other less permanent ways. A protective response on
the part of the host to the presence of microbes on the body's
surfaces may be the release of secretory antibodies onto these
surfaces. Such antibody may neutralize the adhesive properties of the microbe directly or may act in other ways that
prevent the organism from gaining access to or surviving on
the tissue surface (Freter, 1978). Antibodies that neutralize
the adhesive properties of bacteria are produced in response
to pathogens that colonize the urogenital tract (Hanson *et al.*,
1977, Tramont, 1977) and members of the normal flora of the
oral cavity (Williams and Gibbons, 1972). Bacteria may avoid
the consequences of their presence inducing the production of
specific antibody, by selectively destroying the immunoglobu-
lins (Plaut *et al.*, 1975). Theoretically at least, soluble
materials that mimic tissue receptors and which compete with

such receptors for the microbial adhesin should be capable of
reducing the attachment of specific microorganisms. Finally,
the continual shedding of epithelial cells and the flow of
fluids over mucosal surfaces which reduce microbial assemblages
on the surface may be a natural response of the host to the
presence of its microflora. In this respect it may be significant that the turnover of intestinal cells is less in the
germ-free animal than it is in the conventional animal and
that microbial contamination of the small intestine increases
peristalsis, a major antibacterial mechanism of this organ
(Abrams, 1977).

Invasive microbes

    Not all microorganisms are confined to the outer surfaces
of the body. Some organisms are able to invade into and grow
within the living cells of the host whereas others penetrate
into the body and grow in an extracellular state in the body
fluids.

The intracellular habitat   Penetration of the host cell may
range from the superficial to the totally intracellular state.
Partial penetration of the host tissue is achieved by commen-
sal spirochaetes of the human intestinal tract (Lee *et al.*,
1971). The single pole of the organism embedded in the host
cell membrane may function solely as a site of attachment but
feasibly also may provide nutritional advantages to the bacterial cell. This type of association has been reviewed recently by Savage (1980).
    All known truly intracellular microbes of man are patho-
genic. Both professional phagocytic cells, in particular
macrophages, and non-professional phagocytes such as epithelial cells are parasitized. The invasion of the professional
phagocyte appears to occur by the normal phagocytic process
which consists of four identifiable phases; i.e. recognition
of the microbe, attachment of the microbe by the phagocyte,
internalization of the organism within a membrane bound vesicle (phagosome) and finally, the release of the antimicrobial
contents of the lysosomes in the vesicle (phagolysosomes).
Energy is expended by the phagocyte alone during this process
(Stossel, 1974, Silverstein *et al.*, 1977). The mechanism by
which bacteria enter 'non-phagocytic' host cells has remained
unknown until recent studies (Kihlström and Nilsson, 1977,
Hale *et al.*, 1979) which indicates that microbial penetration
of such cells is also an endocytic event perhaps mechanisti-
cally indistinguishable from phagocytosis. However, a distinction does exist. Entry into phagocytes is a host cell
specified event in that recognition, binding and internaliza-

tion is a function of the phagocyte and under appropriate
conditions proceeds independant of the bacteria or particle
being ingested. Internalization of endoparasites by non-
professional phagocytes, in contrast, can be viewed as a parasite specified event in that the recognition and binding phases
appear to be a function of the microbe as may be the initiation of their own internalization (Norrby and Lycke, 1967,
Osada and Ogawa, 1977). Internalization and the rate at which
particles are internalized by phagocytes is dependant upon
continuous circumferential ligand binding between the surfaces
of the particle and the phagocyte throughout the period of
ingestion (Griffin *et al.*, 1975) and upon the strength of such
binding sites (Capo *et al.*, 1978). The Fc portion of immuno-
globulins attached to the particle, interacting with the Fc
receptor of the phagocyte can serve as the ligand in this
respect (Stossel, 1974, Silverstein *et al.*, 1977). A phase
of adhesion does preceed invasion of 'non-phagocytic' cells
by *Salmonella typhimurium*. The binding ligand in this instance is a salmonella protein which possibly recognizes receptors on the host cell (unpublished results); the receptors
for the chlymidial adhesins may be sialic acid residues (Becker,
1978). The extent or type of adhesion also seems to influence
the subsequent invasion by salmonella, for example, in that
adhesion is not necessarily followed by invasion (unpublished
results).

As cited above, the penetration of animal cells by some
(Kihlström and Nilsson, 1977, Hale *et al.*, 1979) but not all
(Gregory *et al.*, 1979) invasive bacteria is prevented by cytochalasin B, an inhibitor of eucaryotic microfilament function
and also phagocytosis (Stossel, 1974, Silverstein *et al.*,
1977). It seems likely, therefore, that intracellular parasites utilize the contractile proteins of the host cell to
effect their own entry. Exactly what this may entail is at
this time unknown but the existance of specific chemical
messages produced by the microbe or selective binding of
microbes to particular receptors that trigger endocytosis are
worthy of consideration.

A typical response of animal cells, particularly phagocytes,
to the internalization of microorganisms is the release of
antimicrobial substances of the lysosomes into the phagosomes
(Stossel, 1974). Some intracellular bacteria such as *Mycobacterium tuberculosis* avoid contact with these potentially
harmful materials by preventing the fusion of the lysosome with
the phagosome (Armstrong and D'Arcy Hart, 1971). Whereas
other bacteria such as *M. lepraemurium* lack this mechanism
and appear to be simply resistant to the lethal effects of
the lysosomal contents (D'Arcy Hart *et al.*, 1972). In the
chlamydiae, the normal invasive phase is a distinct elementary

body which seems responsible for inhibiting lysosome function
thus enabling the intracellular form (the reticulate body) to
replicate (Friis, 1972). When the reticulate body is introduced directly into an animal cell lysosome discharge is not
prevented and death of the bacterium results (Brownridge and
Wyrick, 1979). It is possible, therefore, that host cell
memory of the lysosome/phagosome fusion inhibition message of
the elementary body is retained throughout the replication
phase. Some aspects of the above phenomena have been reviewed (Goren, 1977).

Ecological considerations of the intracellular habitat   Most
of the studies cited above were done in model systems. Although the mechanisms of internalization are probably true re-
flections of those which occur in the natural habitat, ecological considerations of these phenomena are clearly speculative
at this time.

The more apparent advantages of the intracellular state
are that competition from microbes of the tissue surface is
absent and that host cells probably provide a well regulated
and nutritionally rich habitat. Clearly, the potential disadvantages are the antimicrobial mechanisms of the host cell
and the competition with the host cell for available substrates. The means by which microbes may avoid the first disadvantage has been outlined and the second appears not to cur-
tail the replication of the more advanced forms of intracellular parasite. The more highly evolved chlamydial and protozoan parasites have developed ways and means not only of surviving within animal cells but of utilizing parts of the metabolic machinery of the host cell for their own ends (Becker,
1978, Sherman, 1979).

As the microbial species becomes more dependent on the
host cell environment greater dependence on the host cell
metabolism may be expected. Hence, the need for high rates
of intracellular survival and for efficient intercellular
transmission arises. As discussed above, species of the
genus *Chlamydia* are transmitted as a biochemically inert
elementary body that resists environmentally adverse conditions. The elementary body then gives rise intracellularly
to the reticulate body which although osmotically fragile
and not capable of extracellular existance, is biochemically
active and replicates intracellularly (Becker, 1978). DNA
homology studies, however, show the two so-called species of
Chlamydia to be unrelated (Becker, 1978). It seems, therefore, that somewhat similar complex solutions to the same
problem of the intracellular habitat have been found by two
distinct microorganisms. In contrast to the strict intracellular parasites, intracellular pathogens such as *Shigella*

and *Salmonella* species and *Escherichia coli* grow on the simplest of laboratory media. Consequently, it is probable that
in the intracellular state these bacteria function quite independent of the host cell metabolic machinery. Yet the
occupation of the intracellular niche presumably provides
some advantages to the organism. Of the *E.coli*, only some
types are intracellular parasites (DuPont *et al.*, 1971) whereas other types colonize the mucosa of the small intestine
(loc.cit.). The most common forms, however, are those *E.coli*
of the large intestine where the species constitutes but a
minor component of the bacterial flora. It seems possible,
therefore, that the two former types have evolved from populations such as the latter by acquiring the necessary attributes that enable them to colonize the new habitats. However,
the human body finds the occupation of these sites intoler-
able and disease ensues. Indeed, the fact that these two
sites are not commonly colonized by any microorganism does
suggest that host defence mechanisms are directed towards the
exclusion of microbes from such sites.

The intracellular niche is protected against the soluble
substances of the human defence systems such as antibody because such do not readily enter animal cells. However, the
host has other means of coping with this situation. An important response in this respect is cellular immunity (North,
1978).

Invasive extracellular organisms   The colonization of extracellular sites of the body's cavity by microbes that pass
initially through the outer protective surfaces of the body
is invariably inimicable to the host and gives rise to disease
and immune reactions. For pathogens of this type to survive
in the body they must possess the means with which the protective responses of the host may be avoided. Other than
viruses and higher parasites such as the trypanosomes *(loc.
cit.)* growth of most pathogens within the body is of unknown
consequence to the survival of a species, even though these
species may be thought of as primary pathogens. Perhaps
those attributes that assist the microbe in avoiding the host
defence mechanisms have evolved under pressures from the latter but their existence may be also purely fortuitous and
apparent only under unusual circumstances. It is also unclear
if the entry of organisms into the body is a chance event
brought about by their introduction through wounds and the
continual movement of microbes across the apparently intact
mucosal surface of the body (Wolochow *et al.*, 1966) or is due
to inate invasive properties of these microbes. The subsequent survival and growth of some of these bacteria within
the body where their predeliction for particular tissues

(Keppie, 1964) may result from their rate of growth not being exceeded by the rate at which they are removed by the host, leads eventually to disease. The situation is exacerbated by the toxicity of the microbes presence, by defects in the protective mechanisms of the host, by the existence of other pathological states (Mackowiak, 1978) or by the absence of protective antibody (Schneerson *et al.*, 1971). Thus to evoke the concept of microbial invasiveness is perhaps unnecessary although the concept of microbial aggressins or impedins which thwart host protective mechanisms after entry of microbes into the body is useful (Smith, 1977).

Phagocytes, antibody, complement and physiological alterations such as the sequestration of iron (Schlessinger, 1975) may all serve to reduce microbial infections of the tissue. The sequence of events leading to the confrontation of phagocytes and microbe, and the destruction of bacteria by phagocytes has been detailed elsewhere (Stossel, 1974). Bacteria may avoid predation by phagocytes by the elaboration of capsules or surface layers that increase the hydrophilic nature of the surface of the cell, thus rendering the microbe less easily engulfed by the phagocyte (van Oss, 1978). The binding of specific antibody or complement to the microbial surface (i.e. opsonization) increases the hydrophobic characteristic of the capsule and provides recognition and binding sites through which the phagocyte may identify and subsequently attach to the microbe (Stossel, 1974, van Oss, 1978). The M proteins of *S.pyogenes* are such antiphagocytic substances which, upon opsonization, are rendered ineffective as antiphagocytic devices (Fox, 1974). The existence of many antigenic forms of this important virulence factor (Fox, 1974) indicates that the evolution of these substances as antiphagocytic devices occurred in the face of selection by the ever changing specificity of host antibody; alternatively the role of the M protein in the adhesive function of these bacteria may indicate that their action as antiphagocytic devices is purely fortuitous. Of the other means of avoiding phagocytosis, the ability of *M.pneumoniae* to attach to the surfaces of phagocytes without triggering endocytosis is of interest; again antibody negates this phenomenon (Jones and Hirsch, 1971). Some bacteria, not usually considered to be intracellular parasites, nevertheless may resist digestion after phagocytosis (Veale *et al.*, 1976, Smith, 1977).

Antibody can disable microbes in ways other than those noted above. Antibody in conjunction with complement (or in some instances complement alone) may cause lysis of bacterial cells. The strain to strain differences in susceptibility to lysis have in some instances been associated with the extent of encapsulation (Glynn and Howard, 1970). *In vivo* function

for anticomplement systems is indicated by the observations
that bacteria grown in the body are more resistant to such
lytic activity than are bacteria grown in culture (Ward *et al.*,
1970). An intriguing but as yet unexplained means by which
some microbes avoid temporarily the action of antibody is by
the mechanism of continual antigenic change (Coffey and Eveland, 1967, Vickerman, 1978). In the case of *Borrelia* spp,
antigenic forms appear in sequence in response to the production of specific antibody to the existing antigenic type
(Coffey and Eveland, 1967). Ideally however, the survival of
microbes in the body would be enhanced if their presence did
not induce the host protective response. The production of
surface antigens indistinguishable from those of the host
(DeVay and Adler, 1976) appears to be one such mechanism.

Microbial exotoxins

Many microbes, particularly bacteria, are known to elaborate proteinaceous exotoxins which act as enzymes and catalyze
highly specific toxic reactions in the host (Cuatrecasas,
1977, Jeljaszewicz and Wadstrom, 1978). Unlike the chemically different endotoxins (Kass and Wolff, 1973) which are produced by all gram-negative bacteria, a given exotoxin is
characteristic of a given species or even type. Their modes
of action have been extensively studied and in many instances
their precise catalytic role is known in great detail. The
affects of some toxins on the host are known or can be deduced (Arbuthnott, 1978). Thus, diarrhoea caused by *V.cholerae*
can be ascribed to its enterotoxin (reviewed by Jones, 1980),
the spread of *Clostridium perfringens* in the tissues is probably caused by the elaboration of many cytolytic and hydrolytic toxins (Smith, 1979) and the specific killing of leuco-
cytes by the leucocidin toxin of *Staphylococcus aureus*
(Jeljaszewicz *et al.*, 1978) may be responsible for rendering
host defences less effective. Although decidedly involved
in disease, the roles of toxins in the very different process
of colonization is most unclear and sometimes contraindicated.
For example, apparently non-toxigenic *Corynebacterium diphtheriae* can colonize the human pharynx (Jephcott *et al.*, 1975)
albeit the case that the exotoxin is the lethal agent in
diphtherial disease (Barksdale *et al.*, 1960). Moreover,
S.aureus exfoliatin toxin appears to act at sites distal from
sites colonized by the staphylococcus (Rothenberg *et al.*,
1973). Toxins after all, are substances that are perceived
by the body as poisons and then probably only under unusual
circumstances, for example, when a potentially toxigenic
organism is translocated (e.g. translocation of *C.perfringens*
from the intestine to a wound) or when excessive growth of the

organism in its natural habitat occurs (e.g. the toxicity
associated with the overgrowth of clostridial species in the
intestine *(loc.cit.)*). In those organisms that are found in
habitats in addition to that of the human body (e.g. *V.cholerae)*
it may be argued that the evolution of the organism towards
the occupation of a niche in the human body has not coincided
with the elimination of traits (i.e. enterotoxin) only necessary for survival outside the body. Contrary views are held
(Arbuthnott, 1978) and contrary indications of function may
be cited. For example, pathogens that colonize the tissues
and must overcome the full force of host defence mechanisms
and the physical restraints of the tissues would be expected
to, and indeed do, produce many toxins (e.g. *C.perfringens*)
whereas pathogens that colonize less well protected surface
sites (e.g. *V.cholerae*) seldom produce more than a single
toxin. Moreover, toxin production by *Bacillus anthracis*
requires particular environmental conditions such as those
found in the body (Smith *et al.*, 1955) and the sporulation of
C.perfringens coincides with the release of enterotoxin
(Duncan *et al.*, 1972) which induces diarrhoea and hence possibly increases the dispersion of the microbe. From this
point of view, therefore, it may be argued that the range and
specificity of the toxins of particular organisms, the regulation of their production by environmental forces and their
presence as additional acquired traits could be taken to reflect necessary toxin-producing requirements by pathogenic
bacteria of the body which promote growth and/or dispersion.

Other than toxins enabling the microbe to compete with the
host, no role in their competition with other microbes has
been demonstrated. Physical removal of other microbes from
habitats such as the intestine by excess fluid flow caused by
enterotoxins does not appear to occur (Gorbach *et al.*, 1970)
and neither does the modification of the habitat by enterotoxin production enhance the growth of *V.cholerae* (Holmes
et al., 1975). Indeed, the production of enterotoxin by porcine *E.coli* has little influence on colonization (Smith and
Linggood, 1971).

In conclusion, therefore, there is little argument about
the roles some toxins play in the disease process but there
is a paucity of information on whether or not the toxicity
of these substances contributes anything to the process of
colonization or indeed, in what metabolic process they may
be involved when their toxicity is not apparent.

CONCLUDING REMARKS

This subject is too vast and its progress too rapid for one
even to contemplate an up-to-date review. What is written

here is nothing but a brief outline of the aspects that I find
to be of interest. It should be apparent that the majority
of studies are, and always have been, on pathogenic microbes
and those microbial attributes that may account for the virulence of the organism. Virulence is a concept quite different from colonization in that it is in one sense a measure of
the ability of an organism to overcome host defence mechanisms that should be effective in removing the microorganism.
If pathogens are considered allochthonous organisms and their
hold on the habitat tenuous, then the neutralization of any
one of several critically important virulence factors may
cause the organism to be excluded from particular sites of
the body. This has given a distorted view of the interactions
of all microbes with the body in that these are often seen as
consisting of single events rather than the complex series of
interactions, which is surely the case. Indeed, very little
is known about the regulations of the numbers of organisms
associated with the body.

   In many respects the mechanisms of transmission of microbes
is not understood. Is it not possible that the behaviour of
microbes in the body is as much geared to their dispersal as
it is to their continued survival in the body? For example,
does the induction of diarrhoea assist spread of enteric
pathogens? Some organisms live both in the body and elsewhere. How do these populations interact? Studies on adhesive properties consume enormous research effort and a great
deal is known about this phenomenon, although there is a
great deal more to understand. Besides the nature of adhesive substances and their production *in vivo*, what is the
nature of the adhesin receptors and are they normal cell components in all cases? Does the presence of adhesive pathogens
provide the conditions for the selection of receptor-less
individuals? Intracellular organisms probably can find,
attach to and eventually induce their own internalization by
host cells and may subsequently inhibit normal host cell
functions and assume charge of part of the host cell biosynthetic machinery. The earlier phases are known only in the
most rudimentary terms. Is the location of invasive extracellular organisms due in some instances to chance entry and
subsequent survival or are some such microbes equipped to
penetrate through the outer surface of the body without assistance? Exotoxins are enzymes with well characterized activities but what evidence is there that toxin serves the survival of a given species in the body? Where else and under
what circumstances may their activities be important?

   These questions are not applicable in all instances but
in general they are valid and intriguing questions, the answers to which may be arrived at only with great difficulty.

REFERENCES

Abrams, G.D. (1977). Microbial effects on mucosal structure
and function. *American Journal of Clinical Nutrition* 30,
1880-1886.
Aly, R., Maibach, H.I., Shinefield, H.R. and Strauss, W.G.
(1972). Survival of pathogenic microorganisms on human
skin. *Journal of Investigative Dermatology* 58, 205-210.
Arbuthnott, J.P. (1978). Role of exotoxins in bacterial pathogenicity. *Journal of Applied Bacteriology* 44, 329-345.
Armstrong, J.A. and D'Arcy Hart, P. (1971). Response of cultured macrophages to Mycobacterium tuberculosis, with observations on fusion of lysosomes with phagosomes. *Journal
of Experimental Medicine* 134, 713-740.
Barksdale, L., Garmise, L. and Horibata, K. (1960). Virulence, toxinogeny, and lysogeny in Corynebacterium diphtheriae. *Annals of the New York Academy of Science* 88,
1093-1108.
Bartlett, J.G. (1979). Antibiotic-associated pseodomembra-
nous colitis. *Reviews of Infectious Diseases* 1, 530-539.
Beachey, E.H. (1975). Binding a group A Streptococci to
human oral mucosal cells by lipoteichoic acid. *Transactions of the Association of American Physicians* 88, 285-292.
Beachey, E.H. (Ed.). (1980). Bacterial Adherence, Receptors
and Recognition Series B, Vol. 7, Chapman and Hall, London,
In press.
Becker, Y. (1978). The Chlamydia: Molecular biology of procaryotic obligate parasites of eucaryocytes. *Microbiological Reviews* 42, 274-306.
Brownridge, E. and Wyrick, P.B. (1979). Interaction of
Chlamydia psittaci reticulate bodies with mouse peritoneal
macrophages. *Infection and Immunity* 24, 697-700.
Bullen, C.L., Willis, A.T. and Williams, K. (1973). The sig-
nificance of bifidobacteria in the intestinal tract of in-
fants. *In* "Society for Applied Bacteriology Symposium
Series No. 2" (Ed. G. Sylces and F.A. Skinner), pp. 311-
325. Academic Press, London.
Capo, C., Bongrand, P., Benoliel, A.M. and Depieds, R. (1978).
Dependence of phagocytosis on strength of phagocyte-parti-
cle interaction. *Immunology* 35, 117-182.
Cisar, J.O., Kolenbrander, P.E. and McIntire, F.C. (1979).
Specificity of coaggragation reactions between human oral
Streptococci and strains of Actinomyces viscosus or Actinomyces naeslundii. *Infection and Immunity* 24, 742-752.
Coffey, E.M. and Eveland, W.C. (1967). Experimental relap-
sing fever initiated by Borrelia hermsi II. Sequential
appearance of major serotypes in the rat. *Journal of Infectious Diseases* 117, 29-34.

Cooke, E.M., Ewins, S. and Shooter, R.A. (1969). Changing faecal population of Escherichia coli in hospital medical patients. *British Medical Journal* 4, 593-595.

Crowther, J.S. (1971). Sarcina Ventriculi in human faeces. *Journal of Medical Microbiology* 4, 343-350.

Cuatrecasas, P. (Ed.) (1977). The Specificity and Action of Animal, Bacterial and Plant Toxins. Receptors and Recognition, Series B, Vol. 1. Chapman and Hall, London.

D'Arcy Hart, P., Armstrong, J.A., Brown, C.A. and Draper, P. (1972). Ultrastructural study of the behaviour of macrophages toward parasitic mycobacteria. *Infection and Immunity* 5, 803-807.

Davidson, J.N. and Hirsl, D.C. (1975). Use of the K88 antigen for in vivo bacterial competition with porcine strains of enteropathogenic Escherichia coli. *Infection and Immunity* 12, 134-136.

DeVay, J.E. and Adler, H.E. (1976). Antigens common to hosts and parasites. *Annual Review of Microbiology* 30, 147-168.

Drasar, B.S. and Hill, M.J. (1974). Human Intestinal Flora. Academic Press, London, New York, San Francisco.

Drasar, B.S., Shiner, M. and McLeod, G.M. (1969). Studies on the intestinal flora I. The bacterial flora of the gastrointestinal tract in healthy and achlorhydric persons. *Gastroenterology* 56, 71-79.

Duguid, J.P. and Old, D.C. (1980). The adhesive properties of Enterobacteriaceae. *In* Bacterial Adherence (Ed. E.H. Beachey). Receptors and Recognition, Series B, Vol. 7. Chapman and Hall, London. In press.

Duncan, C.L., Strong, D.H. and Sebald, M. (1972). Sporulation and enterotoxin production by mutants of Clostridium perfringens. *Journal of Bacteriology* 110, 378-391.

DuPont, H.L., Formal, S.B., Hornick, R.B., Snyder, M.J., Libonati, J.P., Sheahan, D.G., LaBrec, E.H. and Kalas, J.P. (1971). Pathogenesis of Escherichia coli diarrhea. *New England Journal of Medicine* 285, 1-9.

Ellen, R.P. and Gibbons, R.J. (1974). Parameters affecting the adherence and tissue tropism of Streptococcus pyogenes. *Infection and Immunity* 9, 85-91.

Evans, D.G. and Evans, D.J. (1978). New surface-associated heat-labile colonization factor antigen (CFA/II) produced by enterotoxigenic Escherichia coli of serogroups 06 and 08. *Infection and Immunity* 21, 638-647.

Evans, D.G., Evans, D.J., Tjoa, W.S. and DuPont, H.L. (1978). Detection and characterization of colonization factor of enterotoxigenic Escherichia coli isolated from adults with diarrhea. *Infection and Immunity* 19, 727-736.

Finegold, S.M., Sutter, V.L., Sugihara, P.T., Elder, H.A., Lehmann, S.M. and Phillips, R.L. (1977). Fecal microbial flora in Seventh Day Adventist populations and control subjects. *American Journal of Clinical Nutrition* 30, 1781-1792.

Fowler, J.E. and Stamey, T.A. (1977). Studies of introital colonization in women with recurrent urinary infections. VII. The role of bacterial adherence. *Journal of Urology* 117, 472-476.

Fox, E.N. (1974). M proteins of Group A Streptococci. *Bacteriological Reviews* 38, 57-86.

Freter, R. (1978). Association of enterotoxigenic bacteria with the mucosa of the small intestine - mechanisms and pathogenic implications. *In* Proceedings of Nobel Symposium 43. Cholera and Related Diarrheas-Molecular Aspects of a Global Health Problem. S. Karger , Basel. In press.

Freter, R. and Abrams, G.D. (1972). Function of various intestinal bacteria in converting germfree mice to the normal state. *Infection and Immunity* 6, 119-126.

Freter, R., O'Brien, P.C.M. and Macsai, M.S. (1979). Effect of chemotaxis on the interaction of cholera vibrios with intestinal mucosa. *American Journal of Clinical Nutrition* 32, 128-132.

Friis, R.R. (1972). Interaction of L cells and Chlamydia psittaci: Entry of the parasite and host responses to its development. *Journal of Bacteriology* 110, 706-721.

Fubara, E.S. and Freter, R. (1972). Availability of locally synthesized and systemic antibodies in the intestine. *Infection and Immunity* 6, 965-981.

Gabridge, M.G. and Dee Barden Stahl, Y. (1978). Role of adenine in the pathogenesis of Mycoplasma pneumoniae infections of tracheal epithelium. *Medical Microbiology and Immunology* 165, 43-55.

Garnham, P.C.C. (1966). Malaria Parasites and Other Haemosporidia. Blackwell Scientific Publications, Oxford.

Gemski, P. and Formal, S.B. (1975). Shigellosis: An Invasive infection of the gastrointestinal tract. *In* Microbiology-1975. (Ed. D. Schlessinger), pp. 165-169. American Society for Microbiology, Washington, D.C.

Giannella, R.A., Broitman, S.A. and Zamchek, N. (1972). Gastric acid barrier to ingested microorganisms in man: Studies in vivo and in vitro. *Gut* 13, 251-256.

Gibbons, R.J. and van Houte, J. (1975). Bacterial adherence in oral microbial ecology. *Annual Review of Microbiology* 29, 19-44.

Gibbons, R.J. and van Houte, J. (1980). Bacterial adherence and the formation of dental plaques. *In* Bacterial Adherence (Ed. E.H. Beachey). Receptors and Recognition, Series B, Vol. 7, Chapman and Hall, London. In press.

Glynn, A.A. and Howard, C.J. (1970). The sensitivity to complement of strains of Escherichia coli related to their K
antigen. *Immunology* 18, 331-346.

Gorbach, S.L., Banwell, J.G., Jacobs, B., Chatterjee, B.D.,
Mitra, R., Brigham, K.L. and Neogy, K.N. (1970). Intes-
tinal microflora in Asiatic cholera. II. The small bowel.
Journal of Infectious Diseases 121, 38-45.

Gorbach, S.L., Banwell, J.G., Jacobs, B., Chatterjee, B.D.,
Mitra, R., Sen, N.N. and Mazumder, D.N.G. (1970). Tropical Sprue and malnutrition in west Bengal. I. Intestinal
microflora and absorption. *American Journal of Clinical
Nutrition* 23, 1545-1558.

Gorbach, S.L., Menda, K.B., Thadepalli, H. and Keith, L.
(1973). Anaerobic microflora of the cervix in healthy
women. *American Journal of Obstetrics and Gynecology* 117,
1053-1055.

Gorbach, S.L., Plaut, A.G., Nahas, L., Weinstein, L., Spank-
nebel, G. and Levitan, R. (1967). Studies of intestinal
microflora II. Microorganisms of the small intestine and
their relations to oral and fecal flora. *Gastroenterology*
53, 856-867.

Goren, M.B. (1977). Phagocyte lysosomes: Interactions with
infectious agents, phagosomes, and experimental perturbations in function. *Annual Reviews of Microbiology* 31,
507-533.

Gregory, W.W., Byrne, G.I., Gardner, M. and Moulder, J.W.
(1979). Cytochalasin B does not inhibit ingestion of
Chlamydia psittaci by mouse fibroblasts (L cells) and
mouse peritoneal macrophages. *Infection and Immunity* 25,
463-466.

Griffin, F.M., Griffin, J.A., Leider, J.E. and Silverstein,
S.C. (1975). Studies on the mechanism of phagocytosis I.
Requirements for circumferential attachment of particle-
bound ligands to specific receptors on the macrophage
plasma membrane. *Journal of Experimental Medicine* 142,
1263-1282.

Grinsted, J. and Lacey, R.W. (1973). Ecological and genetic
implications of pigmentation in Staphylococcus aureus.
Journal of General Microbiology 75, 259-267.

Hale, T.L., Morris, R.E. and Bonventre, P.F. (1979). Shigella
infection of Henle intestinal epithelial cells: Role of
the host cell. *Infection and Immunity* 24, 887-894.

Hamada, S. and Slade, H.D. (1980). Mechanisms of adherence
of streptococcus mutans to smooth surfaces in vitro. *In*
Bacterial Adherence (Ed. E.H. Beachey). Receptors and
Recognition, Series B, Vol. 7, Chapman and Hall, London.
In press.

Hanson, L.A., Ahlstedt, S., Fasth, A., Jodal, U., Kaijser, B.,

Larsson, P., Lindberg, U., Olling, S., Sohl-Åkerlund, A. and Svanborg-Edén, C. (1977). Antigens of *Escherichia coli*, human immune response, and the pathogenesis of urinary tract infections. *Journal of Infectious Diseases* 136, S144-S148.

Hardie, J.M. and Bowden, G.H. (1974). The normal microbial flora of the mouth. *In* "The Normal Microbial Flora of Man" (Eds. F.A. Skinner and J.G. Carr), pp. 47-83. Academic Press, London, New York.

Hentges, D.J., Maier, B.R., Burton, G.C., Flynn, M.A. and Tsutakawa, R.K. (1977). Effect of a high beef diet on the fecal bacterial flora of humans. *Cancer Research* 37, 568-571.

Holmes, R.K., Vasil, M.L. and Finkelstein, R.A. (1975). Studies on toxogenesis in Vibrio cholerae III. Characterization of nontoxigenic mutants in vitro and in experimental animals. *Journal of Clinical Investigation* 55, 551-560.

James, J.F. and Swanson, J. (1978). Color/opacity colonial variants of *Neisseria gonorrhoeae* and their relationship to the menstrual cycle. *In* Immunobiology of *Neisseria gonorrhoeae* (Eds. G.F. Brooks, E.C. Gotschlich, K.K. Holmes, W.D. Sawyer and F.E. Young), pp. 338-343. American Society for Microbiology, Washington, D.C.

Jeljaszewicz, J., Szmigielski, S. and Hryniewicz, W. (1978). Biological effects of Staphylococcal and Streptococcal toxins. *In* Bacterial Toxins and Cell Membranes (Eds. J. Jeljaszewicz and T. Wadström), pp. 185-227. Academic Press, London, New York, San Francisco.

Jeljaszewicz, J. and Wadström, T. (Eds.) (1978). Bacterial Toxins and Cell Membranes. Academic Press, London, New York, San Francisco.

Jephcott, A.E., Gillespie, E.H., Davenport, C., Emerson, J.W. and Moroney, P.J. (1975). Non-toxigenic Corynebacterium diphtheriae in a boarding school. *Lancet* 1025-1026.

Jones, G.W. (1977). The attachment of bacteria to the surface of animal cells. *In* "Microbial Interactions" (Ed. J.L. Reissig). Receptors and Recognition, Series B, Vol. 3, pp. 139-176. Chapman and Hall, London.

Jones, G.W. (1980). The adhesive properties of vibrio cholerae and other vibrio species. *In* "Bacterial Adherence" (Ed. E.H. Beachey). Receptors and Recognition, Series B, Vol. 7. Chapman and Hall, London. In press.

Jones, G.W. and Rutter, J.M. (1972). Role of the K88 antigen in the pathogenesis of neonatal diarrhea caused by *Escherichia coli* in piglets. *Infection and Immunity* 6, 918-927.

Jones, T.C. and Hirsch, J.G. (1971). The interaction in vitro of mycoplasma pulmonis with mouse peritoneal macro-

278   G.W. JONES

phages and L-cells. *Journal of Experimental Medicine* 133,
   231-259.
Kaneko, T. and Colwell, R.R. (1978). The annual cycle of
   Vibrio parahaemolyticus in Chesapeake Bay. *Microbial Eco-*
   logy 4, 135-155.
Kaper, J., Lockman, H., Colwell, R.R. and Joseph, S.W. (1979).
   Ecology, serology, and enterotoxin production of Vibrio
   cholera in Chesapeake Bay. *Applied and Environmental Mi-*
   crobiology 37, 91-103.
Kass, E.H. and Wolff, S.M. (Eds.) (1973). Bacterial Lipo-
   polysaccharides: The Chemistry, Biology, and Clinical Significance of Endotoxins. University of Chicago Press,
   Chicago, London.
Keppie, J. (1964). Host and tissue specificity. *In* Microbial Behaviour, in vivo and in vitro. 14th Symposia of
   the Society for General Microbiology 14, 44-63.
Kihlström, E. and Nilsson, L. (1977). Endocytosis of Salmonella typhimurium 395 MS and MR10 by hela cells. *Acta*
   Pathologica Microbiologia Scandinavica Sect. B. 85, 322-
   328.
Kligman, A.M., Leyden, J.J. and McGinley, K.J. (1976). Bacteriology. *Journal of Investigative Dermatology* 67, 160-
   168.
Kornman, K.S. and Loesche, W.J. (1980). The subgingival
   microbial flora during pregnancy. *Journal of Peridontal*
   Research 15. In press.
Krinsky, W.L. (1976). Animal disease agents transmitted by
   horse flies and deer flies (Diptera: Tabanidae). *Journal*
   of Medical Entomology 13, 225-275.
Lee, F.D., Kraszewski, A., Gordon, J., Howie, J.G.R., McSeve-
   ney, D. and Harland, W.A. (1971). Intestinal spirochaeto-
   sis. *Gut* 12, 126-133.
Lippincott, J.A. and Lippincott, B.B. (1980). Microbial ad-
   herence in plants. *In* Bacterial Adherence (Ed. E.H.
   Beachey). Receptors and Recognition, Series B, Vol. 7.
   Chapman and Hall, London. In press.
Mackowiak, P.A. (1978). Microbial synergism in human infec-
   tions. *New England Journal of Medicine* 298, 21-26,83-87.
Mallory, A., Savage, D., Kern, F. and Smith, J.G. (1973).
   Patterns of bile acids and microflora in the human small
   intestine II. Microflora. *Gastroenterology* 64, 34-42.
Marples, J.J. (1965). The Ecology of the Human Skin. Charles
   C. Thomas, Springfield, Illinois.
Mata, L.J., Mejicanos, M.L. and Jiménez, F. (1972). Studies
   on the indigenous gastrointestinal flora of Guatemalan
   children. *American Journal of Clinical Nutrition* 25,
   1380-1390.
McCabe, R.M. and Donkersloot, J.A. (1977). Adherence of

Veillonella species mediated by extracellular glucosyl-
transferase from Streptococcus salivarius. *Infection and
Immunity* 18, 726-734.

Montville, T.J., Cooney, C.L. and Sinskey, A.J. (1978).
Streptococcus mutans dextransucrase: A review. *Advances
in Applied Microbiology* 24, 55-84.

Moon, H.W., Isaacson, R.E. and Pohlenz, J. (1979). Mechanisms of association of enteropathogenic Escherichia coli
with intestinal epithelium. *American Journal of Clinical
Nutrition* 32, 119-127.

Moore, W.E.C. and Holdeman, L.V. (1974a). Human fecal flora:
The normal flora of 20 Japanese-Hawaiians. *Applied Microbiology* 27, 961-979.

Moore, W.E.C. and Holdeman, L.V. (1974b). Special problems
associated with the isolation and identification of intestinal bacteria in fecal flora studies. *American Journal
of Clinical Nutrition* 27, 1450-1455.

Nalin, D.R., Daya, V., Reid, A., Levine, M.M. and Cisneros,
L. (1979). Adsorption and growth of Vibrio cholerae on
chitin. *Infection and Immunity* 25, 768-770.

Nelson, D.P. and Mata, J.L. (1970). Bacterial flora associated with the human gastrointestinal mucosa. *Gastroenterology* 58, 56-61.

Noble, W.C. and Somerville, D.A. (1974). Microbiology of
Human Skin. W.B. Saunders Company Ltd., London, Philadelphia, Toronto.

Nobuo, S., Siegel, J.E. and Savage, D.C. (1979). Ecological
determinants in microbial colonization of the murine gastrointestinal tract: Adherence of Torulopis pintolopesii
to epithelial surfaces. *Infection and Immunity* 25, 139-143.

Norrby, R. and Lycke, E. (1967). Factors enhancing the host-
cell penetration of Toxoplasma gondii. *Journal of Bacteriology* 93, 53-58.

North, R.J. (1978). The concept of the activated macrophage.
Journal of Immunology 121, 806-809.

Ørskov, F. and Sørensen, K.B. (1975). Escherichia coli serogroups in breast-fed and bottle-fed infants. *Acta Pathologica Microbiologica Scandinavica* 83, 25-30.

Osada, Y. and Ogawa, H. (1977). A possible role of glyco-
lipids in epithelial cell penetration by virulent Shigella
flexneri 2a. *Microbiology and Immunology* 21, 405-410.

Patterson, M.J. and Batool Hafeez, A. (1976). Group B
Streptococci in human disease. *Bacteriological Reviews*
40, 774-792.

Plaut, A.G., Gorbach, S.L., Nahas, L., Weinstein, L., Spank-
nebel, G. and Levitan, R. (1967). Studies of intestinal
microflora III. The microbial flora of human small intestinal mucosa and fluids. *Gastroenterology* 53, 868-873.

Plaut, A.G., Gilbert, J.V., Artenstein, M.S. and Capra, J.D. (1975). Neisseria gonorrhoeae and Neisseria meningitidis: Extracellular enzyme cleaves human immunoglobulin A. *Science* 190, 1103-1105.

Roberts, A.P., Linton, J.D., Waterman, A.M., Gower, P.E. and Koutsaimanis, K.G. (1975). Urinary and faecal Escherichia coli O-serogroups in symptomatic urinary-tract infection and asymptomatic bacteriuria. *Journal of Medical Microbiology* 8, 311-318.

Roberts, L.W. and Robinson, D.M. (1977). Efficiency of transovarial transmission of Rickettsia tsutsugamushi in Leptotromibidium arenicola (Acari: Trombiculidae). *Journal of Medical Entomology* 13, 493-496.

Robinet, H.G. (1962). Relationship of host antibody to fluctuations of Escherichia coli serotypes in the human intestine. *Journal of Bacteriology* 84, 896-901.

Rothenberg, R., Renna, F.S., Drew, T.M. and Feingold, D.S. (1973). Staphylococcal scalded skin syndrome in an adult. *Archives of Dermatology* 108, 408-410.

Rutter, J.M., Burrows, M.R., Sellwood, R. and Gibbons, R.A. (1975). A genetic basis for resistance to enteric disease caused by E. coli. *Nature* 257, 135-136.

Salyers, A.A., West, S.E., Vercellotti, J.R. and Wilkins, T.D. (1977). Fermentation of mucins and plant polysaccharides by anaerobic bacteria from the human colon. *Applied and Environmental Microbiology* 34, 529-533.

Satterwhite, T.K., Evans, D.G., DuPont, H.L. and Evans, D.J. (1978). Role of Escherichia coli colonization factor antigen in acute diarrhoea. *Lancet* 181-184.

Savage, D.C. (1977). Microbial ecology of the gastrointestinal tract. *Annual Reviews of Microbiology* 31, 107-133.

Savage, D.C. (1980). Adherence of normal flora to mucosal surfaces. *In* Bacterial Adherence (Ed. E.H. Beachey). Receptors and Recognition, Series B, Vol. 7. Chapman and Hall, London. In press.

Schlessinger, D. (Ed.) (1975). Microbiology 1974 III. Roles of iron in host-parasite interactions. pp. 263-309. American Society for Microbiology, Washington, D.C.

Schneerson, R., Rodrigues, L.P., Parke, J.C. and Robbins, J.B. (1971). Immunity to disease caused by Hemophilus influenzae type b II. Specificity and some biologic characteristics of "natural", infection-acquired, and immunization-induced antibodies to the capsular polysaccharide of Hemophilus influenzae type b. *Journal of Immunology* 107, 1081-1089.

Sherman, I.W. (1979). Plasmodium: Biochemical consequences of life in a red cell. *In* Microbiology-1979. (Ed. D. Schlessinger). pp. 124-129. American Society for Microbio-

logy, Washington, D.C.
Shinefield, H.R., Aly, R., Maibach, H., Ribble, J.C., Boris,
  M. and Eichenwald, H.F. (1975). Factors influencing col-
  onization of mucous membranes and skin surfaces with
  Staphylococcus aureus. *In* Microbiology-1975. (Ed. D.
  Schlessinger). pp. 110-115. American Society for Micro-
  biology, Washington, D.C.
Silverstein, S.C., Steinman, R.M. and Cohn, Z.A. (1977).
  Endocytosis. *Annual Reviews of Biochemistry* 46, 669-722.
Skinner, F.A. and Carr, J.G. (Eds.) (1974). The Normal Mi-
  crobial Flora of Man. Academic Press, London, New York.
Smith, H. (1977). Microbial surfaces in relation to patho-
  genicity. *Bacteriological Reviews* 41, 475-500.
Smith, H., Keppie, J. and Stanley, J.L. (1955). The chemical
  basis of the virulence of Bacillus anthracis. V. The spe-
  cific toxin produced by B. anthracis in vivo. *Journal of
  Experimental Pathology* 36, 460-472.
Smith, W.H. (1962). The development of the bacterial flora
  of the faeces of animals and man: The changes that occur
  during aging. *Journal of Applied Bacteriology* 24, 235-241.
Smith, W.H. (1965). The development of the flora of the alimentary tract in young animals. *Journal of Pathology and
  Bacteriology* 90, 495-513.
Smith, H.W. and Linggood, M.A. (1971). Observations on the
  pathogenic properties of the K88, hly and ent plasmids of
  Escherichia coli with particular reference to porcine
  diarrhea. *Journal of Medical Microbiology* 4, 467.
Smith, L.D.S.(1979). Virulence factors of Chostridium per-
  fringens. *Reviews of Infectious Diseases* 1, 254-260.
Socransky, S.S., Manganiello, A.D., Propas, D., Oram, V. and
  van Houte, J. (1977). Bacteriological studies of develop-
  ing supragingival dental plaque. *Journal of Peridontal
  Research* 12, 90-106.
Stamey, T.A. and Sexton, C.C. (1975). The role of vaginal
  colonization with enterobacteriaceae in recurrent urinary
  infections. *Journal of Urology* 113, 214-217.
Stossel, T.P. (1974). Phagocytosis. *New England Journal of
  Medicine* 290, 717-723,774-780,833-839.
Svanborg Edén, C. (1978). Attachment of Escherichia coli to
  human urinary tract epithelial cells. Scandinavian Journal
  of Infectious Diseases Supplementum 15. Almqvist & Wiksell
  Periodical Company, Stockholm, Sweden.
Svanborg Edén, C. and Hansson, H.A. (1978). Escherichia coli
  pili as possible mediators of attachment to human urinary
  tract epithelial cells. *Infection and Immunity* 21, 229-237.
Tachibana, D.K. (1976). Microbiology of the foot. *Annual
  Review of Microbiology* 30, 351-375.
Tinanoff, N., Gross, A. and Brady, J.M. (1976). Development

of plaque on enamel. *Journal of Peridontal Research* <u>11</u>, 197-209.

Trager, W. (1979). Erythrocyte-malaria parasite interactions. *In* Microbiology-1979. (Ed. D. Schlessinger). pp. 120-123. American Society for Microbiology, Washington, D.C.

Tramont, E.C. (1977). Inhibition of adherence of Neisseria gonorrhoeae by human genital secretions. *Journal of Clinical Investigation* <u>59</u>, 117-124.

van Oss, C.J. (1978). Phagocytosis as a surface phenomenon. *Annual Review of Microbiology* <u>32</u>, 19-39.

Veale, D.R., Finch, H., Smith, H. and Witt, K. (1976). Penetration of penicillin into human phagocytes containing Neisseria gonorrhoea: Intracellular survival and growth at optimum concentrations of antibiotic. *Journal of General Microbiology* <u>95</u>, 353-363.

Vickerman, K. (1978). Antigenic variation in trypanosomes. *Nature* <u>273</u>, 613-617.

von Graevenitz, A. (1977). The role of opportunistic bacteria in human disease. *Annual Reviews of Microbiology* <u>31</u>, 447-471.

Ward, M.F., Watt, P.J. and Glynn, A.A. (1970). Gonococci in urethral exudates possess a virulence factor lost on subculture. *Nature* <u>227</u>, 382-384.

Weinstein, L., Bogin, M., Howard, J.H. and Finkelstone, B.B. (1936). A survey of the vaginal flora at various ages, with special reference to the Döderlein bacillus. *American Journal of Obstetrics and Gynecology* <u>32</u>, 211-218.

Williams, R.C. and Gibbons, R.J. (1972). Inhibition of bacterial adherence by secretory immunoglobulin A: A mechanism of antigen disposal. *Science* <u>177</u>, 697-699.

Wolochow, H., Hildebrand, G.J. and Lamanna, C. (1966). Translocation of microorganisms across the intestinal wall of the rat: Effect of microbial size and concentration. *Journal of Infectious Diseases* <u>116</u>, 523-528.

MISCONCEPTIONS, CONCEPTS AND APPROACHES IN RHIZOSPHERE BIOLOGY

G.D. Bowen

*Division of Soils, C.S.I.R.O., Glen Osmond
Adelaide, Australia*

INTRODUCTION

I believe the last few years have heralded a new era in the study of microbial growth around roots. We have entered a period of increasing concern with population dynamics of the rhizosphere and the interactions which occur there, an increasing concern with the mechanisms involved, and an increase in hypothesis formulation and experiment and in simulation modelling approaches. A large background knowledge of the rhizosphere has been built over many years by direct microscopy and by traditional isolation methods, but these studies tended to be static, not dynamic. As Newman (1978) pointed out, there are a number of questions related to the rhizosphere which conventional studies have not answered. These include questions such as: "What are the specific growth rates of microorganisms in the rhizosphere?", "How do growth rates change with the age of the root?" and "What are the sources of inoculum for roots; what are the quantitative relations between populations of mycorrhizal fungi in soil, infection of the root and plant response?".

The interest in rhizosphere biology originates mainly from effects of soil organisms on plant growth. For example, if we are to realize fully the potential of symbiotic microorganisms to stimulate plant growth we need to ensure that a selected, highly efficient symbiont will establish and persist in the face of competition from possibly less efficient naturally occurring symbionts (Bowen, 1978) and of antagonism from other soil organisms - "biological control" of symbionts is just as real as biological control of plant pathogens. We also need to know appropriate strategies to increase the populations of highly beneficial organisms where they occur naturally, for example, symbionts and organisms suppressive to root diseases. This involves an appreciation of factors which selectively enhance populations of particular microorganisms. We need to learn appropriate management techniques to decrease the impact of root diseases and this requires a knowledge of their epidemiology. The possibility of rhizosphere microbial capture of plant nutrients, of nitrogen fixation by free living organisms and of possible detoxification of allelopathic substances (Newman, 1978), makes an understanding of energy sources available to microorganisms necessary, and how these are affected by soil and plant factors.

Perhaps the most important force in the "renaissance" of rhizosphere ecology is the application of skills derived from other fields. Thus the use of scanning electron microscopy (Campbell and Rovira, 1973; Elliott, Gilmour, Cochran, Coby and Bennett, 1979) and especially transmission electron microscopy (TEM) of the root-soil interface, coupled with histochemical techniques, (Foster and Rovira, 1978; Rovira, Bowen and Foster, 1980) have allowed a more direct examination of the spatial and physiological relations between roots and microorganisms associated with them. These studies have also indicated the possibility of unique types of microorganisms in the rhizoplane (Foster and Rovira, 1978). The use of isotope labelling techniques with plants growing in soil (Barber and Martin, 1976) have allowed a better appreciation of the amounts of substrates coming from roots under almost natural conditions. Conceptual models of complex situations, such as the epidemiology of rhizosphere colonization and the identification of parts which are amenable to simple experimentation in soil (Bowen, 1979; Bowen and Theodorou, 1973), have allowed an evaluation of factors affecting the various components. The application of population dynamics, and of ecological and modelling approaches (Newman and Watson, 1977; Bowen and Rovira, 1976), well developed in other fields, have made rhizosphere biology a more quantitative science.

In this paper I wish to examine some of the more recent advances in rhizosphere biology and how they have moulded our present concepts. The broad conceptual model I shall use is that of Figure 1, which is discussed in detail elsewhere (Bowen, 1979). This is an ecological type model at the plant-soil level and is not appropriate to consideration of physiological mechanisms or of the details of microbial interactions. Figure 1 indicates:

i)   that the plant is the main driving force for the system and thus factors affecting its growth will also affect microbial growth;

ii) that distribution of roots and microbial movement
to them are essential considerations in understanding the
composition of the root microflora;

iii) that microbial growth at the root surface can itself
affect losses from the root and that microbial growth at the
surface may affect substrates available for organisms away
from the root surface.

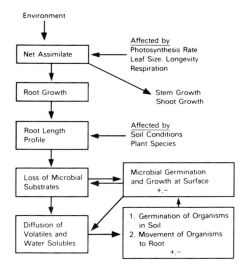

*Fig. 1. A simple conceptual model of plant microorganism
interactions.*

SOME MISCONCEPTIONS

   Old ideas do not die readily and the years have led to a
number of firmly entrenched misconceptions on plant-microbe
interactions and rhizosphere biology.

   The main misconception is that studies of microbial growth,
physiology and interactions in laboratory media or with plants
in culture solutions are highly meaningful for plants growing
in soil. Some of these studies may possibly be relevant, for
example, the production of highly specific chemicals (Krupa
and Fries, 1971), but usually they are not. This erroneous
belief ignores the quite different physical and chemical
nature of soil and of substrates coming from roots growing
in soil compared with solution cultures (which are often
luxurious in nutrients). Theodorou and Bowen (1971) found
that some strains of ectomycorrhizal fungi which grew well
on laboratory media at 16°C could not colonize the rhizoplane

of *Pinus radiata* at this temperature in sterile soil,
although they did at 20°C and 25°C. Also, antibiosis in
laboratory media often has little or no relevance to antibiosis in the rhizoplane, either for plant pathogens
(Broadbent, Baker and Waterworth, 1971) or ectomycorrhizal
fungi (Bowen and Theodorou, 1979) except possibly for
organisms showing consistently strong effects (Marx, 1969).
Antibiosis in laboratory media is influenced markedly by
different media. Therefore there is no real alternative to
studying microbial growth dynamics and interactions in soil
itself. This is not to deny that studies in solution culture
are almost essential (initially) in investigations at a
cellular physiology level, for example, recognition between
roots and symbionts, and the control of transfer of metabolites between microorganisms and roots. The use of sterile
soil in some experimental approaches to epidemiology and
microbial ecology of the root itself introduces artifacts.
However this is closer to reality than solution culture
systems and enables one to isolate effects of soil physical
and chemical variables from those of other soil microorganisms.

A second entrenched misconception is that the dynamics of
microbial growth and interactions in well stirred laboratory
solutions apply to microorganisms in soil and on the root.
As will be seen below, this is not so. This assumption
ignores that these environments are usually comparatively low
in substrates, that solute transfer limitations occur in
such discontinuous matrices as soil, and that growth dynamics
on surfaces are probably quite different from those in
solutions. Prey-predator relations also will be affected by
the spatial restrictions of surfaces and soil.

A third common misconception, particularly relevant to
symbiotic relations of plants, is that natural selection
will have led to the dominance of the most effective strains
of rhizobia or mycorrhizal fungi in an area. However, the
trend of evolution has been for survival, not high productivity, and it is more likely that indigenous strains of
these symbionts have evolved primarily to be compatible with
the existing soil physical chemical and biological environments. There are several instances where the naturally
occurring symbionts are not the most effective (Mosse, 1975).

GROWTH SUBSTRATES

Over the last few years concepts of substrates released
from roots for microbial growth, previously based on leakage
of exudates from solution-grown plants under sterile conditions have changed in three major respects:

i)   It is now recognised that substrates from roots have

many origins. Rovira, Foster and Martin (1979) have classi-
fied these as:
exudates of low molecular weight leaking from intact cells;
secretions released actively from cells;
mucilages of various types;
mucigel consisting of microbial and of plant mucilages
(possibly modified by microbial action) and;
lysates from autolysis of senescing cells.
Sloughed root cells should be added to these (Griffin, Hale
and Shay, 1976).
By using techniques specific for 1,2-diglycol groups in
sugars linked between carbon 1 and 4, Foster has distinguished
between gels produced by roots and gels produced by microorganisms. Furthermore, differential hydrolysis has
indicated fungal gel is more stable than root derived mucilage (see Rovira, Bowen and Foster, 1980). This may be a
reason for the increase in aggregation of soil caused by
vesicular-arbuscular mycorrhizas (v.a.m.) (Sutton and
Sheppard, 1976).

ii) In sharp contrast to losses of substrate from solution
grown plants, losses from soil grown plants can be very
large indeed. Using ^{14}C- carbon dioxide labelling, Barber
and Martin (1976) showed losses into soil of 5-10% of the
assimilated carbon by three-week wheat plants. In four
experiments on wheat and barley by Barber and Martin (1976)
and Martin (1977a), between 3 and 9% of the carbon in roots
was lost as water soluble substances and 17-25% was water
insoluble material. Newman (1978) concluded soluble substrates from roots are frequently 1-10% of the root weight and
may sometimes reach 25%.

iii) The "normal" soil microflora increases losses from
roots by up to 100% (Barber and Martin, 1976). Martin (1977a)
found a loss of 22-24% of the assimilate over the first 6
weeks from wheat plants grown in non-sterile soil via the
roots. The carbohydrate budget of mycorrhizal plants is
receiving increasing study (Harley, 1975; Bowen, 1978) and
it is now apparent that losses of assimilates from roots due
to the non-infecting soil microflora must be considered as
an important factor in carbon balances of soil-grown plants.
This increased loss of assimilates due to microorganisms
raises at least two important questions:

(a) Is the rhizosphere microflora increase of assimilate
loss a significant enough drain on the plant to reduce its
growth or does the plant compensate by photosynthesizing
more? A similar question has been asked with respect to
symbionts (Bowen, 1978) Increases in plant growth following

soil sterilization which are not explicable readily by
control of pests and diseases or by nutrient release, could
be due to the effect of the rhizosphere microflora draining
away organic compounds which otherwise might be utilized by
the plants. In symbiotic systems the benefits normally far
outweigh the assimilate costs but even here, growth depressions may occur either before the plant benefits from the
symbiosis (Cooper, 1975; Gibson, 1966) or where the symbionts
are superfluous (Crush, 1976). It has been calculated that
anything from a few percent to as much as a quarter of a
tree's assimilate may be used by ectomycorrhizas (Newman,
1978).

(b) What are the origins of the increased assimilate loss
in the presence of microoganisms? Martin (1978) considered
the increase was due to microbial lysis of cells. Light
microscopy and TEM studies show heavily colonized individual
senescent cells near the root apex (Bowen and Foster, 1979).
However, we do not know what is cause and effect in this
instance, for no studies have been made on the frequency of
senescent cells with sterile and with non-sterile roots.
Even if sub-clinical pathogens, causing cell senescence
prematurely, are the main reasons for the increased substrate
losses, there are at least two other ways in which substrates
could be lost locally to the advantage of microorganisms
there.

   i) Microbial absorption of exudates could maintain a high
concentration gradient at that point and sustain diffusion
out of the cell in a parallel way to the sustained diffusion
of glucose and fructose from higher plants to mycorrhizal
fungi (Harley, 1975).

   ii) Local production by some microorganisms of compounds
increasing cell permeability would increase rates of exudation at that point. This would give a particular competitive
advantage to the organisms concerned and would be an especially useful property of late arrivals at the root surface.
It would also be useful in colonization of cell surfaces
(away from cell junctions) where exudation appears to be low.
This hypothesis could be tested by electrophysiological
approaches or by examining efflux from roots radioactively
labelled with ions such as chloride which are lost passively.
A number of organisms associated with roots are able to
produce hormone type compounds, for example, some ectomycorrhizal fungi (Slankis, 1973). Highinbotham (1968) found
hormones increased permeability of *Avena* coleoptiles, but
several other compounds may also have this effect. If the
hypothesis above is confirmed, selection of organisms with
such a property might aid in their establishment.

Factors affecting plant growth and composition (Fig. 1) and factors affecting membrane permeability, would be expected to affect substrates coming from the roots and a closer integration of the basic fields of plant physiology and microbiology is necessary. For example, Reid and Mexal (1977) found not only did roots of *Pinus contorta* exude 27-29% more ^{14}C- labelled assimilate at -400 KPa water potential than at -200 KPa, but also 45% more assimilate was translocated to the roots. Some roots exuded over 80% of the total ^{14}C-activity which reached them in 10 days. Similarly Martin (1977b) found wheat roots lost twice as much water-soluble material in water-stress conditions. Phosphorous deficiency increases losses of amino acids and reducing sugars from roots by 2-3 fold (Bowen, 1969; Ratnayake, Leonard and Menge, 1978) and two mechanisms may be responsible. Ratnayake *et al.*, (1978) found evidence for increased permeability of low phosphate roots of 8-10 week old seedlings or sorghum and citrus. Using essentially the same method on 2-4 week old *P. radiata* roots Bowen (1969) found no evidence for increased permeability but found a doubling of free amino acids and amides in phosphate deficient roots. Presumably the large phosphorus supplies of *P. radiata* seed were sufficient to maintain complete root integrity for at least 4 weeks. Barrow (1977) found no growth response of this species to applied phosphate until 9 weeks after application of phosphate. It would be interesting to compare total exudates from closely related C_3 and C_4 plants.

The other basic determinant of substrate distribution at the root soil interface is the diffusion of substrates through soil. Newman and Watson (1977) considered this in a model in which microbial abundance in the rhizosphere was simulated. The growth rate of microorganisms at each point in soil was assumed to be controlled by the concentration of soluble organic substrate. The concentration of substrate changed due to:

i)   its production by the root and diffusion through soil,

ii) its production in the soil by breakdown of insoluble organic matter and

iii) its use by microorganisms.

In the Newman-Watson model diffusion of substrates through wet soil was rapid enough for extensive microbial development as far out as 4 mm. In drier soil (-0.3 bar) some exudate initially diffused into soil thus stimulating microbial growth but as populations developed at the root surface they captured more of the exudate and much less diffused into soil. At -0.3 bar soil matric potential, microbial

abundance at the root surface was 15 times that at 0.1 mm and
430 times that at 0.8 mm at 10 days. With lower diffusion
coefficients of substrates in soil, that is, at -1 bar matric
potential, still a moisture conducive to good root growth,
the gradients of microbial abundance with distance from the
root were even more steep. If this model can be validated,
clearly it has great practical use in predicting the
distances from the roots at which propagules may receive
enough substrate to germinate under various conditions. The
limited diffusion of soluble substrate predicted by the model
suggests that volatiles may be very important in stimulating
microbial activity more than 1 or 2 mm from the roots. In
Newman and Watson's model a 10 fold difference in exudation
led to a 120 fold difference in microbial abundance.

This model, like any model, needs experimental testing
and possibly later refinement in critical areas. The data
of Bowen and Rovira (1973) are remarkably consistent with
it. The steep fall in bacterial numbers with distance from
the root is also confirmed by the TEM data of Foster and
Rovira (1978) in which, at 15-20 μm from the root, the
frequency of bacteria declined to 10% of that at the rhizoplane. TEM techniques should be invaluable in testing
predictions of such models.

It should be remembered that predictions by models are
not proof and they, therefore, need validation. One of
their powers is they indicate the most sensitive components
of the system for further research and testing and the areas
possibly most responsive to manipulation. Some aspects of
modelling in rhizosphere ecology are discussed by Bowen
(1979).

A major limitation of root substrate studies so far is
that they have usually measured gross effects and have not
dealt with heterogeneity between and along roots. Approaches
such as split root techniques and potometer techniques (often
used in ion absorption studies) could be useful in this
respect with soil grown plants. Microbial assays have demonstrated the types of differences which can occur between
and along roots: differences in microbial colonization
of seminal and nodal roots of wheat have been shown by
traditional plating methods (Sivasithamparam, Parker and
Edwards, 1979); that the junctions of root cells with each
other are the major sites of exudation has been shown by
inoculating roots uniformly with organisms and subsequently
observing sites of colony development (Bowen, 1979); that
older parts of roots may provide as much substrate as
younger parts has been shown by studying microbial growth
rates on different parts of washed, sterile roots planted to
non-sterile soil so that all parts were exposed to microorg-

anisms at the same time (Bowen and Rovira, 1973).

RHIZOSPHERE POPULATION DYNAMICS

As Figure 1 shows, the microbial population of the rhizosphere will be determined by factors such as rooting intensity, the germination of propagules in the soil and their
growth to the root, and growth on and along the root. One of
the major deficiencies of rhizosphere biology is the very
small numbers of studies *in soil* on any of these microbial
parameters. Studies on spore germination provide several
good examples of the importance of performing such studies
in soil: Daniels and Duff (1978) found no germination of
spores of three isolates of v.a.m. fungus on agar but 35-45%
germination in soil. Most studies of germination of basidiospores in laboratory media have shown very low rates of
germination, for example, often less than 1.0% (Fries, 1966),
but Theodorou and Bowen (1973) obtained approximately 10%
germination of basidiospores of an ectomycorrhizal fungus
added to planted soil. There appear to be no experiments yet
on germination of spores of ectomycorrhizal fungi on plant
roots in soil. Simple methods are available (or can be
devised) to study the above factors in soil (Bowen, 1979)
without resort to laboratory media or use of buried slides
(or cellophane) which bring their own artifacts (Wong and
Griffin, 1976).

The short generation times of organisms in rich, stirred
laboratory media have not been observed on roots. In what
is one of the only existing studies on growth rates of
bacteria on roots in soil, Bowen and Rovira (1976) found
generation times of 5.2 h and 39 h on *P. radiata* roots in
soil, for a *Pseudomonas* sp. and a *Bacillus* sp. respectively,
compared with 77 h and greater than 100 h in nearby soil.
Barber and Lynch (1977) recorded a "doubling time" of 24 h
for total bacterial counts with barley in solution culture.
Such generation times are by no means uncommon in natural
systems. Generation times should be used in rhizosphere
biology rather than the usual root/soil (R/S) ratios
because :

i)   this is a more dynamic parameter than the static
measurements usually made with rhizosphere studies,

ii)   this brings rhizosphere biology into line with other
fields of population biology and

iii) because the generation time concept is internally
consistent and a more valid measurement than comparing growth
on a surface (root) with that in a volume (soil).

The relatively rapid growth of pseudomonads on the root

in the studies by Bowen and Rovira (1976) could explain why
they are a major group of organisms in the rhizosphere.
However, there are very large growth differences between
closely related strains of pseudomonads. For example, the
relative increase in four strains of *Pseudomonas* sp. inoculated to *Eucalyptus sieberi* were 2, 62, 141 and 380 over a
period of four days. On *E. globulus* the increases were
approximately one sixth of those above (Bowen, 1979 and
unpublished). What are the bases of such large differences
in growth between closely related bacterial strains? Are
they due to different abilities to use common growth substrates or do they reflect the needs of some strains for particular growth factors not supplied by a plant species? Such
a reason seems more likely, at least in the case of the
ectomycorrhizal fungi which showed marked differences between
strains (of the one species) in the rhizosphere at 16°C,
although all grew similarly in laboratory media at this
temperature (Theodorou and Bowen, 1971). The very large
differences in growth in the rhizoplane between mycorrhizal
fungi (Bowen and Theodorou, 1973, 1979) and between rhizobia
sp. (Chatel and Greenwood, 1973) gives appreciable scope for
selection of strains which can colonize particular species
of plants rapidly.

    Contrary to earlier ideas that the rhizoplane is entirely
covered by microorganisms, only 4-10% of the surface is
usually occupied (Bowen and Theodorou, 1973, Rovira
Newman, Bowen and Campbell, 1974), except for such associations as ectomycorrhizas in which short lateral roots may be
completely sheathed by fungus. Ecological sampling methods
developed in other fields and used by Newman and H.J. Bowen
(1974) and by Malajczuk and G.D. Bowen (see Bowen and
Rovira, 1976) have shown a non-random distribution of
organisms on the root and a preferential colonization of
the junctions of cells. There are 50 times as many bacteria
in this position with *E. calophylla* roots as in other areas
(Bowen and Rovira, 1976). The cell junctions are the major
sities of exudation and the major routes of spread of initial
colonization but even so, colonization of them is not
complete.

    The incompetences of root cover by microorganisms is a
reflection of the often limited migration of organisms along
the root and the relatively wide spacing in soil of inoculum
for new root growth. A root will frequently grow faster
than existing colonies can spread (Bowen and Rovira, 1973)
and new root growth may be colonized largely from the soil.
How far do propagules move to the root? This is probably
related to size of propagule and the reserves they contain
(Bowen and Rovira, 1976).

Microbial propagules differ considerably in size from bacteria, 1-4 μm in size, to small fungal propagules of only a few microns diameter to larger spores, such as the spores of v.a. mycorrhizal fungi are commonly 200-400 μm diameter and higher, to propagules such as sclerotia, which are 1-2 mm and larger. Two main reproduction alternatives appear to be followed in fungi, namely the production of many small propagules or of fewer large propagules. This appears reminiscent of r and K strategies respectively in other fields of ecology (see Southwood, 1976). In short, r strategists are basically opportunistic, with rapid growth rates; as the habitats they colonize are often virtual ecological vacuums, high competitive ability may not be required and they are typically small in size; mortality rates may be high but they recover quickly from population reduction. K strategists, by contrast, have large size and, longevity, low mortality rates, high competitive ability and a large investment in each offspring and may recover slowly from reductions in population. (This analogy may be a rewarding one also in considering the biological control of diseases - biological control would be expected to be more feasible on K organisms or the K stage of organisms which can go from K to r, for example, zoosporangia of *Phytophthora* producing zoospores).

Bacterial migration to the root in a film of water is quite limited below field water capacity (Bowen and Rovira, 1976). Therefore initial colonization of a new root surface in the absence of free water will usually be by the chance touching of some of the many small (r type) propagules, namely, bacteria and small spored fungi with rapid germination and growth on simple substrates. Thus while one can predict the main types of organisms which will be the primary colonizers of the rhizoplane, the actual species composition will be largely a probability exercise. Colonization by organisms further from the root may take longer. K strategy organisms with fewer, larger propagules capable of growth over further distances to the root are particularly suited to grow longer distances and may be the major component of the "late" colonizers, especially with plant species with low rooting intensity. K strategy organisms in other fields of ecology often have a high competitive ability and this might be expected also with the K type microorganisms suggested here.

Initial colonization of roots is often associated with organic matter and as pseudomonads are often in the organic matter their frequent occurrence in the rhizosphere is not surprising. The importance of organic matter as a source of inoculum suggests a close link between organisms coloniz-

ing one season's roots and the next. The judicious selection
of previous crops therefore may be one avenue of increasing
the inoculum potential of symbionts (for example, rhizobia
with non-legumes, Diatloff, 1969) or of root disease suppre-
ssing organisms. Thus grass roots stimulate high populations
of *Phialophora radicola* var. *graminis* which is antagonistic
to *Gauemannomyces graminis* var. *tritici* (Slope, Sah, Broom
and Gutteridge, 1978). Also changes of the microbial
composition of soil are likely to be rapid with crop sequences with large root biomasses in the first few years, for
example, with grasses rather than with trees (Bowen, 1979).

Although there is some evidence (Dazzo, Napoli and Hubbell,
1976) that adsorption of rhizobia to roots may be related to
host specificity in *Rhizobium*-clover symbioses, it is not
necessary to invoke lectins (or other recognition phenomena
related to *infection*) in growth specificity around roots.

Many of us are used to thinking about monocultures of
plants, yet most natural systems and many agricultural
pasture systems are mixed. It is important, therefore,
that we recognise that biological interactions at the plant-
plant-microorganism level also occur. Thus Iqbal and
Qureshi (1971) reported that the crucifer *Brassica campestris*
reduced v.a.m. infections of field grown wheat and reduced
wheat yield by 25%. Christie, Newman and Campbell (1978)
found *Lolium perenne* to decrease v.a.m. infection of
Plantago lanceolata by almost 40% and a pasture mixture
increased bacterial and fungal growth on *Plantago* by over
200%. Such important effects warrant further study especially with interactions between plants and "economic" microorganisms. As Christie, Newman and Campbell (1978) pointed out,
some of the effects may have been indirect, arising from
changes in plant nutrient status, assimilation and root
exudation of a plant when grown with other plants. However
not all effects were explicable in this way and direct
effects might occur, for example, a substance from *Calluna
vulgaris* can inhibit ectomycorrhizal fungi of *Picea abies*
in heathland (Robinson, 1972).

Models for plant response

The epidemiological considerations above are of
particular importance in predicting the impact of known
populations of diseases, (that is, predicting the necessity
of employing expensive control measures) and in predicting
situations in which appreciable responses would be obtained
by inoculation with mycorrhizal fungi (possibly with some
added fertilizer) even when naturally occurring fungi are of
an efficient type. Empirical or mechanistic models could be
developed for these situations. I have made the simplifying

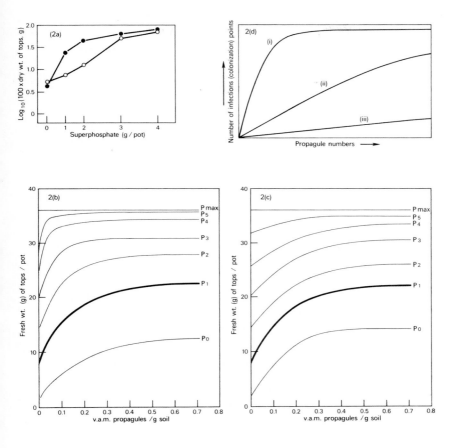

Fig. 2.   *(a): Response of subterranean clover to v.a.*
mycorrhiza infection at 5 soil phosphate levels; mycorrhizal
(●), non-mycorrhizal (○). (b) and (c): Possible response
curves to varying mycorrhizal infection with different phos-
phate levels; P_1 is an experimentally derived curve on
Medicago truncatula (Bowen and Smith, unpublished observations)
(d): hypothesized number of infection or colonization points
with varying populations of propagules which move (i) over
long distances to the root, (ii) moderate distances and (iii)
small distances.

assumption that root diseases and mycorrhiza are "seen" by
the plant as decreases or increases respectively in effective
"root" length for absorption of nutrients (assuming no
water deficits). I shall refer particularly to the mycorrhizal response.

An empirical model could be based on the type of response
to mycorrhizal infection as given in Fig. 2a, from Abbott
and Robson (1977) on *Trifolium subterraneum* with 50-100% of
the root mycorrhizal. As the limiting nutrient, phosphate,
increased, the infection became redundant. The question then
becomes one of the infection needed to obtain the full mycorrhizal response at any level of added phosphate. The
two possibilities are indicated in Fig. 2b and 2c, in which
the dark line was experimentally obtained on *Medicago
truncatula* at one level of phosphate. The infection-response
curve and the number of propagules of the v.a. mycorrhizal
fungus in soil were obtained by diluting soil with the same
soil which had been sterilized, and later counting the
numbers of infections (Smith and Bowen, 1979, and unpublished
observations). The upper line (P_{max}) is the level of phosphate at which no response would be obtained from inoculation.

A more general mechanistic model demands a model of
plant nutrition which generates rooting intensity. At
present such a model would be generated by a mixture of
mechanistic and empirical sub-models. The infection levels,
as a function of numbers of propagules and the distance they
move to roots, is suggested in Fig. 2d. The generation of
plant response from this infection then depends on the
extent of growth of mycelial into soil infection then
extending the rooting system; considering the poor state of
this subject, it would be necessary to use empirical values
(for example, from Sanders, Tinker, Black and Palmerley,
1977). The more widely spaced the root the more dependent
the plant is likely to be on mycorrhizal infection (Baylis,
1975).

MICROBIAL INTERACTIONS

Microbial interactions such as antagonism, competition
and synergism in the soil and the rhizoplane are probably
the most important, yet the least understood, phenomena in
rhizosphere biology. These can determine the success or
failure of introduced inoculants and the biological "control"
of diseases. As the rhizoplane is occupied incompletely by
microorganisms we have a number of spatially discrete small
communities rather than one large interacting community.
Despite this incomplete cover of the root, interactions
between organisms do occur because they tend to occupy the
same favoured sites, that is, the longtitudinal junctions of

cells, and a general microflora has been shown experimentally
to decrease the rate of spread of an ectomycorrhizal fungus
along the root by approximately half (Bowen and Theodorou,
1973). That is, microbial communities in soil and on the
root are best considered as being in many micro-islands and
they tend to interact only when their "territory" is crossed,
for example, lysis of germ tubes of fungi as they grow through
soil. The many different types of organisms which occur on
roots cannot all be considered to occupy different "physiological niches" as is sometimes thought; many may be occupy-
ing the same niche but at different parts of the root.

The types of interaction which can occur on the root are
indicated in Table 1, derived from inoculation studies on
plants in sterilized soils (Bowen and Theodorou, 1979):

i)   Very large differences occurred between the mycorrhizal fungi (as they do between strains of the one fungus
species) in colonization of the root. These differences may
be important ecologically, for example, the slow growth of
Pisolithus tinctorius (only one strain was studied) may be
the reason why it appears to be the dominant ectomycorrhizal
fungus only in extreme situations, such as mine wastes.

ii)   Large depressions of growth can be caused by antagonism, for example, *P. fluorescens* in Table 1. Although all
fungi were depressed, some, for example, *T. terrestris*, were
not as severely affected as others. However, antagonisms
in simple mixtures may not be expressed in more complex
situations for in other studies (Bowen and Theodorou, 1979)
a *Bacillus* sp. with no effect on colonization by *R. luteolus*
overcame the marked inhibition by *P. fluorescens* toward
this fungus even though it did not reduce the numbers of *P.
fluorescens*.

iii) Reductions in growth may be caused also by competition for substrates in short supply. Other organisms are
neutral and some, for example, *Bacillus* sp.2 may be stimu-
lative. TEM studies (for example, Foster and Rovira, 1978)
show communities consisting of microcolonies of different
organisms. Are these necessarily competitive? Could they
sometimes be mutualistic? Although competition is the
conceptual backbone of much ecological thought it is extremely difficult to study under natural conditions. However
there are experimental approaches to this which have developed in other fields. For example, replacement series and
diallel analyses (Williamson, 1972) and microbiologists could
use such approaches with profit.

The possibility of synergistic associations opens up
interesting avenues for further study, not only with regard
to the mechanisms of synergism but also with regard to applied aspects such as establishment of symbionts and their

TABLE 1

Rhizoplane Interactions between Mycorrhizal Fungi
and Bacteria (Bowen and Theodorou, 1979)

	Length of Root colonized by fungi (mm)				
		Bacteria present			
		Exp. 1			Exp. 2
	0	Ps. fluorescens	Bacillus sp. #1	0	Bacillus sp. #2
Rhizopogon luteolus	27.6	8.3	23.7	30.7	34.1
Suillus luteus	16.2	3.2	8.7	19.1	26.2
Thelephthora terrestris	22.4	12.9	18.1	12.1	11.2
Corticium bicolor	17.2	3.2	15.3	15.8	27.1
Pisolithus tinctorius	6.9	5.7	4.2	6.9	1.9

| | L.S.D. | $p < 0.05$ | 5.4 | | 6.2 |
| | | $p < 0.01$ | 7.1 | | 8.2 |

Experiments were conducted in sterile soil for 4 weeks at $20^{\circ}C$ day
(12 hr), $16^{\circ}C$ night.

effects on plant growth. Azcon, Barea and Hayman (1976)
reported an instance in which lavender plants inoculated
with v.a.m. fungi and phosphate-dissolving bacteria absorbed
more phosphate from soil fertilized with rock phosphate than
those inoculated with v.a.m. fungi or bacteria alone.

Soil microfauna

The possible role of soil microfauna is an important
omission from most studies on microorganism-plant interactions. However the impact of soil microfauna may be large:
nematodes may enhance root disease severity (Meagher, Brown
and Rovira, 1978), pests may attack legume nodules - a
situation in which remaining nodules may sometimes compensate
by being more active - and v.a.m. fungi and nematodes can be
mutually suppressive (see Bowen, 1978).

Protozoans may play a particularly important part in
controlling microbial activities in the rhizosphere, and in
adjacent soil. Coleman, Anderson, Cole, Elliott, Woods
and Campion (1978) have adopted an experimental approach
to studying changes of bacteria, carbon and phosphorus
in soil microcosms inoculated with bacteria, nematodes and
amoebae and a related more rapid recycling of phosphorus in
bacterial biomass. However amoebae do not decrease bacteria
populations in soil to very low numbers and this is partly
because of the ability of the bacteria to reproduce and
replace cells consumed by predation (Habte and Alexander,
1978). The roles of protozoans in the rhizosphere await more

study and the prey-predator relations may be quite different
in the two dimensional matrix of the root surface from the
three dimensional framework of soil. There are, indeed,
suggestions from solution culture studies that amoebae
may directly affect substrate availability from roots. Darbyshire and Greaves (1970) found addition of an amoeba led to
a large increase in the population of a *Pseudomonas* sp. on
the roots of the grass *Phleum pratense* in solution culture
although the reasons for this were not studied.

CONCLUSION

There is a need, not merely to understand systems and
their controlling factors ("descriptive" science) but also
to manage them if possible ("manipulative" science).
Microbial ecologists are now beginning to come to grips with
the task of understanding microbial growth around roots, but
this is just a beginning and there is much more to be done
on each of the aspects I have referred to above. Perhaps the
major advance of the last 5 to 10 years has not been the
application of new skills, important though this has been,
but the questioning concepts of microbial growth derived
from solution cultures which have been learnt by scientists
during their traditional training. Instead, they are now
examining what really happens in soil and on plant roots,
that is, with 2-3 dimensional discontinuous systems, growth
on surfaces, and situations in which diffusion of substrates
into (and out of) the system may control growth.

There seems little point in trying to control the entire
microbial composition of the rhizosphere and, given the largely stochastic nature of infection of new root growth, such
an aim would be almost unattainable. However, there is
great point in controlling the population of particular
microorganisms and especially to increase the populations
of beneficial symbionts. We have two general variables
which could be manipulated, namely, plants and microorganisms,
and in selection programmes we should take advantage of the
variation in many properties of *each* of these. More avenues
of management also arise because of the impact of non-host
plants on colonization and infection of the root by symbionts.

High soil populations, high growth rates, resistance to
antibiosis and high competitive ability appear to be key
characters in successful establishment of particular microorganisms on the root. Organisms which locally increase
permeability of the root cell may have a further competitive
advantage. Establishment would be further enhanced by a
selective or specific food source for the organism and this
could be obtained by breeding (or engineering) both plants
and symbionts for production and use respectively of unusual

substrates. In fact the ultimate in "competition" would be
highly specific host-microorganism combinations, for example,
shortfalls in legume nitrogen fixation due to competition
to inoculated rhizobia from poorly effective, naturally
occurring rhizobia could be eliminated by breeding legumes
so specific in their *Rhizobium* requirements that they nodulate
only with the inoculated strain.

Although very significant advances have been made in microbial ecology of the rhizosphere in recent years, we still
have much to learn. The new dimensions this field offers
in plant physiology, physical chemistry of soil, growth at
surfaces and population dynamics are challenging, exciting
and highly relevant to plant-microbe interactions and plant
productivity.

REFERENCES

Abbott, L.K., and Robson, A.D. (1977). Growth stimulation
    of subterranean clover with vesicular arbuscular mycorr-
    hizas, *Australian Journal of Agricultural Research* 28,
    639-649.
Azcon, R., Barea, J.M. and Hayman, D.S. (1976). Utilization
    of rock phosphate in alkaline soils by plants inoculated
    with mycorrhizal fungi and phosphate-solubilizing bact-
    eria, *Soil Biology and Biochemistry* 8, 135-138.
Barber, D.A. and Lynch, J.M. (1977). Microbial growth in
    the rhizosphere, *Soil Biology and Biochemistry* 9, 305-308.
Barber, D.A. and Martin, J.K. (1976). The release of organic
    substances by cereal roots in soil, *New Phytologist* 76,
    69-80.
Barrow, N.J. (1977). Phosphorus uptake and utilization by
    tree seedlings, *Australian Journal of Botany* 25, 571-584.
Bayliss, G.T.S. (1975). The Magnolid mycorrhiza and myco-
    trophy in root systems derived from it. *In* "Endomycorr-
    hizas" (Ed. F.E.T. Sanders, B. Mosse and P.B.H. Tinker)
    pp.
Bowen, G.D. (1969). Nutrient status effects on loss of
    amides and amino acids from pine roots, *Plant and Soil*
    30, 139-142.
Bowen, G.D. (1978). Dysfunction and shortfalls in symbiotic
    responses. *In* "Plant Disease" (Ed. J.G. Horsfall and
    E.B. Cowling), Vol. 3, pp. 231-256. Academic Press,
    New York.
Bowen, G.D. and Foster, R.C. (1979). Dynamics of microbial
    colonization of plant roots. *In* "Symposium on Soil
    Microbiology and Plant Nutrition" (Ed. W.J. Broughton
    and C.K. John) pp. 14-31. University of Malaya, Kuala
    Lumpur.

Bowen, G.D. and Rovira, A.D. (1973). Are modelling approaches useful in rhizosphere microbiology? *Bulletin of Ecological Research Communications*, Stockholm 17, 443-450.

Bowen, G.D. and Rovira, A.D. (1976). Microbial colonization of plant roots, *Annual Review of Phytopathology* 14, 121-144.

Bowen, G.D. and Theodorou, C. (1973). Growth of ectomycorrhizal fungi around seeds and roots. *In* "Ectomycorrhiza : Their Ecology and Physiology" (Ed. G.C. Marks and T.T. Kozlowski), pp. 107-150. Academic Press, New York.

Bowen, G.D. and Theodorou, C. (1979). Interactions between bacteria and ectomycorrhizal fungi, *Soil Biology and Biochemistry* 11, 119-126.

Broadbent, P., Baker, K.F. and Waterworth, Y. (1971). Bacteria and actinomycetes antagonistic to funal root pathogens in Australian soils, *Australian Journal of Biological Science* 24, 925-944.

Campbell, R. and Rovira, A.D. (1973). A study of the rhizosphere by scanning electron microscopy, *Soil Biology and Biochemistry* 5, 747-752.

Chatel, D.L. and Greenwood, R.M. (1973). Differences between strains of Rhizobium trifolii in ability to colonize soil and plant roots in the absence of their specific host plants, *Soil Biology and Biochemistry* 5, 809-813.

Christie, P., Newman, E.I. and Campbell, R. (1978). The influence of neighbouring grassland plants on each others endomycorrhizas and root-surface microorganisms, *Soil Biology and Biochemistry* 10, 521-527.

Coleman, D.C., Anderson, R.V., Cole, C.V., Elliott, E.T., Woods, L. and Campion, M.K. (1978). Trophic interactions in soils as they affect energy and nutrient dynamics. IV. Flows of metabolic and biomass carbon, *Microbial Ecology* 4, 373-380.

Crush, J.R. (1976). Endomycorrhizas and legume growth in some soils of The Mackenzie Basin, Canterbury, New Zealand, *New Zealand Journal of Agricultural Research* 19, 473-476.

Daniels, B.A. and Duff, D.M. (1978). Variation in germination and spore morphology among four isolates of Glomus mosseae, *Mycologia* 70, 1261-1267.

Darbyshire, J.F. and Greaves, M.P. (1970). An improved method for the study of the interrelationships of soil microorganisms and plant roots, *Soil Biology and Biochemistry* 2, 63-71.

Dazzo, F.B., Napoli, C.A. and Hubbell, D.H. (1976). Adsorption of bacteria to roots as related to host specificity in the Rhizobium - clover symbiosis, *Applied and Environmental Microbiology* 32, 166-171.

Diatloff, A. (1969). The introduction of Rhizobium japonicum
    to soil by seed inoculation of non-host legumes and
    cereals, *Australian Journal of Experimental Agricultural
    and Animal Husbandry* 9, 357-360.
Elliott, L.F., Gilmour, C.M., Cochran, V.L., Coby, C. and
    Bennett, D. (1979). Influence of tillage and residues on
    wheat root microflora and root colonization by nitrogen-
    fixing bacteria. *In* "The Soil-Root Interface". (Ed.
    J.L. Harley and R. Scott-Russell), pp. 242-258. Academic
    Press, London.
Foster, R.C. and Rovira, A.D. (1978). The ultrastructure of
    the rhizosphere of Trifolium subterraneum L. *In* "Micro-
    bial Ecology" (Ed. M.W. Loutit and J.A.R. Miles), pp.
    278-290. Springer-Verlag, Berlin.
Fries, N. (1966). Chemical factors in the germination of
    spores of Basidiomycetes. *In* "The Fungus Spore" (Ed.
    M.F. Madelin), pp. 189-199. Butterworth, London.
Gibson, A.H. (1966). The carbohydrate requirements for
    symbiotic nitrogen fixation: A "whole plant" growth anal-
    ysis approach, *Australian Journal of Biological Science*
    19, 499-515.
Griffin, G.J., Hale, M.G. and Shay, F.J. (1976). Nature
    and quantity of sloughed organic matter produced by roots
    of axenic peanut plants, *Soil Biology and Biochemistry*
    8, 29-32.
Habte, M. and Alexander, M. (1978). Mechanisms of persist-
    ence of low numbers of bacteria preyed upon by protozoa,
    Soil Biology and Biochemistry 10, 1-6.
Harley, J.L. (1975). Problems of mycotrophy. *In* "Endomy-
    corrhizas" (Ed. F.E. Sanders, B. Mosse and P.B. Tinker),
    pp. 1-24. Academic Press, London.
Highinbotham, N. (1968). Cell electropotential and ion
    transport in higher plants. *In* "Transport and Distribut-
    ion of Matter in Cells of Higher Plants" (Ed. K. Mothes,
    I. Muller, A. Nelles and D. Neumann), pp. 167-177.
    Akademie-Verlag, Berlin.
Iqbal, S.H. and Qureshi, K.S. (1976). The influence of
    mixed sowing (cereals and crucifers) and crop rotation on
    the development of mycorrhiza and subsequent growth of
    crops under field ocnditions, *Biologia* 22, 287-298.
Krupa, S. and Fries, N. (1971). Studies on the ectomycorr-
    hizae of pine 1. Production of volatile organic compounds,
    Canadian Journal of Botany 49, 1425-1431.
Martin, J.K. (1977a). Factors influencing the loss of
    organic and carbon from wheat roots, *Soil Biology and
    Biochemistry* 9, 1-9.

Martin, J.K. (1978). The variation with plant age of root carbon available to soil microflora. *In* "Microbial Ecology" (Ed. M.W. Loutit and J.A.R. Miles), pp. 299-302. Springer-Verlag, Berlin.

Marx, D.H. (1969). The influence of ectotrophic mycorrhizal fungi on the resistance of pine roots to pathogenic infections. II. Production, identification, and biological activity of antibiotics produced by Leucopaxillus cereals, var. piceina, *Phytopathology* 59, 411-417.

Meagher, J.W., Brown, R.H. and Rovira, A.D. (1978). The effects of cereal cyst nematode (Heterodera avenae) and Rhizoctonia solani on the growth and yield of wheat. *Australian Journal of Agricultural Research* 29, 1129-1137.

Mosse, B. (1975). Specificity in VA mycorrhizas. *In* "Endomycorrhizas" (Ed. F.E.T. Sanders, B. Mosse and P.B.H. Tinker), pp. 469-484, Academic Press, London.

Newman, E.I. (1978). Root microorganisms : their significance in the ecosystem, *Biological Review* 53, 511-554.

Newman, E.I. and Bowen, H.J. (1974). Patterns of distribution of bacteria on root surfaces, *Soil Biology and Biochemistry* 6, 205-209.

Newman, E.I. and Watson, A. (1977). Microbial abundance in the rhizosphere. A computer model, *Plant and Soil* 48, 17-56.

Ratnayake, M., Leonard, R.T. and Menge, J.A. (1978). Root exudation in relation to the supply of phosphorus and its possible relevance to mycorrhizal formation, *New Phytologist* 81, 543-552.

Reid, C.P.P. and Mexal, J.G. (1977). Water stress effects on root exudation by lodgepole pine, *Soil Biology and Biochemistry* 9, 417-421.

Robinson, R.K. (1972). The production by roots of Calluna vulgaris of a factor inhibitory to growth of some mycorrhizal fungi, *Journal of Ecology* 60, 219-224.

Rovira, A.D., Bowen, G.D. and Foster, R.C. (1980) The nature of the rhizosphere and the influence of the rhizosphere microflora and mycorrhizas on plant nutrition. *In* "Encyclopedia of Plant Physiology, New Series" (Ed. A. Lauchi and R.L. Bieleski), Vol. 12, In Press. Springer-Verlag, Berlin.

Rovira, A.D., Foster, R.C. and Martin, J.K. (1979). Note on terminology: origin, nature and nomenclature of the organic materials in the rhizosphere. *In* "The Soil-Root Interface" (Ed. J.C. Harley and R.S. Russell), pp. 1-4. Academic Press, London.

Rovira, A.D., Newman, E.I., Bowen, H.J. and Campbell, R. (1974). Quantitative assessment of the rhizoplane microflora by direct microscopy, *Soil Biology and Biochemistry*

6, 211-216.

Sanders, F.E., Tinker, P.B., Black, R.L.B. and Palmerley, S.M. (1977). The development of endomycorrhizal roots systems 1. Spread of infection and growth-promoting effects with four species of vesicular-arbuscular endophyte, *New Phytologist* 78, 257-268.

Sivasithamparam, K. and Parker, C.A. (1979). Rhizosphere microorganisms of seminal and nodal roots of wheat grown in pots, *Soil Biology and Biochemistry* 11, 155-160.

Slankis, V. (1973). Hormonal relationships in mycorrhizal development. *In* "Ectomycorrhizae" (Ed. G.C. Marks and T.T. Kozlowski), pp. 231-298. Academic Press, London.

Slope, D.B., Sah, G.A., Broom, E.W. and Gutteridge, R.J. (1978). Occurrence of *Phialophora radicicola* var. *graminicola* and *Gaeumannomyces graminis* var. *tritici* on roots of wheat in field crops, *Annals of Applied Biology* 88, 239-246.

Southwood, T.R.E. (1976). Bionic strategies and population parameters. *In* "Theoretical Ecology, Principles and Applications" (Ed. R.M. May), pp. 26-48. Blackwell, Oxford.

Smith, S.E. and Bowen, G.D. (1979). Soil temperature, mycorrhizal infection and nodulation of Medicago truncatula and Trifolium subterraneum, *Soil Biology and Biochemistry*, In Press.

Sutton, J.C. and Sheppard, B.R. (1976). Aggregation of sand dune soil by endomycorrhizal fungi, *Canadian Journal of Botany* 54, 326-333.

Theodorou, C. and Bowen, G.D. (1971). Influence of temperature on the mycorrhizal associations of Pinus radiata, *Australian Journal of Botany* 19, 13-20.

Theodorou, C. and Bowen, G.D. (1973). Inoculation of seeds and soil with basidiospores of mycorrhizal fungi, *Soil Biology and Biochemistry* 5, 765-771.

Williamson, M. (1972). The analysis of biological populations. Arnold, London.

Wong, P.T.W. and Griffin, D.M. (1976). Bacterial movement at high matric potentials. II. In fungal colonies, *Soil Biology and Biochemistry* 8, 219-223.

SCANNING ELECTRON MICROSCOPY IN THE STUDY OF MICROBIAL DIGESTION OF PLANT FRAGMENTS IN THE GUT

T. Bauchop

*Applied Biochemistry Division
Department of Scientific and Industrial Research
Palmerston North, New Zealand*

INTRODUCTION

In studying microbes associated with animals we are concerned with the nature of the microbe-host interaction and its magnitude. The task of the ecologist in approaching a microbial habitat in an animal, is to discover by *direct observation* the major microbes present and, in the first instance at least, to determine the major physicochemical factors in the environment, as they govern the size and composition of the microbial population. The growth of micro-organisms, frequently taken up as a primary task, can only be developed rationally when the fundamental nature of the ecosystem has been appreciated. This has been well exemplified in many studies on the microbiota of the alimentary tract. Despite the advantages of a relatively constant environment, microbiologists have made only limited progress in understanding this ecosystem. As Savage (1978) pointed out at the First International Microbial Ecology Symposium, "for most animal types including humans, the gastrointestinal microbiology is not understood". The rumen of domestic ruminants was without doubt the major exception to the above statement. Yet even there, fundamental aspects of rumen microbiology have recently been shown to have been overlooked (Bauchop, 1979c).

In recent years it has come to be appreciated that decades of study on human faecal flora were ill-founded, bascially through lack of appreciation of the physical conditions prevailing in the hindgut. Important chemical factors in the ecosystem, such as branched-chain volatile fatty acids (VFA), were also overlooked and were not supplied in culture media used. The value of habitat-simulating media, including use

of critical anaerobic techniques, can now clearly be seen in
improved understanding of the bacteriology. Better understanding of the chemistry of the hindgut environment should
result in additional advances in this knowledge.

Study of the rumen ecosystem possessed many advantages
compared with the human hindgut. Use of surgically-modified
animals permitted ready access to rumen contents at different
stages of digestion. Furthermore, the nature of the major
substrates for the microbes could be readily determined by
chemical analysis of feedstuffs. The overall character of
the fermentation was firmly established by the demonstration
of the composition of the VFA end products (Elsden, Hitchcock,
Marshall and Phillipson, 1946). In addition, the value of
habitat-simulating media, including the use of critical
anaerobic techniques, for culture of rumen microbes was
appreciated early (Hungate, 1950). The overall result was
that the rumen microbiota, comprising protozoa and bacteria
became the most widely researched and best understood of the
complex microbial populations of the gut. Yet even with this
well researched ecosystem, it has been suggested recently
(Bauchop, 1979a) that the accepted concepts of microbiol
digestion in the rumen require reappraisal. This view
resulted from the demonstration that the rumen anaerobic fungi
colonized in large populations on plant fragments in cattle
and sheep (Bauchop, 1979 a,c,d). The extent of this rumen
flora was indicated by the following finding (Bauchop, 1979c):
"With every piece of these plant fragments suspended in the
rumen, extensive fungal colonization was obtained". Much
has been achieved in rumen microbiology and in understanding
of ruminant digestion but for many years direct microscopical
approaches have been neglected, particularly in relation to
the rumen digesta plant solids.

DIRECT MICROSCOPIC OBSERVATIONS

In discussing the role of the microflora of the alimentary
tract, Baker and Harriss (1947) stressed "the necessity for
the control of cultural investigations by direct microscopic
examination of materials taken from the living animal".
Partly in reaction to claims made for the over-riding importance
of the purely microscopical approach, but mainly due to
successes in culturing key rumen bacteria, direct microscopical
approaches were largely ignored for many years. However,
direct observations remain the cornerstone of any ecological
study, even in the ecology of microorganisms. Despite the
problems resulting from limited morphological characteristics,
every effort must be made to extract the fullest possible
information by means of direct observations.

Any or all of the tools of microscopic observation can be exploited, including the full range of techniques in light microscopy, transmission and scanning electron microscopy (TEM and SEM).

The use of SEM in the discovery of bacterial colonization of the rumen epithelium (Bauchop, Clarke and Newhook, 1975) led to consideration of its application to the study of fibre digestion in the rumen. Although the role of the anaerobic cellulolytic bacteria as the main agents of fibre digestion was widely accepted, SEM examination of bacteria on digesta plant fragments seemed worthwhile. As a result of early findings on protozoa (Bauchop and Clarke, 1976) this work developed into a much broader study of microbial colonization of plant fragments in the gut. Although this work is still at an early stage of development, several interesting findings have already emerged and the following examples are presented to illustrate the application of SEM in the study of microbial digestion in the gut.

SEM studies of gut microbes

The surfaces of the gut epithelia and of digesta solids comprise the two types of surfaces available for microbial colonization. SEM has found ready and valuable application in studies on adherence phenomena involving microorganisms on gut surfaces in a wide range of animals but digesta solids have not been widely studied. In the rumen, extensive colonization of plant fragments by protozoa was demonstrated (Bauchop and Clarke, 1976) as well as protozoal involvement in physical degradation of plant tissues (Bauchop, 1979b). SEM also revealed the large populations of anaerobic fungi colonizing plant fragments in the rumen, stomach and hindgut of a wide range of herbivorous animals (Bauchop, 1979 c,d). In other *in vivo* studies SEM has been used to examine the degree of microbial digestion of starch grains in the rumen (Davis and Harbers, 1974), and the degree of digestion of the roots of pasture plants by a soil-inhabiting insect larva (Bauchop and Clarke, 1977). Other SEM studies on digestion of plant materials have involved mainly *in vitro* incubations of plant materials with strained rumen contents (Akin, 1976; Akin and Amos, 1979). This work revealed details of microbial attack under these conditions, although long incubations used (up to 4 days) mean that conditions may be far removed from those prevailing in the rumen. SEM has been used also to study adhesion of rumen bacteria to plant fragments in anaerobic cultures (Latham, Brooker, Pettipher and Harris, 1978; Minato and Suto, 1979).

Rumen protozoa

Most of the rumen protozoa are ciliates, holotrichs and
entodiniomorphs, although a few species of flagellates are
frequently present also. Rumen protozoa have been extensively
studied in rumen liquor by light microscopy and are generally
envisaged as swimming free in the liquid fraction of rumen
contents. Although it is well known that some of the larger
protozoa ingest plant fragments, it had been believed that
only a small fraction of the fragments in the rumen were of a
size small enough to be ingested by even the larger ciliates
(Baker and Harriss, 1947). More recent studies with SEM and
light microscopy have established that large numbers of some
of the important members of the rumen fauna can colonize large
plant fragments in the rumen (Bauchop and Clarke, 1979b). The
that the entodiniomorph, *Epidinium*, can ingest plant tissues
directly from large plant fragments (Bauchop, 1979b). The
holotrich, *Dasytricha ruminantium*, also has been found to
colonize plant fragments in large numbers, but in this case
plant tissues were not degraded (T. Bauchop, in preparation)
and presumably only soluble compounds are utilized.

Epidinium   In experiments where pieces of lucerne stem
(Medicago sativa L.) in wide-mesh nylon bags were suspended
in the rumen of sheep, *Epidinium* was found to colonize damaged
tissues in high numbers (Bauchop and Clarke, 1976). The
highest numbers were found concentrated on exposed cortex
tissues between the epidermis and the vascular cylinder (Fig.
1a) and epidinia were found attached to stem samples suspended
for periods from 1 to 40 h in the rumen. By means of light
microscopy large numbers of epidinia were found on stem and
leaf fragments in rumen digesta taken from animals given
different fresh diets. Also in sheep fed dried feeds, such
as chaffed lucerne, high numbers of protozoa were found
associated with plant fragments, particularly with leaves.
Further details of the colonization were revealed from SEM
studies of lucerne stem suspended in the rumen for shorter
periods of time. Even by 15 min large populations were
attached to damaged regions of stem (Fig. 1b) forming a
complete ring around the cut end of the stem between the
epidermis and the vascular cylinder. There was considerable
degradation of phloem and cortex tissues resulting in exposure
of the vascular cylinder although the epidermis did not appear
to be degraded. Degradation of tissues at the cut ends
continued in this pattern until by 2 h considerable pieces of
vascular cylinder were exposed due to degradation of the
adjacent thin-walled tissues. Examination at higher magni-
fications of the attached protozoa revealed that they were
ingesting plant tissues directly on the large stem pieces (Fig.2)

Fig. 1. Epidinium on the cut end of a piece of lucerne stem suspended in a sheep rumen. (a) Epidinium densely packed between the epidermis and the vascular cylinder after 30 min exposure. (b) Epidinium forming a complete ring around the cut end between the epidermis and the vascular cylinder after 15 min exposure.

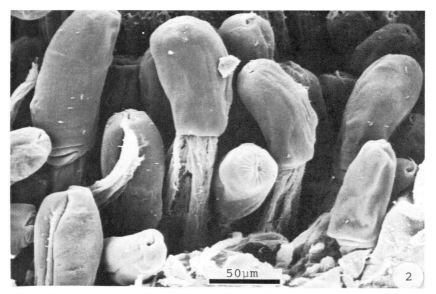

Fig. 2. Epidinium ingesting phloem elements on lucerne stem suspended in a sheep rumen for 15 min.

This extremely rapid colonization exhibited high specificity in two forms. Firstly, only a single protozoan genus was attached to the tissues, despite the fact that a complex protozoal fauna was present in the rumen. Secondly, the colonization occurred mainly on the cortex and phloem tissues indicating the existence of a powerful chemotaxis. The firm attachment of these protozoa to the stem pieces was also remarkable in view of the many washings involved in processing tissues for SEM, but could be explained by the deep engulfment of ingested plant tissues by the protozoa. Thus these findings revealed that epidinia participates in the physical degradation of forage, thereby exposing additional internal plant tissues for further microbial colonization and digestion.

In *in vitro* experiments with rumen contents from cattle, results were obtained indicating that *Epidinium* also degraded intact leaf blades of grasses (Amos and Akin, 1978). From results of TEM studies it was concluded recently (Akin and Amos, 1979) that *Epidinium* partially hydrolysed cell walls of grass tissues by extracellular enzymes. It was believed that this resulted in the separation of tissues into individual cells that were then ingested by the protozoa. Such a complex process would appear unlikely in view of the large quantities of plant tissues shown to be ingested directly by epidinia. The ingestion of such large quantities of cellulosic plant

material raises the question of the digestive capabilities of
Epidinium. Views on cellulolytic activity by rumen protozoa
are still in conflict and although cellulolytic bacteria may
function within protozoa, further study is needed on intrinsic
digestion of cellulose by rumen protozoa.

Dasytricha   The composition of the rumen fauna differs with
host species and diet as well as with geographical region.
In addition, individual animal differences also occur, and
within a single flock of sheep one or two animals may be found
to be lacking *Epidinium* even although it is present in other
members of the flock (Clarke, 1977). The extensive colonization of plant fragments by *Epidinium* raised the question of
which member of the rumen microbiota might take its place in
these animals. A sheep was found that lacked *Epidinium* and
the previous experiments were repeated. In this case a large
population of the holotrich protozoon, *Dasytricha ruminantium*,
colonized the cut ends of lucerne stem by 15 min (T. Bauchop,
in preparation). The main colonization region differed in this
case and the central area of pith was specifically selected
(Fig. 3a). Although in the previous experiments epidinia were
found on pith also (Bauchop, 1979b), colonization at that site
was always much less than on the cortex and phloem regions.
The *Dasytricha* colonization differed also in that plant tissues
were not physically degraded as found with *Epidinium* although
many *Dasytricha* penetrated deeply within damaged parenchymatous
cells (Fig. 3b). Attachment to plant particles by the rumen
large holotrich, *Isotricha*, has been demonstrated *in vitro* also
(Orpin and Letcher, 1978). The rumen holotrichs are known to
utilize soluble sugars and chemotactic responses by *Isotricha*
to sucrose, glucose and fructose have been demonstrated
(Orpin and Letcher, 1978). The dense and highly specific
colonization by *Dasytricha* may be explained by a similar
chemotactic response, and information on the carbohydrate
composition of lucerne stem pith would be of great interest.
In experiments with stem pieces from chaffed lucerne, examples
have been found of *Dasytricha* colonizing in rows on the
epidermal surface (T. Bauchop, unpublished observations).
Stem from dried feeds contains many epidermal lesions where
bacteria also colonize, and the colonization by *Dasytricha* in
this case may be the sites of such lesions.
     These observations make clear that considerable numbers of
protozoa may populate large fragments of plant materials in the
rumen. Their sequestration in this way may help to explain many
anomalies concerning the protozoa, including their survival
despite generation times lower than the turnover time of rumen
liquor. (Bauchop and Clarke, 1976).

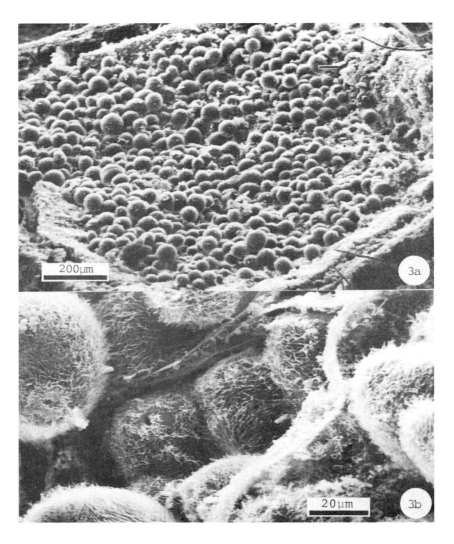

Fig. 3. Dasytricha on the cut end of a piece of lucerne stem suspended in a sheep rumen. (a) Heavy colonization of central pith region after 2 h. (b) Deep penetration into damaged cells of parenchyma by Dasytricha after 15 min exposure.

Gut anaerobic fungi

The discovery of large populations of anaerobic fungi colonizing plant fragments in the gut of ruminants and other herbivorous animals (Bauchop, 1979 c,d) resulted from a study using SEM. When the nature of the fungal colonization sites was appreciated, fungi were readily demonstrated by light microscopy.

Many fungal spores present on fresh and dried herbage pass into the rumen and aerobic fungi have been isolated from rumen contents. These fungi are considered to be merely transient, however, because of their continual entry into the rumen and their ability to grow anaerobically.

During a study of flagellate protozoa in the rumen of sheep, Orpin, (1975) isolated a microbe that appeared similar to certain types of aquatic phycomycetes. Rumen flagellates had not been studied extensively due mainly to the fact that their small mass and low numbers indicated that they were relatively unimportant members of the rumen microbiota. Study of this new "flagellate" revealed that it possessed a life-cycle involving a motile flagellated form and a "non-motile vegetative reproductive form". However its classification as a fungus was approached with caution mainly due to its obligately anaerobic nature, and the multi-flagellated form of the motile stage, two characters not readily reconcilable with fungal classification. Two additional "flagellates" were isolated later by Orpin and the demonstration of chitin in the cell walls of these three organisms, together with the knowledge of their morphology, led to the conclusion that they were true fungi (Orpin, 1977). However, only low numbers of zoospores ("flagellates") or sporangia had been found in the rumen and it was concluded (Orpin, 1977) that "the overall effect on rumen metabolism is probably not great, due to the relatively low population density (up to 3.6×10^4 ml^{-1}) of the chitin synthesizing vegetative stages of the three organisms". Thus the importance of Orpin's discoveries was not immediately appreciated and his findings were not taken up by other workers.

In his studies Orpin had followed established procedures used in rumen microbiology, and worked mainly with strained rumen fluid, discarding the rumen solids fraction. The finding that the commonly discarded rumen digesta plant fragments were heavily colonized by large populations of the anaerobic fungi indicated that the fungi formed a significant part of the rumen microbiota (Bauchop, 1979c).

Initial observations During an investigation by SEM of microbial digestion of plant materials suspended in a nylon bag in the sheep rumen, large numbers of spherical or ovoid

bodies (6-10 μm) were found attached to exposed vascular
cylinders of partially degraded lucerne stem (Fig.4). Their
identity was revealed by the occasional finding on the plant
pieces of similar but flagellated forms that corresponded with
the descriptions of some of the zoospores of the rumen anaerobic
phycomcetous fungi given by Orpin (1975, 1977). The attached
non-flagellated form was then identified as an early developmental stage in the life-cycle of these fungi. Apparently the
flagellae were lost soon after attachment of the zoospores to
plant tissues. The zoospores attached rapidly, within 15 min
of the lucerne being suspended in the rumen, and within 2 h
large numbers had colonized vascular tissues.

In similar experiments with sisal twine (vascular bundles
surrounded by sclerenchymatous sheaths from leaves of *Agave
sisalona*) large numbers of sporangia were found on the surface
of this hard fibre after 24-48 h in the rumen (Fig.5).

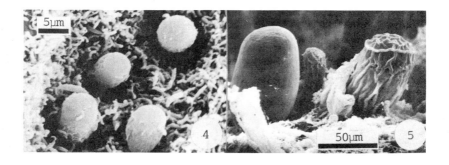

*Fig. 4. Early stage in the development of sporangia from
zoospores of rumen phycomcetes attached to vascular
cylinder of lucerne stem suspended in a sheep rumen for 2 h.
Fig. 5. Sporangia of rumen phycomycetes on cut end of sisal
twine (vascular tissue) suspended in a sheep rumen for 48 h.*

The sporangia were similar to those described in cultures
of the rumen phycomycetes. The discovery of large numbers
of two stages (zoospores and sporangia) in the life cycle
of the rumen phycomycetes on fibrous tissues from different plant

materials, together with the large increase in size of sporangia
on sisal during rumen incubation, gave rise to the idea that
significant amounts of fungal hyphae might underlie them, and
that this might be important in rumen digestion. These
findings also suggested a central role for fibrous plant
materials in the life cycle of these fungi in the rumen, and
prompted a search for fungi on natural digesta in the rumen
of cattle and sheep.

Rumen digesta plant fragments In all large animals consuming
appreciable amounts of plant material, large numbers of fungi
with sporangia at various stages of development were readily
found (Figs. 6 and 7). Highest populations were found with
stalky fibrous diets, for example, meadow hay in cattle and
chaffed lucerne in sheep. High populations of fungi were
present in the rumen of animals grazing stalky pastures
also: in cattle feeding on pasture of predominantly perennial ryegrass (*Lolium perenne* L.), and in sheep grazing
pure perennial ryegrass. Lower populations of fungi were
present in the rumen of sheep grazing stalky, pure lotus
(*Lotus pedunculatus* L.). Anaerobic fungi were not detected
in rumen contents of sheep on diets relatively low in fibre,
for example, barley grain (85%) ration, continuously grazed
pasture that did not have the opportunity to develop stalks
or seedheads, or pure stands of young lucerne, red clover
(*Trifolium pratense* L.) or white clover (*Trifolium ripens* L.)
    The presence of fungi at different stages of development
on single pieces of plant materials demonstrated that fungal
infection could occur at widely differing times. By the
time fungi were fully developed, after approximately 24 h in
the rumen, thin-walled tissues of the stem generally were
digested and the fungi were attached to stem vascular
cylinder. Examples were found also where extensive portions
of fungal rhizoids were exposed following digestion of thin-
walled plant tissues (Fig. 8).
    The initial discoveries of these large populations of
fungi on digesta plant fragments were made by means of SEM.
It was then discovered that fungi could be readily detected
by light microscopy with digesta plant fragments stained with
trypan blue (Fig. 11a,b) (Bauchop, 1979a). With relatively
thin plant fragments such as pieces of grass stems, fungi
could be observed directly on unstained material also by
light microscopy. These findings permitted work on fungal
colonization to proceed more rapidly.
    Fungi were found principally colonizing stems, perhaps
because leaves are degraded too rapidly during digestion.
With more slowly digested leaves, such as those from wheat
straw, extensive fungal colonization also occurred. In

Fig. 6. Rumen phycomycetes attached to vascular cylinder of lucerne stem from rumen digesta of a sheep. Samples were obtained before feeding.

Fig. 7. Sporangia of rumen phycomycetes attached to the inside surface of the hollow stem of a grass from rumen digesta of a cow. Sample obtained before feeding.

legumes, colonization was mainly at regions of epidermal damage. With grasses fungi were found chiefly on the inside surface of the hollow stems. Leaves were colonized principally in areas close to vascular bundles. Stomatal invasion, found in some samples, appeared of minor significance.

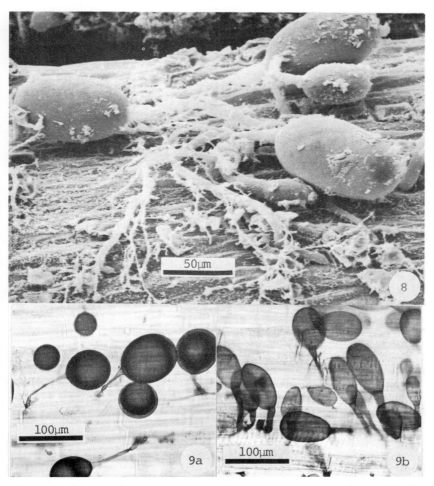

Fig. 8. Rumen phycomycetes on lucerne stem vascular cylinder in rumen digesta from a sheep. Part of rhizoids remain after thin-walled stem tissues have been digested.
Fig. 9. Light micrograph of sporangia of rumen phycomycetes attached to plant fragments in rumen digesta. (a) Grass stem from a cow. (b) Lucerne stem from a sheep.

Plant materials suspended in the rumen   Extensive fungal
colonization has been demonstrated with many different plant
materials suspended in nylon bags in the rumen of cattle and
sheep (Bauchop, 1979c).   In all of these experiments every
piece of plant material was extensively colonized, an
indication of the importance of the fungal flora in the rumen.
In studying details of the development of these fungi in the
rumen the value of wheat straw leaf as a substrate was
appreciated early.   The thinness of the leaf and the absence
of chlorophyll, that would mask staining, permitted details
of fungal development to be observed readily using trypan blue
stain.   SEM had revealed that zoospores rapidly attached to
plant surfaces and that the flagellae were lost soon afterwards
(Bauchop, 1979c).   In experiments with wheat straw leaf, by 3 h
hyphae were detected in mesophyll tissue.   Subsequent fungal
proliferation resulted in development of an extensive mycelium.
Development of sporangia commenced at approximately 12 h and
by 24 h had grown to mature size.   The fungal life-cycle
recommenced with release of zoospores from sporangia apparently
around 24-30 h.

Growth on solid substrates in vitro   The rumen anaerobic fungi
colonize and grow readily on plant fragments from all of the
common fibrous feedstuffs.   The use of sisal twine as substrate
proved particularly useful for culture of the fungi as they grow
more slowly and survive longer on sisal.   Wheat straw leaf was
valuable also as a substrate in cultures, and mycelial development was easily monitored by staining pieces of leaf.   As the
mycelium developed, the mesophyll and epidermal cells were
digested from the inside.   Epidermal long cells were found to
be digested whereas the silicified short cells resisted
digestion (Bauchop, 1979d).   Demonstration of plant cell wall
digestion raised the question of cellulolytic activity.
Cellulase was demonstrated in cultures containing strips of
filter paper.   The rumen phycomycetes grew as distinct colonies
and the paper was digested in areas of fungal development.   The
mat of paper fibres was eventually replaced by a mat of fungal
rhizoids and sporangia.

Distribution of gut anaerobic fungi   Gut anaerobic fungi have
now been found to occur in several herbivorous animals
consuming fibrous diets.   Similar fungi were found attached
to plant fragments from the rumen of feral red deer, reindeer
and impala (Fig. 10a).   Six species of macropod marsupials,
the kangaroo tribe, also had similar fungi attached to plant
fragments obtained from their ruminant-like stomachs (Fig. 10b).
In two examples of herbivores possessing a major fermentation
in the hind gut, the horse and the African elephant, large

numbers of anaerobic fungi were found colonizing plant material in fresh faecal samples (Fig. 10c,d). Fungi from both of these latter sources were cultured anaerobically although not from the other sources where only formalin-treated samples were available. Results from these animals and domestic ruminants indicate that the requirements for development of gut fungi include a large organ of fermentative digestion and a high fibre diet. Similar fungi were not detected in faeces from the rabbit, possum or man.

Importance of gut fungi The extent of attachment, colonization, and growth of the gut fungi on fibrous plant fragments suggests a significant role in fibre digestion, perhaps as initial colonizers in ligno-cellulose digestion (Bauchop, 1979c). Furthermore, the broad distribution of anaerobic fungi in widely differing herbivores indicates that they are well adapted to the gut environment. In the rumen and other gut sites the relative importance of the cellulolytic activities of fungi and bacteria remain to be assessed. The cellulase activities of the known rumen cellulolytic bacteria have not adequately explained cell wall digestion in ruminants (Dinsdale, Morris and Bacon, 1978); details of the enzymic activities of the rumen anaerobic fungi will thus be awaited with considerable interest.

Rumen bacteria on plant fragments

Bacteria associated with rumen solids have been little studied, even although results of *in vitro* fermentation measurements indicated that more microorganisms were associated with solids than with the liquid fraction (Hungate, 1966). By means of electron microscopy, many tissues of plant fragments have been found to be heavily populated by bacteria (Akin, 1979). Bacteria adhering to plant fragments came not simply from random contamination, but from highly specific colonization by a limited range of bacteria to particular tissues. Bacterial colonization can also be extremely rapid (T. Bauchop, unpublished observations). Fig. 11a shows a high population of bacteria attached to the vascular cylinder of lucerne stem after 15 min in a sheep rumen. A population of predominantly curved rods on the pith of lucerne stem is shown in Fig. 11b. During digestion of lucerne stem, the epidermis did not appear to be degraded but merely slid down the stem as underlying tissues were degraded at sites of physical damage (Bauchop, 1979b). Fig. 11c shows bacteria adhering to the inside surface of a section of epidermis that has folded over. The resemblance to a cultured colony of bacteria is belied only by the 15 min period required for its development. Mixed bacterial populations were also found on

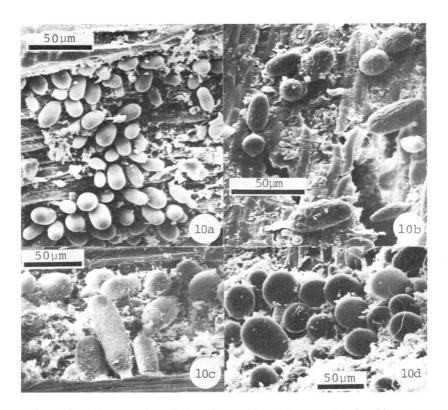

Fig. 10.   Sporangia of fungi on plant fragments in digesta
taken from different sites and animals.   (a) Rumen of an impala.
(b) Stomach of a grey kangaroo.   (c) Fresh faeces from a horse.
(d) Fresh faeces from an African elephant.

fragments of different feedstuffs examined after various periods
in the rumen (Fig. 12).   The interrelationships of individual
plant tissues and bacterial flora, including bacterial succession during plant tissue degradation, remain to be studied in
detail.

Physical damage of plant materials and microbial digestion

The highly resistant nature of plant cutinized surfaces to
microbial digestion, and the effect of lignin in retarding
tissue breakdown were recognised from light microscope studies
(Baker and Harriss, 1947).   Use of SEM has extended these

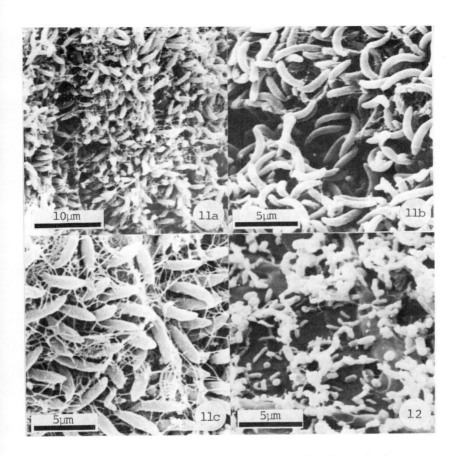

Fig. 11.   Bacterial populations at cut end of fresh lucerne stem suspended in a sheep rumen.   (a) Vascular cylinder after 15 min.   (b) Central pith region after 30 min.   (c) Inside surface of a piece of folded over epidermis after 15 min.
Fig. 12.   Mixed bacterial population on inside surface of the hollow stem of a grass after 6 h in the rumen of a cow.

observations.

All major colonization of plant fragments by gut microbes appears to occur at sites of physical damage or on internal tissues, such as the inside surfaces of the hollow stems of grasses.   Intact cutinized plant surfaces had relatively few microorganisms present (Fig. 13a; cf. Fig. 11a,b).   SEM examination of the surface of fresh lucerne stem pieces, suspended in a sheep rumen even for as long as 40 h, revealed no obvious

changes.   Seemingly delicate structures such as stem hairs
(also cutinized) also revealed no signs of degradation after
40 h (T. Bauchop, unpublished observations).   During this
period however, extensive colonization by bacteria and protozoa,
together with digestion and degradation, occurred at the cut
ends of the stem.   With dried feeds also, such as chaffed lucerne
intact cutinized surfaces had few associated bacteria.   However
many minor lesions exist on these stem surfaces and primary
colonization by bacteria occurred at these sites, in addition
to major areas of damage.

It is possible that some fungi may be able to penetrate
stem cuticular layers (although the existence of small lesions
could not be ruled out) but numerically they were unimportant
compared with the fungal population colonizing internal
tissues (Bauchop, 1979c).   However fungi were capable of
penetrating minor lesions on plant surfaces that were not
colonized by protozoa or bacteria (Fig. 13b).   In addition
fungi can gain access to internal tissues by stomatal invasion
(Fig. 13c), although this mechanism also appeared of minor
significance.

In addition to cutinized surfaces, the surfaces of highly
lignified vessel elements also have few associated bacteria
(Fig. 13d).   These elements pass out of the animal apparently
unchanged and can constitute a large proportion of the
indigestible material in poor quality diets.   Silica also may
influence digestion in some cases although it is probably of
minor importance in most diets.   The rumen anaerobic fungi
appeared unable to digest the silicified short cells of
wheat straw leaf epidermis (Bauchop, 1979d).

Importance of chewing in plant digestion   The extent of
digestion of plant materials in the gut results from the
combined effects of chewing, microbial fermentation, the
animal's digestive enzymes, and detrition in the gut.   The
importance of physical damage of plant materials due to
chewing in relation to microbial fermentation has been little
studied even in ruminants, although prevention of rumination
is well known markedly to increase retention time and turnover
time of feed.   Considerable nutritional advantages accrue
from foregut fermentative digestion compared with that in the
hindgut and are expressed in the highly successful colonization
of the high-fibre, low nitrogen environments by ruminants
(Moir, 1968).   While several factors contribute to the
nutritional superiority of foregut fermentation, the foregoing
results suggest that the importance of physical damage of
plant materials produced by chewing (rumination) may have been
underestimated.   Examination of faecal plant fragments from
animals with major hindgut fermentations (rabbit, horse,

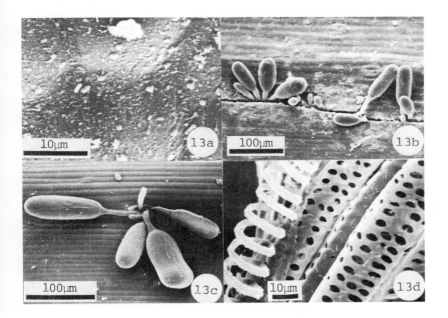

Fig. 13. Influence of plant surface on microbial colonization.
(a) Few bacteria are present on intact cuticle of fresh lucerne
stem suspended in a sheep rumen for 15 min (cf. Fig. 11a,b).
(b) Fungal colonization at minor crack in surface of grass
stem taken from rumen digesta of a cow. (c) Fungal coloniz-
ation via stomata in lucerne stem taken from rumen digesta of
a sheep. (d) Few bacteria are present on highly lignified
vessel elements at cut end of lucerne stem after digestion
for 40 h in a sheep rumen.

elephant), indicated that very limited degradation of plant
materials occurred compared with that in ruminants (T. Bauchop,
unpublished observations). These observations are in accord
with digestibility data on horses and rabbits. The extent of
physical damage due to chewing could be one important difference
between the two groups. Although food ingested by both groups
is exposed to chewing, only during rumination is chewing
repeated and extended over long periods.

CONCLUSION

Direct observation is only one of the many approaches to be
employed in elucidating a problem in microbial ecology, but it
does provide a solid framework for the development of such a

study. SEM is only one of the techniques available for direct
observation but it has proved to be a powerful one and already
has yielded much useful information on microbial colonization
of gut epithelia. In herbivorous animals, digestion involves
mainly degradation of insoluble substrates. Thus we are
concerned chiefly with microbial colonization at solid-liquid
interfaces, a situation well suited to investigation by SEM.
This constitutes a high area for research and the present
findings represent merely a beginning. Investigation of even
a few of the many possible combinations of animal, diet and
plant tissue will surely yield much new information on microbial
digestion in the gut. The highly organized structure of plant
materials imposes spatial relationships on microbial colonization, as shown here. In the rumen ecosystem, understanding of
these relationships is required for the integration of the
accumulated knowledge of the microbiota involved in plant
digestion. With so many avenues available for exploration,
microbial ecologists can look forward to exciting discoveries
and a better understanding of the complex interrelationships
in the gut ecosystem.

ACKNOWLEDGEMENTS

I thank D. Hopcroft for preparation of samples and for
assistance with operation of the SEM.

REFERENCES

Akin, D.E. (1976). Ultrastructure of rumen bacterial
    attachment to forage cell walls, *Applied and Environmental
    Microbiology* 31, 562-568.
Amos, H.E. and Akin, D.E. (1978). Rumen protozoal degradation
    of structurally intact forage tissues, *Applied and
    Environmental Microbiology* 36, 513-522.
Akin, D.E. (1979). Microscopic evaluation of forage digestion
    by rumen microorganisms - a review, *Journal of Animal Science*
    48, 701-710.
Akin, D.E. and Amos, H.E. (1979). Mode of attack on orchard
    grass leaf blades by rumen protozoa, *Applied and
    Environmental Microbiology* 37, 332-338.
Baker, F. and Harriss, S.T. (1947). Microbial digestion in
    the rumen (and caecum) with special reference to the
    decomposition of structural cellulose, *Nutrition Abstracts
    and Reviews* 17, 3-12.
Bauchop, T. (1979a). Rumen anaerobic fungi, *Abstracts.
    Annual Meeting of the American Society for Microbiology*,
    Abstract RT 24, 382.
Bauchop, T. (1976b). The rumen cilate *Epidinium* in primary
    degradation of plant tissues, *Applied and Environmental
    Microbiology*, 37, 1217-1223.

Bauchop, T. (1979c). Rumen anaerobic fungi of cattle and sheep, *Applied and Environmental Microbiology* 38, 148-158.

Bauchop, T. (1979d). The rumen anaerobic fungi : colonizers of plant fibre, *Annales de Recherches Veterinaires* 10, 246-248.

Bauchop, T., Clarke, R.T.J. and Newhook, J.C. (1975). Scanning electron microscope study of bacteria associated with the rumen epithelium of sheep, *Applied Microbiology* 30, 668-675.

Bauchop, T. and Clarke, R.T.J. (1976). Attachment of the cilate *Epidinium* Crawley to plant fragments in the sheep rumen, *Applied and Environmental Microbiology* 32, 417-422.

Bauchop, T. and Clarke, R.T.J. (1977). Degree of plant root digestion by the larva of the beetle *Costelytra zealandica*, *Journal of Insect Physiology* 23, 65-71.

Clarke, R.T.J. (1977). Protozoa in the rumen ecosystem. *In* "Microbial Ecology of the Gut" (Ed. R.T.J. Clarke and T. Bauchop) pp. 251-275. Academic Press, London.

Davis, A.B. and Harbers, L.H. (1974). Hydrolysis of sorghum grain starch by rumen microorganisms and purified porcine α-amylase as observed by scanning electron microscopy, *Journal of Animal Science* 38, 900-907.

Dinsdale, D., Morris, E.J. and Bacon, J.S.D. (1978). Electron microscopy of the microbial populations present and their modes of attack on various cellulosic substrates undergoing digestion in the sheep rumen, *Applied and Environmental Microbiology* 36, 160-168.

Elsden, S.R., Hitchcock, M.W.S., Marshall, R.A. and Phillipson, A.T. (1946). Volatile acid in the digesta of ruminants and other animals, *Journal of Experimental Biology* 22, 191-202.

Hungate, R.E. (1950). The anerobic mesophilic cellulolytic bacteria, *Bacteriological Reviews* 14, 1-49.

Hungate, R.E. (1966). *"The Rumen and Its Microbes"*. Academic Press, New York.

Latham, M.J., Brooker, B.E., Pettipher, G.L. and Harris, P.J. (1978). *Ruminococcus flavefaciens* cell coat and adhesion to cotton cellulose and to cell walls in leaves of perennial ryegrass (*Lolium perenne*), *Applied and Environmental Microbiology* 35, 156-165.

Minato, H. and Suto, T. (1979). Technique for fractionation of bacteria in rumen microbial ecosystem. III. Attachment of bacteria isolated from bovine rumen to starch granules. *in vitro* and elution of bacteria attached therefrom, *Journal of General and Applied Microbiology* 25, 71-93.

Moir, R.J. (1968). Ruminant digestion and evolution. *In* "Handbook of Physiology" (Ed. C.F. Code), Sect.6, Vol.5, pp. 2673-2694. American Physiological Society, Washington.

Orpin, C.G. (1975). Studies on the rumen flagellate, *Neocallimastix frontalis*, *Journal of General Microbiology* 91, 249-262.

Orpin, C.G. (1977). The occurrence of chitin in the cell walls of the rumen organisms *Neocallimastix frontalis, Piromonas communis* and *Sphaeromonas communis, Journal of General Microbiology* 99, 215-218.

Orpin, C.G. and Letcher, A.J. (1978). Some factors controlling the attachment of the rumen holotrich protozoa *Isotricha intestinalis* and *I. prostoma* to plant particles *in vitro*. *Journal of General Microbiology* 106, 33-40.

Savage, D.C. (1978). Gastrointestinal microecology: one opinion. *In* "Microbial Ecology" (Ed. M.W. Loutit and J.A.R. Miles), pp. 234-239. Springer-Verlag, Berlin.

MICROBIAL ANTAGONISM - THE POTENTIAL FOR
BIOLOGICAL CONTROL

Kenneth F. Baker

Collaborator, U.S. Department of Agriculture SEA-AR
Ornamental Plants Research Laboratory
Oregon State University, Corvallis, Oregon 97330.

INTRODUCTION

In the early developmental period of pure-culture techniques
for microorganisms, William Roberts (1874) observed an inhibitory
effect between "*Penicillium glaucum*" and bacteria. He was one of
the first to note antibiotic action and introduced the term
antagonism into microbiology. Thus began the subject of microbial
interaction, laying the groundwork for the rise of biological
control.

It was almost fifty years after the recognition of microbial
interaction and inhibition that serious attempts were made to
utilize the phenomenon in control of plant or animal diseases.
Hartley (1921), Sanford (1926), Millard and Taylor (1927),
Sanford and Broadfoot (1931), Henry (1932) and Weindling (1932,
1934) had shown by 1934 that both introduced and resident
antagonists could diminish disease caused by plant pathogens,
that they could produce antibiotics and that the biological
balance could be manipulated by changes in the soil environment
(organic matter content, temperature, and pH).

However, it was the discovery by A. Fleming in 1928 of the
production of penicillin by *Penicillium notatum* and its
purification and demonstrated medical use by H.W. Florey and
E.B. Chain in 1938 that started a period of intensive antibiotic
research (Hare, 1970). More than 100 antibiotics had been
described by 1949, most of them from saprophytic soil microorganisms. The single greatest stimulus to studies of antagonists
to plant pathogens probably was the success story of penicillin.
Some plant pathologists were sceptical but hopeful that biological
control of plant diseases would be more than a novelty of
Petri-dish biologists and would prove successful under field
conditions. The numerous reasons for this attitude have been

detailed elsewhere (Baker and Cook, 1974; Garrett, 1956).
However, by 1965 interest in biocontrol had increased markedly
among plant pathologists. This is shown by the symposia on
the subject held in Berkeley, California in 1963, London in
1968, Philadelphia in 1971, Burnley, Australia in 1972,
Minneapolis, Minnesota and Lausanne, Switzerland in 1973,
Tokyo in 1974, Rydalmere, Australia in 1977 and in Munich in
1978.
     Perhaps the change in the past decade has resulted from
man's tardy realisation that it is better to work with natural
biological forces than to ignore or attempt to override them.
The widespread concern over chemical pollution of air and
water may have also been involved. Multiple controls (integrated
control), involving the restricted use of chemicals in support
of ecological and biological controls, seem to be the probable
method of the future.
     Examples of biocontrol discussed in this paper include only
those that have been successfully demonstrated under field or
commercial conditions, and most of which are commercially used
today. More extensive treatments of biocontrol are provided by
Baker and Snyder (1965), Baker and Cook (1974), Cook (1977),
and Baker (1980).

BIOLOGICAL BALANCE

     Microbial populations have a resilient dynamic stability
produced by biological buffering from the competition within
the group. Microorganisms compete for nutrients, oxygen and
favourable ecological niches and are selected for tolerance
of ambient conditions of pH, carbon dioxide, water and microbial
toxins. Microorganisms secrete metabolites, some of which
inhibit other microorganisms (antibiotics), while others
stimulate some microorganisms to form essential stages of
their life cycle (probiotics). *Pseudomonas* spp. thus stimulate
Phytophthora cinnamomi to form zoosporangia and zoospores
(Ayres, 1971), the chief agents for infecting roots, and a
Chromobacterium sp. induces formation of chlamydospore
resting structures by the fungus (Broadbent and Baker, 1974a).
A negative microbial interaction can therefore directly
inhibit a pathogen or it can, by inhibiting a stimulatory
microorganism, indirectly inhibit the pathogen. Both positive
and negative interactions, in the sense of this symposium,
are thus involved in biological control.
     The fluctuating population density of a microorganism is
maintained within certain definite limits by starvation and by
antagonists. Established microbial associations have developed
interlocking, flexibly buffered ecological niches. An alien
microorganism can establish in such a stable ecosystem if:
     i)   it is better adapted than are some of the residents,

ii) it is introduced in such numbers as to temporarily or permanently swamp the residents,

iii) if it modifies the environment to make it favourable to itself, or

iv) if the natural balance has been so disturbed by man that the environment is more favourable to the alien than to the residents.

Manipulation of the Biological Balance

The balance of the resident microorganisms can be altered through "nudging" by slight environmental changes, or by a "shock" such as chemical or heat treatment of soil or the addition of organic amendments. Unfortunately, man has emphasised the shock approach through the use of chemicals, and has long exceeded the tolerance limits of biological systems. It is now becoming clear that subtle nudging may achieve the same results, more slowly but also more permanently. This is exemplified by soil treatment with aerated steam at 60°C for 30 minutes in commercial nurseries, instead of the older 100°C for 30 minutes or more. Creation of a biological vacuum is thus avoided and the surviving saprophytes inhibit any chance pathogen that may be introduced (Baker, quoted in Toussoun, Bega and Nelson, 1970; Baker, 1971).

Microorganisms are adaptable and have incredibly complex, resourceful means of survival, both factors which are important evolutionary requirements. To successfully manipulate the microbiota for his own purposes, man has to be inventive and sagacious. Plant pathologists have found that they cannot depend on a single procedure to control a pathogen. Biocontrol must be viewed, not as a single-shot, all-purpose treatment, but as part of an integrated control programme and obviously, it must work within the limits of biological balance, of which it is a specialised application.

Environmental factors operate selectively on the crop, pathogen and antagonists. Soil bacteria antagonistic to *Fusarium roseum* 'Culmorum', cause of wheat foot rot, cease growth in moderately dry soil of -10 to -15 bars water potential, but the pathogen can grow under drier conditions (-75 to -85 bars). The wheat plant, drawing water from deep in the soil, continues to grow after the surface soil dries, but is increasingly attacked by the *Fusarium* sp. as the bacteria become inactive. The disease is controlled in Washington by using a dust mulch, seeding late and keeping soil fertility at low levels to produce smaller plants, and selecting varieties for high water efficiency – factors that delay soil-water depletion and prolong antagonism by the bacteria (Cook and Papendick, 1970; Papendick and Cook, 1974).

Atkinson, Neal, and Larson (quoted in Bruehl, 1975) found

that "the genotype of the host governs the magnitude and
composition of bacterial populations in the rhizosphere with
surprising specificity ... presumably through its control of
the quantity or quality of root exudates, or both". This
opens an exciting new area of biological control by genetic
engineering, acting through the effect of root exudates on
the rhizosphere microflora.

Biological balance of microorganisms is achieved through
antagonism, "the balance wheel of nature". There are three
mechanisms of antagonism:

    i)   competition for nutrients, favourable sites and
oxygen;
    ii) antibiosis by metabolites of other microorganisms
and
    iii) parasitism or predation by other microorganisms.

Soils that are biologically "live" (that is, have a
heterogeneous, abundant, and active microbiota) are more
likely to exert biological control of an introduced plant
pathogen than are infertile, biologically depauperate soils.
Live soils have a fairly high content of organic matter and
mineral nutrients, and are well aerated and moist. The soil
is more stable in almost all respects than the aerial environment, and soil-borne microorganisms have become adapted to
these stable conditions. For this reason slight changes in
one or more factors of the soil environment may exert a
profound effect on soil microorganisms. This is especially
true for plant pathogens, which are less tolerant of
unfavourable conditons (high temperature, antibiotics) and
are less adaptable nutritionally than are saprophytes (Baker,
quoted in Toussoun, Bega and Nelson, 1970).

Interactions between Microorganisms

Many types of interactions between microorganisms are
involved in achieving a biological balance:

    i)   A microorganism may have no effect on the associate.
    ii) It may stimulate growth or development of the
associate. Thus, *Melanospora damnosa* is stimulated by *Fusarium
roseum* 'Culmorum' to form ascospores, *Mycena citricolor* is
stimulated by *Penicillium oxalicum* to form basidiocarps, and
Agaricus bisporus is stimulated by bacteria to form sporophores.
    iii) It may inhibit growth or development of the associate
(Linderman and Gilbert, quoted in Bruehl, 1975). Mycelial
growth of *Phytophthora cryptogea* may be inhibited by thiamine-
requiring bacteria, and enhanced by thiamine-synthesising
bacteria (Erwin and Katznelson, 1961). As is well known,
penicillin produced by *Penicillium notatum* inhibits formation
of the cell wall of staphylococci.
    iv) It may stimulate formation of resting structures by

the associate. *Fusarium solani* f. sp. *phaseoli* may be
stimulated to form chlamydospores by species of *Protaminobacter*,
Arthrobacter or *Bacillus* (Ford, Gold and Snyder, 1970).
Sclerotium rolfsii may be stimulated by *Streptomyces* sp. to
form sclerotia (Baker and Cook, 1974).

v)   It may inhibit the formation of resting spores by the
associate. *Sclerotium rolfsii* is stimulated by some isolates
of *Bacillus subtilis* to increase mycelium production and
decrease sclerotium formation (Baker and Cook, 1974).   The
stimulation of *Phytophthora cinnamomi* by *Pseudomonas* spp. to
form zoosporangia also decreases production of chlamydospores.

vi)   It may enforce dormancy on the associate.   Thus
fungistasis often is associated with sites of high microbiological
activity.

vii) It may cause lysis of the associate.   Bacteria in
soils suppressive to *Phytophthora cinnamomi* may attach to the
surface of mycelia and zoosporangia, causing endolysis
(Broadbent and Baker, 1974b).

It is becoming clear that the progression through the
stages of the life cycle of a soil microorganism is determined
at least as much by the associated microflora and abiotic
environment as by the genes of the microorganism.

Microbial populations in undisturbed nature have great
resistance to change because only the well-integrated, well-
adapted, and competitive biota have found an ecological niche
in which to develop.   Those not well adjusted have simply been
replaced by those that are.   It follows that, to be enmeshed
in such an association is to gain strength and protection.
An alien microorganism has great difficulty in invading such
a microflora whose stability is a product of its biological
complexity.   However, man sometimes has been able to manipulate
the microbial balance by altering some factor of the environment
and thus control some plant diseases.   The occurrence of a
plant disease indicates that the biological balance is not
in equilibrium and, the greater the imbalance, the more severe
the disease is apt to be.   Elimination of a pathogen from a
site is not necessary for control, indeed is rarely achieved
in any disease control.   Rather, the inhibition of pathogenic
activities without overkill of other microorganisms is the
objective; a biological vacuum is to be avoided, because the
first microorganism to return will then flourish.   If this is
a pathogen, severe disease loss will result.   Thus, when soil
that has been steamed at 100°C becomes contaminated by a
pathogen, the result can be disastrous.   Soil treatment at 60°C
for 30 minutes by aerated steam (Baker, 1971) is widely used in
nurseries because pathogens are killed, but many saprophytes
survive and buffer any pathogens inadvertently introduced.
This is selective manipulation of the soil microflora.

There are other means of selective treatment of soil to favour antagonists over the pathogen. Orchard soils have been successfully treated for many years in California for the control of *Armillaria mellea*. Diseased trees are removed and the soil fumigated with carbon disulphide, a weak fungicide. Exposed mycelium is killed, and that in the roots is weakened, and then killed by the mycoparasite, *Trichoderma viride*. Treatment with methyl bromide or with aerated steam (43°C for 2 hours) produces a similar control (Munnecke, Wilbur and Darley, 1976). *Armillaria* sp. hyphae normally produce an antibiotic that protects them from *Trichoderma* sp. but methyl bromide prevents its formation (Ohr and Munnecke, 1974). Carbon disulphide treatment of soil was also used for control of wilt of asparagus caused by *Fusarium oxysporum* f. sp. *asparagi* in California before the advent of resistant varieties.

In natural ecosystems disastrous plant-disease epidemics are unknown or rare and pathogen suppression is usual; health rather than disease is the normal state of life. Man can upset the balance by abrupt, violent ecological changes in agricultural practices and diseases then intensify. Biological control, in this sense, is the attempt to retain or restore such a disease-suppressive balance. This usually is not the original balance, but a new one that is stable and compatible with the changed conditions. It is unlikely that agriculture could have developed if such biocontrol was not common, even under the disturbed conditions of cultivation.

If biocontrol is common in nature, how is it recognised? This can only be done when a susceptible crop and a pathogen are brought together. When the biological balance of the host-environment-pathogen-antagonist complex favours disease production, it is recognised by the symptoms produced -- a positive observation. When the antagonists suppress the activity of the pathogen there is no disease to see -- a negative observation. Man only has time to cope with the problems he recognises and rarely investigates those that give him no trouble.

Man has therefore, unknowingly benefited from naturally occurring biological control and has attributed the healthy state to absence of the pathogen or to unfavourable environment. This has been the case with fusarium wilt diseases in the Salinas Valley of California. *Fusarium oxysporum* f. sp. *pisi*, cause of pea wilt, has long been abundantly introduced with seed of susceptible pea varieties grown for canning, freezing and seed production. However, wilt has appeared in only two small areas now subdivided for homesites. The soil is naturally suppressive to this wilt fungus, but after the soil is steamed or fumigated the disease develops when infested pea seed is planted in it (Tousson, quoted in Bruehl, 1975).

The first step in applying biocontrol is to find instances where it is working. These sites should then be studied to determine the mechanism of action and to devise means of transfer of suppressiveness to other localities.

BIOLOGICAL CONTROL OF SOIL-BORNE PLANT PATHOGENS

Resident Antagonists and Suppressive Soils

Composite microorganism populations antagonistic to a given pathogen or disease occur only in soils biologically suppressive to the pathogen, but individual microorganisms antagonistic to it may occur in many soils. On the basis of present evidence, it appears unlikely that a naturally suppressive soil is due to the presence of a single antagonist.

It is suggested that a direct approach to biocontrol is best at present. Seek antagonists (either individuals or populations) where the disease does not occur, declines with time, or cannot establish. Try transferring antagonist populations or mixed isolates of them, to conducive soil that is either nontreated or so treated as to decrease resident microorganisms.

Resident microfloras of soils can be manipulated so as to control plant pathogens. *Fomes lignosus, Ganoderma pseudo-ferreum,* and *F. noxious* cause the white, red, and brown root diseases of plantation rubber. Old jungle or rubber trees are poisoned before replanting; the quick death allows rapid invasion by saprophytes. Old stumps are creosoted to prevent infection by airborne spores. Creeping legumes are planted between the rows of new trees, but kept back three feet from them. Old stumps rapidly decay under the legume cover, and the food bases of the fungi are depleted by enhanced fructification, and by rhizomorph formation by *F. lignosus.* Actinomycetes and other saprophytes apparently contain the pathogens within infected roots and diminish spread. Infected trees are removed at each quarterly inspection, and the collars of neighbouring trees are exposed and painted with PCNB (pentachloronitrobenzene) in wax to prevent spread along the roots (Fox, quoted in Toussoun, *et al.,* 1970). This is the successful commercial control procedure for root rot of rubber trees in Malaysia.

Ectomycorrhizae have been shown (Marx, quoted in Bruehl, 1975) to decrease infection of feeder roots by pathogens such as *Phytophthora cinnamomi.* The mycorrhizal fungus must infect before the pathogen and apparently is effective by the possession principle, antibiotic production, and increase of the host's vigour.

Examples of suppressive soils will now be considered, arranged in three categories, based on the way they are recognised in the field.

The pathogen is unable to establish The example of the
inability of *Fusarium oxysporum* f. sp. *pisi* to establish in
the suppressive soils of the Salinas Valley of California has
already been discussed, and it was pointed out that sterilisa-
tion of the soil destroyed the suppressiveness. This is one
of the hazards of overkill treatments of field soil. Apparently
soil suppressive to this *Fusarium* sp. also occurs in Illinois
and in New South Wales, Australia, but not in Wisconsin or
the Palouse area of Washington. In the Palouse fields fusarium
wilt of peas is an important disease, but *F. roseum* 'Culmorum'
causes almost no concern in this, one of the world's most
fertile wheat-growing areas. This shows the high specificity
common to suppressive soils.

Smith and Snyder (1972) compared the behaviour of *F. oxysporum*
f. spp. *batatas, cubense,* and *lycopersici* with closely similar
saprophytic *F. oxysporum* isolates from the suppressive soil.
When the pathogens were inoculated into suppressive and
conducive soils, chlamydospore germination was less and germ
tubes were shorter in the suppressive soil. The saprophytes
were less inhibited in each case, showing the specificity
of the interaction. An *Arthrobacter* sp. increased more
rapidly in the suppressive than in the conducive soil. Smith
(1977) found that germination of chlamydospores of *F. oxysporum*
f. spp. *vasinfectum* and *tracheiphilum* was less in suppressive
than in conducive soil, and ceased in 24 as compared with 60
hours. An *Arthrobacter* sp. was associated with germlings in
suppressive but not conducive soil.

Reinking and Manns (1933) showed the clear-cut existence
of soils "tolerant" and "intolerant" to *F. oxysporum* f. sp.
cubense, cause of Panama disease of banana in Central America.
In "short-life" sandy soils bananas died in three to four
years after planting and developed high populations of the
pathogen, but in "long-life" clay soils, often only a few
yards away, plants remained healthy for up to twenty years,
and developed only low populations of the pathogen. This
distinction was important in selecting fields for new plantings
until resistant varieties were used in the early 1960's. Stotsky
and associates in 1961-1967 showed that montmorillonoid-type
clays were present in long-life soils and absent in short-life
soils; the proportion of montmorillonite present was directly
related to the degree of suppression. This clay stimulated
bacterial activity, perhaps checking growth of *Fusarium*
(Stotzky and Rem, 1966).

A strain of *Rhizoctonia solani* that causes a bare-patch
disease of wheat in the Eyre Peninsula, South Australia, was
massively introduced into a field in Adelaide. It caused

disease for a short time, but the pathogen disappeared from
the suppressive soil in four months (Baker, Flentje, Olsen
and Stretton, 1967). However, another (crucifer) strain of
the fungus occurs commonly in both soils, again demonstrating
the specificity of the relationship. That the disease
appeared briefly in the Adelaide field illustrates the
important point that addition of abnormally large amounts of
a pathogen into a suppressive soil may temporarily swamp the
antagonists. This shows the significance of the pathogen-
antagonist balance in the incidence of plant disease.

R.S. Smith, (reported by Toussoun, quoted in Bruehl, 1975)
found that soils of California pine forests are suppressive
to *Fusarium* spp., but that forest nursery soils are not.
Seedlings transplanted to the forest soil showed a decline
in populations of *Fusarium oxysporum* within one year, and
its disappearance in three years. It was thought that suppressiveness of pine forest soil resulted from the absence of
annual plants that supported development of fusaria, and from
the effect of needle litter, suberized roots, and rhizosphere
microorganisms. This type of suppression lacks specificity
in that it apparently applies to all fusaria.

Louvet, Rouxel and Alabouvette (1976) and Rouxel, Alabouvette
and Louvet (1977) studied a soil in France that was biologically
suppressive to *F. oxysporum* f. sp. *melonis*. They transferred
the suppressive microflora *in toto* by inoculating conducive
soil with 10% suppressive soil. The pathogen survived but did
not multiply in non-treated suppressive soil; it multiplied
as the suppression progressively declined after aerated steam
treatment temperatures above 55°C. Because populations of
F. oxysporum and *F. solani* in the soil roughly paralleled the
soil suppressiveness following the thermal treatments, they
were thought to be involved in the suppression by competition
with the pathogen.

The evidence suggests that a soil suppressive to one forma
specialis of *F. oxysporum* will be suppressive to other formae
speciales as well, but not to *F. roseum, F. solani,* or the
saprophytic *F. oxysporum*. This indicates that these formae
speciales are closely similar in characteristics that affect
their interaction with the suppressive microbiota. On the
other hand, the host strains of *Rhizoctonia solani* appear to
be sufficiently distinct that microbiota in soil suppressive
to the wheat strain are not suppressive to the crucifer strain.

*The pathogen establishes but causes no disease Phytophthora
cinnamomi* has been present for nearly forty years in an
avocado grove in Queensland, Australia, under 250 cm or more
of annual rain, without causing root rot. Steaming the soil
from this grove at 100°C for 30 minutes makes it fully conducive

when inoculated with the pathogen, but after treatment with aerated steam at 60°C for 30 minutes it is still suppressive. These facts show that the antagonists are thermotolerant spore-forming bacteria or actinomycetes. Multiple antagonists are involved. Suppression may be temporarily lost after water-logging or flooding, presumably due to a shift of microbial balance. The suppressive soil is high in organic matter, calcium and ammonium nitrogen tied up in the organic cycle typical of tropical rain forests. The microbiota of the soil is both abundant and varied (Broadbent and Baker, 1974a,b; Broadbent, Baker and Waterworth, 1971). Intensive cover-cropping throughout the year, heavy applications of poultry manure, use of superphosphate and addition of dolomite limestone to keep the pH near neutrality has protected the grove.

Zoospores are the principal infective structures of *P. cinnamomi*, although mycelia may also infect, apparently mainly through wounds. Stimulation by material produced by *Pseudomonas* spp. is required for the pathogen to produce abundant zoospores (Zentmyer, 1965; Ayers, 1971). These bacteria occur in most soils, whether *P. cinnamomi* is present or not, and apparently develop abundantly on the surface of mycelia of the fungus when it is present; the stimulatory compound then reaches effective concentrations. The concentration of the compound, either in soil steamed at 40-50°C for 10 minutes or the soil extract passed through Millipore filters to eliminate live bacterial cells, apparently is too low for stimulation. There are inhibitory agents in suppressive soil that either inhibit the stimulatory bacteria or break down their stimulatory compound; these agents seem not to occur in conducive soils. The few zoosporangie formed in suppressive soils are attractive to antagonistic bacteria, which attach and cause lysis and discharge of the undifferentiated contents (Broadbent and Baker, 1974b).

The high organic matter, calcium and ammonium and nitrate content of suppressive soils
     i)   create conditions which favour antagonists and inhibitors and
     ii)   improve soil aeration and drainage.

Pegg (1977) has shown that in some groves badly injured by root rot, replanted and given the above culture for three years, the pathogen and root rot became difficult to detect. The Queensland avocado industry has extensively adopted this procedure and root rot is no longer important there. Other soils suppressive to *P. cinnamomi* have since been found in Australia (Broadbent, Baker and Malajczuk, unpublished observations).

Pegg (1977) found that biocontrol of *P. cinnamomi* root rot of pineapple is possible, but by a different technique.

Maintaining soil at neutrality causes chlorosis in pineapple
and sulphur is therefore added to bring the pH below 5.4. At
this pH, antagonist bacteria and nitrifying bacteria are
inhibited and nitrogen remains in the ammonium form somewhat
inhibitory to *Phytophthora* sp. *Trichoderma viride* is
favoured by the acid conditions and its antibiotics are most
effective under these conditions. Effective control of
pineapple root rot is thus commercially accomplished in
Queensland.

In Western Australia, *P. cinnamomi* causes severe root rot
losses to the valuable timber tree, *Eucalyptus marginata*,
and its proteaceous understory shrubs. Shea and Malajczuk,
(1977) have shown that a change from the highly susceptible
Banksia grandis understory to resistant legumes such as
Acacia pulchella has provided good control of the pathogen.
This change can be achieved by high-intensity prescription
burning which kills the *Banksia* induces seed germination of
legumes and favours accumulation of litter. Low-intensity
burning has the opposite effect and favours root rot. *Acacia*
roots and leaf debris inhibit sporulation of the pathogen.
Since high-intensity burning may injure the Eucalyptus, it
may be necessary to use low-intensity fire to weaken the
Banksia and then sow the areas with seed of *Acacia* that have
been heat-treated to break dormancy.

*The pathogen establishes but the disease diminishes with
continued culture* The take-all disease (caused by *Gaeumannomyces
graminis* var. *tritici*) has been found in the wheat-growing
areas of the world to reach maximum severity in two to three
years of continuous wheat and to subsequently decline to an
economically tolerable level. When such suppressive soil is
sterilized and reinfested with the pathogen it becomes
conducive; treatment with aerated steam at 40°C for 30
minutes diminishes and at 60°C eliminates the suppressiveness.
This indicates that the antagonists are nonspore-forming
bacteria, fungi, or actinomycetes; nonspore-forming fluorescent
bacteria are the present suspects. Multiple antagonists
apparently are involved. If a single crop not susceptible to
take-all is planted in rotation, the disease decline may be
reversed and severe disease develops in the next wheat crop.
Apparently the live virulent pathogen must be present for the
suppressive microflora to develop (Gerlagh, 1968), suggesting
an intimate relationship between the pathogen and antagonists.
Addition of rich organic matter to the soil controls take-all,
probably by stimulating development of antagonists. Shipton,
Cook and Sitton (1973) showed that it is possible to transfer
the antagonist population *in toto* by transferring 0.5-1.0% (w/w)
of suppressive soil to a 12-15 cm depth of conducive soil that
has been fumigated with methyl bromide and reinoculated with

the pathogen. Take-all was diminished in the first year and almost completely controlled in the second and subsequent years.

The potato common scab pathogen (*Streptomyces scabies*) declines in California and Washington after eight years of monoculture, but seems to take longer when potatoes are grown in rotation with other crops. Incorporation of green barley or pea residue into soil before each potato crop thus permits an annual increase in scab severity over that in potato monoculture. However, a green soybean cover crop ploughed under between potato crops and decomposed by *Bacillus subtilis* produces compounds inhibitory to the pathogen, and prevents an increase in scab (Weinhold, Oswald, Bowman, Bishop and Wright, 1964). Suppressiveness is destroyed by autoclaving, indicating that antagonists are involved. A suppressive soil microflora was first transferred *in toto* by Menzies (1959) for common scab of potato. Transfer of 10% suppressive soil plus 1% alfalfa meal to nontreated conducive soil greatly reduced scab. *Streptomyces* sp. infects potato tubers through stomata and unsuberized lenticels. Lapwood and Hering (1970) found that maintaining soil moisture at about field capacity during tuber formation largely prevented scab, due to the enhanced competitive advantage of bacteria in the stomata and lenticels over the pathogen under these conditions. If the soil was allowed to dry during tuber formation, scab was greatly increased. However, in axenic culture of potato tubers, scab infection was favoured, not suppressed, by moisture. Bacteria were dominant in lenticels of immature tubers in wet soil, actinomycetes in tubers in dry soil. Soil bacteria seem clearly to be involved, not just water alone (Lewis, 1970).

Although *Fusarium solani* f. sp. *phaseoli* causes severe root rot of beans in irrigated lands of the Columbia River Basin, some fields have remained free of the disease for twenty years. These suppressive soils are water-deposited, limy, fine sandy loams. Autoclaving the soil destroys suppressiveness, indicating a biological cause, but this cannot be transferred to conducive soils. The soils are conducive at first, but become suppressive with bean monoculture. In suppressive soil hyphal growth is extensive, with few small chlamydospores slowly formed. In conducive soil hyphal growth is limited, with numerous large chlamydospores quickly formed (Burke, 1965; Burke, quoted in Cook and Watson, 1969).

Introduced Antagonists

In the examples of effective biological control by an introduced antagonist, the infection site is occupied by the antagonist before arrival of the pathogen, or seeds or other propagules provide a favourable food base for antibiotic

production. Introduction of an antagonist into soil with a
greatly simplified microflora (for example, steamed, fumigated,
desert or polder soils) or on to freshly cut host surfaces
enables it to luxuriate and possibly become the dominant
organism.

Nontreated soil Introduction of an antagonist into nontreated
soil is difficult, because it attempts to establish an alien
microorganism into a biologically buffered community. Although
difficult, it can be done with the right organism properly
applied, even as new weeds or pathogens appear in man's
cultivated fields. The host can serve as a selective substrate
for the pathogen, greatly reducing or nullifying the biological
buffering effect. It is as difficult to introduce a pathogen
into nontreated soil without its host as it is to introduce
an antagonist.

Treated soil Introduction of an antagonist into treated
soil is much easier because the organism essentially is
invading a biological vacuum. Soil steamed at $100^{\circ}C$ for 30
minutes, as it often is in commercial nurseries and glasshouses,
may be inoculated with an antagonist to provide effective
biocontrol. Ferguson (1958) used various saprophytic fungi
and Olsen and Baker (1968) used *Bacillus subtilis* to protect
seedlings from *Rhizoctonia solani*; the antagonists were
applied to the treated soil prior to or simultaneously with
the pathogen. Broadbent *et al.*, (1971) introduced species
of *Bacillus* or *Streptomyces* into soil and reduced damping-off
by *Pythium ultimum* and *R. solani*. However, some microorganisms
so introduced were found to increase the amount of disease.

Pseudomonas tolaasii the cause of brown blotch of the
cultivated mushroom, *Agaricus bisporus*, is omnipresent in
casing soils and is spread during watering. Peat cultures of
P. multivorans, *P. fluorescens* or *Enterobacter aerogenes*, in
concentration ratios of at least 80:1 over the pathogen, may
be added to the treated casing soil after it is applied to
the compost. This provides effective biological control of
the disease, apparently through competition (Nair and Fahy,
1976). No antibiosis is observed in culture.

Host propagules Antagonists inoculated on seeds or other
propagules prior to planting have provided effective biocontrol
to the crop produced. Corn seed inoculated with *B. subtilis*
or *Chaetomium globosum* and planted in Minnesota fields that
had a moderate inoculum density of *F. roseum* 'Graminearum',
gave about as good control of seedling blight as captan or thiram
seed treatment (Chang and Kommedahl, 1968; Kommedahl and Mew,
1975). This was thought to result from antibiotic production
by the antagonist that persisted in the rhizosphere for the

short period that protection was needed.

Biological control of crown gall (caused by *Agrobacterium tumefaciens*) by inoculating the seeds before planting or the nursery plants at time of root pruning with *Agrobacterium radiobacter* strain 84 in numbers at least equal to those of the pathogen, has been spectacularly effective in many areas of the world (Kerr, 1972; New and Kerr, 1972; Moore and Warren, 1979). Strains of the pathogen are prevented from transferring their Ti plasmid to the wounded host, apparently because a bacteriocin produced by the antagonist either kills or prevents attachment of the pathogen to the host receptor site. This is the most widely successful biological control of a plant disease to date.

BIOLOGICAL CONTROL OF PATHOGENS OF AERIAL PARTS

Fungus and Bacterial Antagonists

Commercial control of *Heterobasidion annosum (Fomes annosus)* of pine on 62,000 hectares of forest in England has been achieved by inoculating freshly cut stumps with oidia of the low-grade pioneer pathogen, *Peniophora gigantea*. *Heterobasidion annosum* causes root rot of pines in many parts of the world. Wind-borne spores infect newly cut stumps, and mycelium grows through the roots to adjacent living trees. Rishbeth (quoted in Bruehl, 1975) studied this disease for fifteen years and devised the technique for culturing and inoculating the antagonist in this successful control. This apparently is an example of the possession principle (a microorganism already in a substrate will retain possession of it), operating through a short-range antibiotic effect by a primary antagonist.

Eutypa canker of apricots is controlled in South Australia by inoculating fresh pruning wounds with a spore suspension of *Fusarium lateritium* plus benomyl (methyl 1-(butyl-carbamoyl) -2-benzimidazole carbamate). The antagonist is insensitive to benomyl which protects the wounds against *Eutypa armeniacae* until the antagonist has established (Carter and Price, 1974). This is a good example of integrated control using a combination of a nonsensitive antagonist and a chemical toxic to the pathogen.

Transient or casual microorganisms on aerial plant parts settle there from the air and may multiply in foreign debris and plant exudates. Such a microflora affects the biological balance and infection by plant pathogens. In a series of papers, Fokkema and associates reported that pollen grains on plants increased after flowering, and that *Cladosporium herbarum* increased on them. This had no effect on the number of infections of rye by the obligate parasite, *Puccinia recondita* f. sp. *recondita*, but under some conditions increased the number of infections by *Helminthosporium sativum* and *Septoria nodorum*. Normal accumulation of pollen apparently promotes an antagonistic microflora that inhibits pathogens, but pollen applied with facultative pathogens may provide a good base for infection, and in the absence of inhibitory microorganisms, cause severe disease (Fokkema, 1978; Fokkema, van de Laar, Nelis-Blomberg and Schippers, 1975). The possession principle seems clearly to be involved here. Blakeman and Fraser (1971) showed that unidentified bacteria on the surface of leaves of some chrysanthemum varieties were responsible for the failure of *Botrytis cinerea* to germinate in water drops and to infect. Infection would occur, however, if dextrose was added to the drops or the leaves were senescent. Blakeman (1972) found that the total number of bacteria and the proportion of a yellow *Pseudomonas* sp. that inhibited germination of *B. cinerea* spores, increased as the age of a beet leaf exceeded nine weeks. Increased leakage of amino acids with leaf age was thought to selectively favour the *Pseudomonas* sp. which inhibited *Botrytis* spore germination.

Virus-like Antagonists

The recent discovery of hypovirulent strains of certain plant pathogens presents a potential for biological control of undetermined general value. However, the virus-like entities associated with hypovirulence in the fungus *Endothia parasitica*, cause of the destructive chestnut blight, apparently are spontaneously reducing this disease to unimportance in Italy (Mittempergher, quoted in MacDonald, Cech, Luchock and Smith, 1979). The fungus reached Italy about 1938, causing extensive damage, and in 1950 A Biraghi observed the spontaneous healing of cankers on chestnut stump sprouts. It was found that the cankers were infected with a fungus strain of decreased virulence that had gradually replaced the virulent strain. Because of this effective control of the disease, the situation has been extensively investigated in France, Italy and the United States. It has been found that the hypovirulent strains have a double-stranded ribonucleic acid only very rarely present in virulent strains. This transmissible hypovirulence is a symptom of a disease or group of diseases of the pathogen that causes reduced pathogenicity, but not necessarily reduced saprophytic vigour. Hypovirulence can be transferred to a virulent strain by hyphal anastomosis *in vitro* or in cankers, especially if the lesion is wounded. Established cankers heal after they are invaded by a hypovirulent strain because of strong callus formation by the host. The hypovirulence factor does not spread in the U.S. because of the many compatability factors there that affect anastomosis. Because of the rarity of perithecia in France and Italy, apparently there is a single anastomosis group there. Ascospores in the U.S. are

the main means of spread of the fungus and provide the
mechanism for development of the many anastomosis groups.
Hypovirulence is not carried by ascospores nor by all conidia.
Hypovirulent strains tend not to produce perithecia and form
a decreased number of pycnidia. The many anastomosis groups
in the U.S. decrease the probability of hyphal fusion and it
is therefore thought best to inoculate with mixed cultures of
several anastomosis groups. Because of the prevalence of
ascospore infections in the U.S., cankers contain the virulent
strains. It is thought that an insect vector is needed in
the United States to more effectively carry the hypovirulent
strains to the cankers, because of the many anastomosis
groups (Anagnostakis, Day, Dodds, Elliston, Jaynes and Van
Alfen, Bowman and Simmons, quoted in MacDonald *et al.*, 1979).

BACTERIZATION

The inoculation of seeds with specific bacteria before
planting in nontreated soil has increased plant growth in
numerous tests. Early work in Russia, partially confirmed
in studies at Rothamsted Experimental Station in England and
by the Commonwealth Scientific and Industrial Research
Organisation in Australia, emphasised the effect of hormones
and increased nutrients (solubilisation and nitrogen fixation)
as causes of increased growth. Broadbent *et al.*, (1971) in
studies on biological control of damping-off of seedlings,
found that *Bacillus subtilis* A13 increased growth of seedlings
over the controls, even in the absence of *Rhizoctonia solani*
or *Pythium ultimum*. This was confirmed and extended by
Broadbent, Baker, Franks and Holland (1977), who found no
nitrogen fixation, phosphate solubilisation or production
of indoleacetic acid and only moderate amounts of gibberellin,
as well as no increase in nutrients in the seedlings. The
antagonists studied were selected for their wide-spectrum
antibiotic effect in culture media on a group of plant
pathogens. There was a direct relationship between antibiotic
efficacy and effectiveness in increasing growth. Their
suggestion that the effect might be due to inhibition of
nonparasitic root pathogens was strongly supported by later
studies and by Baker and Cook (1974). Kloepper and Schroth
(1978) showed that non -antibiotic-producing mutants of
antibiotic-producing and growth-stimulating isolates of
Pseudomonas spp. had also lost their ability to induce an
increased growth response. Their bacteria apparently inhibit
rhizosphere microorganisms (nonparasitic pathogens) that
have an adverse effect on plant growth. Burr, Schroth and
Suslow (1978) and Merriman, Price, Baker, Kollmorgen, Piggott
and Ridge (quoted in Bruehl, 1975) reported outstanding results
in field trials on a variety of crops with *Pseudomonas* spp.

Bacillus spp. and *Streptomyces griseus*. Growth of the plant may sometimes be reduced by soil or seed inoculation with a microorganism, presumably acting as a nonparasitic pathogen. The increased growth response from bacterization may thus prove to be another example of biological control. In any case, this subject is expanding the scope of plant pathology beyond the traditional concern with parasites that penetrate the host and produce disease. The field must now also include microorganisms (nonparasitic pathogens or exopathogens) in the rhizosphere that decrease plant growth but rarely or never penetrate the roots.

REFERENCES

Ayers, W.A. (1971). Induction of sporangia in *Phytophthoria cinnamomi* by a substance from bacteria and soil, *Canadian Journal of Microbiology* 17, 1517-1523.

Baker, K.F. (1971). Soil treatment with steam or chemicals. *In* "Geraniums" (Ed. J.W. Mastalerz), pp. 72-93. Pennsylvania Flower Growers, University Park, Pennsylvania.

Baker, K.F. (1980). Biological control. *In* "Fungal Wilt Diseases of Plants" (Ed. M.E. Mace, A.A. Bell and C.H. Beckman), in press. Academic Press, New York.

Baker, K.F. and Snyder, W.C. (1965). "Ecology of Soil-borne Plant Pathogens . Prelude to Biological Control" University of California, Press, Berkeley.

Baker, K.F., Flentje, N.T., Olsen, C.M. and Stretton, H.M. (1967). Effect of antagonists on growth and survival of *Rhizoctonia solani* in soil, *Phytopathology* 57, 591-597.

Baker, K.F. and Cook, R.J. (1974). "Biological Control of Plant Pathogens" W.H. Freeman & Co., San Francisco.

Blakeman, J.P. (1972). Effect of plant age on inhibition of *Botrytis cinerea* spores by bacteria on beetroot leaves, *Physiological Plant Pathology* 2, 143-152.

Blakeman, J.P. and Fraser, A.K. (1971). Inhibition of *Botrytis cinerea* spores by bacteria on the surface of chrysanthemum leaves, *Physiological Plant Pathology* 1, 45-54.

Broadbent, P. and Baker, K.F. (1974a). Behaviour of *Phytophthora cinnamomi* in soils suppressive and conducive to root rot, *Australian Journal of Agricultural Research* 25, 121-137.

Broadbent, P. and Baker, K.F. (1974b). Association of bacteria with sporangium formation and breakdown of sporangia in *Phytophthora* spp., *Australian Journal of Agricultural Research* 25, 139-145.

Broadbent, P., Baker, K.F. and Waterworth, Y. (1971). Bacteria and actinomycetes antagonistic to fungal root pathogens in Australian soils, *Australian Journal of*

Biological Science 24, 925–944.

Broadbent, P., Baker, K.F., Franks, N. and Holland, J.
(1977). Effect of *Bacillus* spp. on increased growth
seedlings in steamed and in nontreated soil, *Phytopathology*
67, 1027–1034.

Bruehl, G.W. (1975). "Biology and Control of Soil–Borne
Plant Pathogens", American Phytopathological Society,
St. Paul, Minnesota.

Burke, D.W. (1965). Fusarium root rot of beans and behaviour
of the pathogen in different soils, *Phytopathology* 55,
1122–1126.

Burr, T.J., Schroth, M.N. and Suslow, T. (1978). Increased
potato yields by treatment of seed pieces with specific
strains of *Pseudomonas fluorescens* and *P. putida*,
Phytopathology 68, 1377–1383.

Carter, M.V. and Price, T.V. (1974). Biological control of
Eutypa armeniacae. II. Studies of the interaction between
E. armeniacae and *Fusarium lateritium* and their relative
sensitivities to benzimidazole chemicals, *Australian
Journal of Agricultural Research* 25, 105–119.

Chang, I.P. and Kommedahl, T. (1968). Biological control
of seedling blight of corn by coating kernels with
antagonistic microorganisms, *Phytopathology* 58, 1395–1401.

Cook, R.J. (1977). Management of the associated microbiota.
In "Plant Disease – and Advanced Treatise" (Ed. J.G.
Horsfall and E.B. Cowling), Vol. 1, pp. 145–166.
Academic Press, New York.

Cook, R.J. and Papendick, R.I. (1970). Soil water potential
as a factor in the ecology of *Fusarium roseum* f. sp.
cerealis 'Culmorum', *Plant and Soil* 32, 131–145.

Cook, R.J. and Watson, R.D. (1969). "Nature of the
influence of crop residues on fungus–induced root diseases"
Washington Agricultural Experiment Station Bulletin 716,
1–32.

Erwin, D.C. and Katznelson, H. (1961). Suppression and
stimulation of mycelial growth by phytophthora cryptogea
synthesising bacteria, *Canadian Journal of Microbiology*
7, 945–950.

Ferguson, J. (1958). "Reducing plant disease with fungicidal
soil treatments, pathogen–free stock, and controlled
microbial colonization", Ph.D. Thesis, University of
California, Berkeley.

Fokkema, N.J. (1978). Fungal antagonists in the phyllosphere,
Annals of Applied Biology 89, 115–119.

Fokkema, N.J., van de Laar, J.A.J., Nelis–Blomberg, A.L.
and Schippers, B. (1975). The buffering capacity of the
natural micoflora rye leaves to infection by *Coch-
liobolus sativus*, and its susceptibility to benomyl,

Netherlands Journal of Plant Pathology 81, 176-186.

Ford, E.J., Gold, A.H. and Snyder, W.C. (1970). Induction of chlamydospore formation in *Fusarium solani* by soil bacteria, *Phytopathology* 60, 479-484.

Garrett, S.D. (1956). "Biology of Root-infecting Fungi" Cambridge University Press, London.

Gerlagh, M. (1968). Introduction of *Ophiobolus graminis* into new polders and its decline, *Netherlands Journal of Plant Pathology* 74, (Supplement 2), 1-97.

Hare, R. (1970). "The Birth of Penicillin and the Disarming of Microbes" George Allen and Unwin, London.

Hartley, C. (1921). Damping-off in forest nurseries, *U.S. Department of Agriculture Department Bulletin* 934, 1-99.

Henry, A.W. (1932). Influence of soil temperature and soil sterilization on the reaction of wheat seedlings to *Ophiobolus graminis* Sacc., *Canadian Journal of Research* 7, 198-203.

Kerr, A. (1972). Biological control of crown gall: Seed inoculation, *Journal of Applied Bacteriology* 35, 493-497.

Kloepper, J.W. and Schroth, M.N. (1978). Association of *in vitro* antibiosis with inducibility of increased plant growth by *Pseudomonas* spp., *Phytopathological News* 12, 136.

Kommedahl, T. and Mew, I.C. (1975). Biocontrol of corn root infection in the field by seed treatment with antagonists, *Phytopathology* 65, 296-300.

Lapwood, D.H. and Hering, T.F. (1970). Soil moisture and the infection of young potato tubers by *Streptomyces scabies* (common scab), *Potato Research* 13, 296-304.

Lewis, B.G. (1970). Effects of water potential on the infection of potato tubers by *Streptomyces scabies* in soil, *Annals of Applied Biology* 66, 83-88.

Louvet, J., Rouxel, F. and Alabouvette, C. (1976). Recherches sur la résistance des sol aux maladies. I. Mise en évidence de la nature microbiologique de la résistance d'un sol au developpement de la Fusariose vasculaire du Melon, *Annales Phytopathologie* 8, 425-436.

MacDonald, W.L., Cech, F.C., Luchok, J. and Smith, C. (1979). "Proceedings of the American Chestnut Symposium" West Virginia University Books, Morgantown.

Menzies, J.D. (1959). Occurrence and transfer of biological factor in soil that suppresses potato scab, *Phytopathology* 49, 648-652.

Millard, W.A. and Taylor, C.B. (1927). Antagonism of microorganisms as the controlling factor in the inhibition of scab by green manuring, *Annals of Applied Biology* 14, 202-215.

Moore, L.W. and Warren, G. (1979). *Agrobacterium radiobacter* strain 84 and biological control of crown gall, *Annual Review of Phytopathology* 17, 163-179.

Munnecke, D.E., Wilbur, W. and Darley, E.F. (1976). Effect of heating or drying on *Armillaria mellea* or *Trichoderma viride* and the relation to survival of *A. mellea* in soil, *Phytopathology* 66, 1363-1368.

Nair, N.G. and Fahy, P.C. (1976). Commercial application of biological control of mushroom bacterial blotch, *Australian Journal of Agricultural Research* 27, 415-422.

New, P.B. and Kerr, A. (1972). Biological control of crown gall: Field measurements and glasshouse experiments, *Journal of Applied Bacteriology* 35, 279-287.

Ohr, H.D. and Munnecke, D.E. (1974). Effects of methyl bromide on antibiotic production by *Armillaria mellea*, *Transactions of the British Mycological Society* 62, 65-72.

Olsen, C.M. and Baker, K.F. (1968). Selective heat treatment of soil, and its effect on the inhibition of *Rhizoctonia solani* by *Bacillus subtilis*, *Phytopathology* 58, 79-87.

Papendick, R.I. and Cook, R.J. (1974). Plant water stress and the development of Fusarium foot rot in wheat subjected to different cultural practices, *Phytopathology* 64, 358-363.

Pegg, K.G. (1977). Biological control of *Phtophthora cinnamomi* root rot of avocado and pineapple in Queensland, *Australian Nurserymen's Association Ltd., Annual Conference Seminar Papers* 1977, 7-12.

Reinking, O.A. and Manns, M.M. (1933). Parasitic and other fusaria counted in tropical soils, *Zeitschrift für Parasitenkunde* 6, 23-75.

Roberts, W. (1874). Studies on biogenesis, *Philosophical Transactions of the Royal Society, London* 164, 457-477.

Rouxel, F., Alabouvette, C. and Louvet, J. (1977). Recherches sur la résistance des sols aux maladies. II. Incidence de traitements thermiques sur la résistance microbiologique d'un sol à la Fusariose vasculaire du Melon, *Annales Phytopathologie* 9, 183-192.

Sanford, G.B. (1926). Some factors affecting the pathogenicity of *Actinomyces scabies*, *Phytopathology* 16, 525-547.

Sanford, G.B. and Broadfoot, W.C. (1931). Studies of the effects of other soil-inhabiting microorganisms on the virulence of *Ophiobolus graminis*, *Scientific Agriculture* 11, 512-528.

Shea, S.R. and Malajczuk, N. (1977). Potential for control of eucalypt dieback in Western Australia, *Australian Nurserymen's Association Ltd., Annual Conference Seminar Papers* 1977, 13-19.

Shipton, P.J., Cook, R.J. and Sitton, J.W. (1973). Occurrence and transfer of a biological factor in soil that suppresses take-all of wheat in eastern Washington, *Phytopathology* 63, 511-517.

Smith, S.N. (1977). Comparison of germination of pathogenic *Fusarium oxysporum* chlamydospores in host rhizosphere soils conducive and suppressive to wilts, *Phytopathology* 67, 502-510.

Smith, S.N. and Snyder, W.C. (1972). Germination of *Fusarium oxysporum* chlamydospores in soils favourable and unfavourable to wilt establishment, *Phytopathology* 62, 273-277.

Stotzky, G. and Rem, L.T. (1966). Influence of clay minerals on microorganisms. I. Montmorillonite and kaolinite on bacteria, *Canadian Journal of Microbiology* 12, 547-563.

Toussoun, T.A. Bega, R.V. and Nelson, P.E. (1970). "Root Diseases and Soil-Borne Pathogens" University of California Press, Berkeley.

Weindling, R. (1932). *Trichoderma lignorum* as a parasite of other soil fungi, *Phytopathology* 22, 837-845.

Weindling, R. (1934). Studies on a lethal principle effective in the parasitic action of *Trichoderma lignorum* on *Rhizoctonia solani* and other soil fungi, *Phytopathology* 24, 1153-1179.

Weinhold, A.R., Oswald, J.W., Bowman, T., Bishop, J. and Wright, D. (1964). Influence of green manures and crop rotation on common scab of potato, *American Potato Journal* 41, 265-273.

Zentmyer, G.A. (1965). Bacterial stimulation of sporangium production in *Phtophthora cinnamomi*, *Science* 150, 1178-1179.

ON UNDERSTANDING PREDATOR-PREY INTERACTIONS

F. M. Williams

Department of Biology, and Ecology Program
The Pennsylvania State University
University Park, PA   16802, U.S.A.

INTRODUCTION

The ways to understanding predator-prey interactions have been rocky ones, with the 'microbiologists' and 'ecologists' taking high and low roads. They have not often met, and they certainly have not arrived. I shall try to trudge a few more steps, with one foot on each road. My approach will be selective and idiosyncratic, with a destination of coherence rather than completeness.

Predator-prey dynamics should, by its quantitative nature, be approachable by the theory-observation paradigm that supposedly characterizes a maturing science. Progress has been far from satisfactory, however, partly because of early methodological errors and partly because of what we now perceive to be the inherent complexity of the dynamical interactions.

Historical Background

It all began, for our present purposes, with the more or less independent publication of the same model by Alfred J. Lotka (1925) and Vito Volterra (1926). The pair of equations comprising this model are called the "Lotka-Volterra predator-prey equations."

Both Lotka and Volterra made explicit assumptions that stripped the dynamics to their bare minima - birth, death, and a random predatory encounter.

These familiar equations are

$$\frac{dN_1}{dt} = r_1 N_1 - p_1 N_1 N_2 \qquad (1)$$

and $\dfrac{dN_2}{dt} = p_2 N_1 N_2 - d_2 N_2,$ (2)

where N_1 is prey population density,
        N_2 is predator population density,
        r_1 is prey intrinsic rate of increase,
        d_2 is predator mortality rate,
and $p_1 > p_2$ the predation interaction constants.

The equations' predictions are well known, occurring in
every elementary textbook (e.g., Ricklefs, 1979). Both populations oscillate with periods and amplitudes determined
by their initial displacement from equilibrium; the equations
are neutrally stable.

Early 'confirmation' attempts

Now oscillations are exciting stuff, and a lot of early
effort went into 'confirming' the model in laboratory populations that approximated the simplified conditions envi-
sioned by Lotka and Volterra. Obvious examples of this
effort are the classic and monumental experiments of
G. F. Gause (1934) on yeast and protozoa and of C. B.
Huffaker (1958) on herbivorous and carnivorous mites. The
experiments failed to yield oscillatory behavior - in fact
failed to yield coexistance of prey and predator at all.

A conceptual error

This is where the ecologists' methodological error crept
in. They tried to patch up the experiments - e.g., with
environmental heterogeneity - to achieve the predicted
oscillations. But what was inadequate was the theory: A
'good' theory should have predicted the experimental outcome, even if it was extinction.

I shall return to this point later, after a more formal
development of theory to lay the groundwork for assessing
the present state of the art. Because Lotka and Volterra
developed their predation models in a more general framework including competition, mutualism, etc., their simplifying assumptions tend to get lost unless explicitly re-
stated. As we shall see, the validity of these assumptions
may be critical.

In the following four sections I shall state the assumptions of the most common predator-prey models, describe
some critical predictions, describe a few experimental results at variance with the predictions, and finally discuss

which assumptions may be in error. The last point is not
easy to determine.

ASSUMPTIONS

I start by an explicit and rigorous statement of the
assumptions implicit in the Lotka-Volterra equations presented above. I shall then modify the assumptions to include most current predation models. In this way I hope to
facilitate the assessment of the models by the experimental
results. A test of the model is nothing more than a test of
the assumptions. The logic of the development is identical
to that I have used for single species populations
(Williams, 1972, 1973).

I recognize three major classes of assumption, each with
a distinct and different purpose: *simplifying environmental,
simplifying biological,* and *explanatory* (anacalyptic)
assumptions (Williams 1972, 1973). I explain these by
example for the Lotka-Volterra equations:

Simplifying environmental (S.E.) assumptions

S.E. 1.   The environment, with respect to all
          properties perceptibly affecting the
          organisms, is uniform (or random) in
          space.

S.E. 2.   The environment, with respect to all
          properties perceptibly affecting the
          organisms, is constant in time.

Here we use the term 'environment' in the restrictive sense,
not including the prey and predator populations, since they
are the two state variables of the system.

Clearly we do not wish to state here that the real world
is so constituted (although see Odum, 1957, for a natural
environment that is very close). These simplifying environmental assumptions do, however, lay down the conditions
that one would specify for a well-controlled laboratory experiment. We make the environmental background constant so
that the biological mechanisms under study may more clearly
be seen and tested. Once we are satisfied that our understanding of the mechanisms is adequate, we may relax the
assumptions in the hope that predicted behavior in a variable - perhaps even a natural - environment is also satisfactory. This is also subject to empirical verification.

Simplifying Biological (S.B.) *assumptions*

      S.B. 1.   All organisms, with respect to their
                 impact on the environment and on each
                 other, are identical throughout the
                 populations (and thus throughout space).

      S.B. 2.   All organisms, with respect to their
                 impact on the environment and on each
                 other, are identical throughout time.

Although similar in form to the S.E. assumptions, the S.B.
assumptions are totally different in purpose and relationship to reality. The simplifying biological assumptions
cannot be approximated in any laboratory; they are simply
outright lies about the nature of any biological population,
whether in nature or laboratory.

The purpose of these S.B. assumptions is to limit the
number of state variables needed to characterize the populations. As they are written above, the assumptions tell
us that it makes no difference whether we measure predator
numbers and prey biomass or predator protein and prey ATP.
All quantities will be related to each other as simple pro-
portionalities. For examples of the demonstrated inadequacy
of these assumptions and their subsequent relaxation, see
Williams (1971, 1972, 1973) and Hunt (1977).

Two further simplifying biological assumptions are needed
to satisfy the Lotka-Volterra equations:

      S.B. 3.   Predation loss is the only limitation
                 on the prey population - in the absence
                 of predation the prey increases exponen-
                 tially.

      S.B. 4.   Predation input determines completely
                 the predator growth rate - in the
                 absence of predation the predator dies
                 exponentially.

The purpose of these assumptions is to isolate the phenomenon
of predator-prey dynamics for study. We do not wish to cloud
the picture with intraspecific competition or other confound-
ing influences.

Anacalyptic (A) *assumptions*

    A. 1 The predatory encounter is random in
          time and space - like a bimolecular
          collision it is proportional to the
          product of predator and prey populations.

Anacalyptic (Williams, 1971) assumptions are the assumptions
that embody the mechanisms we propose - they are the hypotheses to be tested.

Functions of assumptions

    I summarize the differences between the three classes of
assumption by the action required in the event of their
failure. Failure of a simplifying environmental assumption
means that we have not designed our observations in as controlled a manner as required - we must redesign to ensure
unobscured observation of the phenomena of interest. Failure of a simplifying biological assumption means that we
have not included sufficient complexity in our representation of the biological systems - we must reformulate the
models to include more ancillary biological sophistication
and complexity. Thus failure of either class of simplifying assumption is a nuisance that interferes with our understanding of the subject at hand. But failure of an anacalyptic assumption means that we have *mis*understood the nature
of the phenomenon - we must reformulate our notions about
the mechanisms involved. They are the 'back to the drawing
board' assumptions.

Lotka-Volterra invalidation

    The inadequacy of the Lotka-Volterra equations has been
apparent for some time. In part it is sufficient to interpret correctly the early experimental results of Gause
(1934) and Huffaker (1958) cited above. In addition, non-
oscillatory steady states and/or apparent limit cycles (as
opposed to neutrally stable oscillations) have been observed
in mixed species chemostats (Boraas, 1979, 1980; Bazin and
Saunders, 1978; reviews by Curds and Bazin, 1977; and
Fredrickson, 1977). These behaviors are all at variance
with the predictions, and hence assumptions of the Lotka-
Volterra equations.
    To this I add emphasis by singling out experiments in my
laboratory by Dr. Martin Boraas that constitute about as
rigorous a test as is possible. Using the green alga

Chlorella pyrenoidosa as prey for the rotifer *Brachionus calicyflous* in well-controlled chemostat cultures (Boraas, 1979; 1980), Dr. Boraas was able to establish stable non-oscillatory steady states for both populations. The combination of chemostat washout and predatory removal of the *Chlorella* required that they be growing at their maximal growth rate for that medium and temperature. Thus assumption S.B.3., unimpeded exponential increase of the prey population, is satisfied as nearly as is possible for any real biological system. Now when the steady state was severely perturbed by an added pulse of algal limiting nutrient (nitrate) or by an added pulse of the algal prey themselves, restoration of the original pre-pulse steady state values was both rapid and non-oscillatory (Boraas, 1979). Since all of the simplifying assumptions are satisfied as much as is possible, the anacalyptic assumption A.1. concerning the mechanism of predator-prey encounter must be in error. The Lotka-Volterra equations are incorrect.

Functional and numerical responses

Direct short-term measurements on predation rate ('functional response' in ecology) and/or specific growth rate ('numerical response') as a function of prey density have shown that the appropriate predation function is some sort of saturation relationship, approaching a maximum value at some level of prey population availability (e.g., Boraas, 1979; Canale, *et al*, 1973; Garver, 1966; Salt, 1974; and for many metazoans: Holling 1959; Ricklefs, 1979). These results are of course contrary to the linear predation function in the Lotka-Volterra equations.

Therefore, although there may be in some of the cases cited reasons to suspect violation of simplifying assumptions, there is abundant evidence that the anacalyptic assumption of Lotka-Volterra dynamics is in error.

The most common, but by no means the only, function currently used to represent the predatory encounter mechanism is the Monod (1942) rectangular hyperbola, which I shall here write as

$$\frac{F_2 \; M_1}{K_2 + M_1} \tag{3}$$

where F_2 is either the predator's maximum feeding rate or maximum specific growth rate, depending on context,

K_2 is the predator's 'half-saturation constant',

and M_1 is the prey population density, now expressed
       as biomass.

Presumably this function was settled on initially by analogy
with its use to represent bacterial growth on dissolved
nutrients (Monod, 1942, 1950; Novick and Szilard, 1950;
derivation by Williams, 1972; review by Fredrickson and
Tsuchiya, 1977). It is thus interesting to note that this
same rectangular hyperbola was independently derived for a
metazoan predator's feeding rate by Holling (1965). His
derivation is more appropriate to the predation act than the
usual analogy drawn between the Monod equation and enzyme
kinetics (Williams, 1972). Holling's derivation involves a
'search time' and a 'handling time' needed by the predator
for each prey item consumed. A maximum predation rate occurs
when all available time is occupied in 'handling'. In the
Appendix I present his plus some original alternative non-
enzyme-kinetic derivations, one for predation rate controlled
by satiation, and also one for filter feeding.
    It seems it is not difficult conceptually to arrive at a
rectangular hyperbola, but that does not guarantee its
correctness. I have shown that the use of the rectangular
hyperbola even for dissolved nutrient uptake systems involves a conceptual error (Williams, 1972); however, since
there are no data to estimate quantitatively the magnitude
of the error involved, I will remain with the simple Monod
function until we question it later in the paper.
    Holling's and my derivations are for what the ecologist
calls the 'functional response', that function relating
feeding rate of an individual predator to prey density. In
this context the F_2 in Formula (3) is interpreted to be the
individual predator's maximum feeding rate. Typically the
microbiologist has used the formula directly as the 'numerical response', that function relating the predator population's specific growth rate to prey density. In that context
the F_2 is the predator population's maximum specific growth
rate.
    The ecologist usually envisions a large, even qualitative,
difference between functional and numerical responses (e.g.,
May, 1977). Although it is obvious that there are time lags
and material losses involved in the translation of a feeding
rate into growth and reproduction (Williams, 1971; 1973), I
shall adhere here to the usual microbiologist's convention:
the functional responses and numerical responses are related
to each other by a constant of proportionality that will be
absorbed into the numerical value of F_2.

Saturation kinetics equations

It is usual to write the equations for the predatory relationship in an open system with material inputs and out-
puts, a chemostat in which the prey population is nutrient
limited.   Thus for the predator population (M_2),

$$\frac{dM_2}{dt} = \frac{F_2 M_1 M_2}{K_2 + M_1} - R_2 M_2 - k_o M_2 \qquad (4)$$

For the prey population (M_1),

$$\frac{dM_1}{dt} = \frac{F_1 C\, M_1}{K_1 + C} - \frac{F_2 M_1 M_2}{Y_2\,(K_2 + M_1)} - R_1 M_1 - k_o M_1. \qquad (5)$$

For the prey limiting nutrient (C),

$$\frac{dC}{dt} = k_o C_o - k_o C - \frac{F_1 C\, M_1}{Y_1\,(K_1 + C)}. \qquad (6)$$

And for the combined respiratory & excretory products (W),

$$\frac{dW}{dt} = R_1 M_1 + R_2 M_2 - k_o W. \qquad (7)$$

In these equations,

  Y_1, Y_2 are the yield constants,
  F_1, F_2 are the maximum specific growth rates,
  K_1, K_2 are the half-saturation constants,
and R_1, R_2 are the pooled respiratory, excretory and
        mortality loss rates,

for the prey and predator populations respectively.   The
chemostat turnover (dilution) rate is k_o and the input concentration of prey limiting nutrient is C_o.   The numerical
response of prey or predator is shown in Fig. 1.
    The assumptions corresponding to this saturation kinetics
model of predator-prey interaction are as follows:

Saturation kinetics assumptions

  S.E.1.   The environment...is uniform (or random)
           in space (unchanged).
  S.E.2.   The environment...is constant in time,
           except for limiting nutrient concentration
           (C) and waste materials (W).
  S.E.3.   The system is either closed ($k_o = 0$) or,
           if it is open, it is a chemostat ($k_o > 0$).

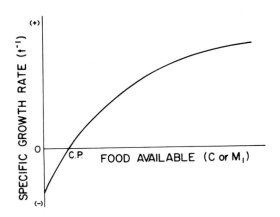

Fig. 1. The saturation kinetic numerical responses. 'C.P.' indicates 'compensation point', where inputs just balance losses.

S.B.1.   All organisms...are identical throughout the populations - and thus throughout space (unchanged).

S.B.2.   All organisms...are identical throughout time.

S.B.3.   Nutrient limitation and predation loss are the only limitations on the prey population.

S.B.4.   Predation input determines completely the predator growth rate...(unchanged).

S.B.5.   Functional and numerical responses are identical in form, related by a constant of proportionality.

S.B.6.   Respiratory, excretory, and mortality loss rates (R_1, R_2) are constant.

S.B.7.   There is no biological effect on the combined waste materials (W).

S.B.8.   The ecological efficiencies (yields) of predator and prey are constant (Y_1, Y_2).

A.1.   Prey growth rate is a simple saturation kinetics function of its limiting nutrient concentration.

A.2.   Predator growth rate is a simple saturation kinetics function of the prey density.

I believe this is the minimal set of assumptions needed
to specify even the very simple set of equations (4) - (7),
embodying only the simplest and most general ideas about
predatory interactions. With a dozen assumptions, most of
which fall in the simplifying biological (S.B.) category,
it becomes apparent that it might be difficult to determine
which assumption(s) might be in error should experiment and
prediction be at variance with one another. This is indeed
the case, as we shall see.

PREDICTIONS

There are a number of variations on the basic theme of
the model represented by Equations (4)-(7), many of which
are simpler by virtue of lacking the respiratory loss terms
(e.g., Canale, 1969; 1970; Tsuchiya et al, 1972; Rosenzweig,
1971) and many of which are more complicated for a variety
of reasons (Hunt, 1977; Canale et al, 1977; Curds and Bazin,
1977; Fredrickson, 1977). I shall not elaborate these variations at this point because (i) with few notable exceptions
the dynamical behaviors of the models are qualitatively
similar, (ii) As yet there are no data sets complete enough
to test the details of different models, and (iii) for
present purposes I suggest the use of a technique I believe
to be simpler, more general, and more powerful - a graphical
technique originally proposed by Rosenzweig and MacArthur
(1963).

Predictions of Equations

The saturation kinetics models predict steady states,
approached either smoothly or via damped oscillations, or
else oscillatory limit cycle behavior. All three have been
observed (Boraas, 1979, 1980; Bazin and Saunders, 1978;
reviews by Curds and Bazin, 1977 and Fredrickson, 1977).
Once the species have been chosen, i.e., the biological
parameter values selected, the mode of behavior is entirely
a function of the two non-biological parameters of the
system, the turnover (dilution) rate k_o and the input prey
nutrient level C_o. The system is 'driven' by those input
parameters. The relationship of the steady state behavior
to these two parameters is displayed graphically in what is
known as an 'operating diagram' (Fredrickson and Tsuchiya,
1977). In general, high values of input nutrient and low
values of turnover rate tend to predict oscillatory behavior,
whereas high values of turnover rate and low values of input
nutrient tend - within the limits that the species can

survive - to damp into steady states.

Rosenzweig and MacArthur (1963) proposed, as an alternative to the Lotka-Volterra equations, a graphical model of predator-prey interactions that gave qualitatively different predictions. Their basic model is shown in Fig. 2: On a phase plane plot of predator vs prey density are drawn the locus of points (M_1, M_2) at which the prey population has a zero growth rate and also that at which the predator population has a zero growth rate. These are called the 'predator (zero growth) isocline' and the 'prey (zero growth) isocline', respectively.

Typically in its simplest form the predator isocline is a vertical line whose abscissal value equals that prey population density at which the predator population could just maintain itself, neither increasing nor decreasing. The extent to which the line is vertical and straight is the extent to which assumption S.B.4, that food completely determines the predator growth rate, is satisfied.

The prey-isocline, by contrast, has a 'hump' - an ascending phase, a maximum ordinate value, and then a descending phase to intercept at the prey population density ('carrying capacity') as it would be in the absence of predators.

It is the intersection of the prey and predator isoclines that determines the dynamical character of the system. In Fig. 2 three possible locations are shown for the intersection of prey and predator isoclines. If intersection occurs on the ascending side (a), the population densities spiral outward in time until one or both become extinct. If intersection occurs at the peak (b), the population densities oscillate stably around the intersection. If intersection occurs on the descending side (c), the population densities approach steady states at the intersection, either smoothly or via damped oscillations. Thus the graphical model seems capable of producing all the behaviors that are observed.

Rosenzweig and MacArthur (1963) and Rosenzweig (1977) have analyzed a number of data sets from the ecological literature and have shown in general that the experimental results were in accord with the shapes and locations of the isoclines. But it has always been a bit obscure to ecologists whence came the shapes of the isoclines, especially the prey isocline with its 'hump'. Even Rosenzweig's (1969) attempt to clarify, in his paper entitled "Why the prey curve has a hump", did not help much; it has remained a $Just-So-$ $Story$.

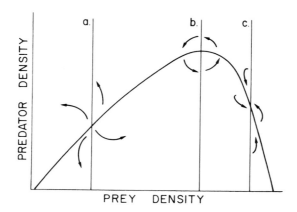

Fig. 2. The predator-prey isoclines with three possible behaviors dependent on predator isocline location (After Rosenzweig and MacArthur, 1963).

Equations and isoclines

It is now possible to shed a bit more light on the shapes of the Rosenzweig-MacArthur isoclines: They are just the shapes that one would calculate from Eqs. (4)-(7). Setting Eq. (4) equal to zero we get a vertical line, the predator isocline. Setting Eq. (5) equal to zero and making some messy substitutions to eliminate C, we get a 'hump' shaped quadratic, the prey isocline. Thus the saturation kinetics equations and the graphical solutions are qualitatively isomorphic; they imply the same underlying mechanisms to the predator-prey dynamics.

Looking at both the equations and the isoclines together, it is perhaps possible to get a clearer picture of the im-plied dynamics.

The predator isocline

As stated earlier the predator isocline is a vertical straight line to the extent that assumption S.B.4. is satis-fied: each predator behaves essentially independently of each other predator in its response to prey density. In the ecologists' terminology, competition among predators is purely the 'exploitation' or 'scramble' type. If there is

mutual 'interference' (e.g., collision or chemical warfare),
then the isocline will slope to the right. If there is
mutual reinforcement (e.g., medium conditioning or hunting
packs), then the predator isocline will slope to the left.

The prey isocline: enrichment

For our present purposes, however, the behavior of the
prey isocline is of more interest. Suppose we have the
stable predator-prey interaction depicted in Fig. 3a, where
the predator isocline falls to the right of the hump. An
increase in input nutrient concentration (C_0) of about 40%
has no effect at all on the predator isocline. But as seen
in Fig. 3b, the nutrient increase shifts the prey isocline
dramatically to the right such that the predator isocline
now falls to the left of the hump. What had been a stable
predator-prey interaction is made unstable by the enrichment of the input nutrient concentration to the system.

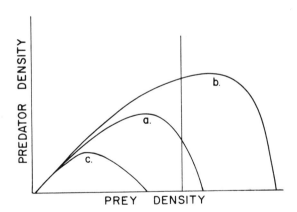

*Fig. 3. The effects of prey resource enrichment (b) and
prey resource impoverishment (c) on a previously stable
system (a).*

This destablizing effect is what Rosenzweig (1971) called
the "paradox of enrichment". Its implication for the
eutrophication process is obvious.
    Now consider the situation shown in Fig. 3c. There the
input nutrient concentration has been reduced by about 30%
from the original. The prey isocline has been shifted to
the left such that it no longer intersects the predator

isocline; prey may exist at a low density but predators
must go extinct in the nutrient-impoverished environment.
These predictions are qualitatively identical to those
gotten by adjusting input nutrient ('feed') concentrations
in the 'operating diagram' (Fredrickson and Tsuchiya, 1977).

Turnover rate

Now let us consider the other driving parameter, the
turnover rate k_o. Fig. 4a depicts a stable predator-prey
system with predator isocline to the right of the hump. In
Fig. 4b the turnover rate has been decreased by about 50%;
this shifts both isoclines, the predator isocline to the
left, and the prey isocline upward to the right.
The predator isocline is by far more sensitive.

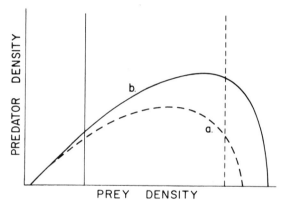

*Fig. 4. The effect of lowering the dilution rate (b) on a
previously stable system (a).*

The effect of decreasing the dilution or turnover rate of
the system is thus predicted to be destabilization. This is
again consistent with predictions from the operating dia-
grams. Ironically the predator can get by with fewer prey
at lower turnover rates, but it is less stable as a result.

Predator affinity

One further prediction concerns the predator efficiency –
or affinity for prey at low prey densities. This is measured
by the half-saturation constant K_2 in Eqs. 3-5. If the stable
predator-prey system depicted in Fig. 5a were to be altered

such that the predator became more effective at low prey
densities, i.e., the value of K_2 were decreased, then the
predator could sustain itself at lower prey densities than
before. The effect of evolving the more efficient predator

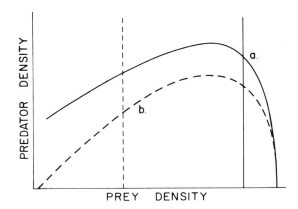

*Fig. 5. The effect of increasing the predators' efficiency
(b; decreasing K_2) at low densities on a previously stable
system (a).*

is shown in Fig. 5b; the prey isocline is lowered somewhat —
fewer predators can control the prey — and the predator
isocline is significantly decreased — fewer prey can sustain
the predators. The result is destabilization: a more effective predator does not necessarily ensure his own liveli-
hood.

Summary

   In this section I have described the simplest and most
general set of equations — based on saturation kinetics —
that are commonly used to describe predator-prey systems.
I then showed that those equations translate directly into
the graphical techniques introduced by Rosenzweig. Several
qualitative predictions were made, consistent with both
models. Briefly, these were: *enrichment, low turnover, and
predator efficiency all tend to destabilize interactions.*
In the next section I shall examine some selected results
and argue that, in our present state of understanding, the
more general and qualitative graphical models may be more
useful and insightful.

SELECTED EXPERIMENTS

   A very limited number of results will be discussed in
this section as they bear upon the general predictions ex-
tracted from the models of the previous section. Some results accord with the predictions and some do not. In general the discord is more eloquent, belonging typically to
the larger data sets. The first two studies are non-chemostat studies, with methods similar to those of Gause (1934).
The second group of six studies involves carefully controlled chemostat experiments.

Non-chemostat confirmations

   The first to be mentioned is the experiment, elegant in
its predictive decisiveness, by Maly (1969), using the
metazoan rotifer *Asplanchna* as predator on *Paramecium*.
Maly performed numerous short-term experiments at different
prey and predator densities, just long enough to establish
vector directions on the predator-prey phase plane. Among
these vectors he interpolated the prey and predator isoclines. The predator isocline fell very far to the left of
the maximum. He thus predicted extinction, and the experimental results decisively confirmed the prediction, looking
very much like Gause's *Didinium-Paramecium* results. The
graphical prediction was adequate for extinctions.
   Luckinbill (1973, 1974) repeated the *Didinium-Paramecium*
experiments with the same outcome as Gause (1934), extinction in less than two generations. He then reduced the
predator capture efficiency by growing the populations in a
medium laced with methylcellulose; this allowed the predator
and prey to coexist with increasing amplitude oscillations
for 40-50 generations before the *predator* became extinct and
only prey remained. Enrichment of this system produced
similar results, except that both predator and prey became
extinct. But *impoverishment* of *Paramecium* food by 50% resulted in continued limit cycle coexistence for a couple
hundred generations, the duration of the experiment.
   We interpret Luckinbill's experiments as follows: Introduction of methylcellulose reduced the predator capture
efficiency (parameter a in derivations in appendices),
thereby increasing K_2, the predator half-saturation constant;
this stabilized by shifting the predator isocline to the
right near the maximum. Subsequent enrichment shifted the
prey isocline to the right, destabilizing the system slightly. On the other hand, impoverishment shifted the prey isocline to the left, finally creating limit cycle stability.

I mention these two examples in detail because they are
not very familiar in the microbiological literature and because they seem to be unequivocally (although sparsely) consonant with the theoretical predictions.

Single stage chemostat studies

The following experimental sets are all single stage
chemostat studies. Want of space precludes considering two
stage studies.

Tsuchiya *et al* (1972) studied the dynamics of the slime
mold *Dictyostelium* feeding on glucose limited *E. coli*.
The observed sustained oscillations were reasonably similar
to those predicted by the saturation kinetics model (Eqs. 4-
7) using independently measured parameters. An increase in
dilution rate (k_o) produced an increase in oscillatory frequency, consistent with the model (Curds, 1971). An observed abrupt change from oscillations to steady state has been
interpreted as a developmental change (Curds and Bazin, 1977),
but observation in my laboratory of a similar phenomenon in
a totally different system (Boraas, 1979) may cast this
interpretation in doubt; this will be discussed below.

Canale *et al* (1973) studied the ciliate *Tetrahymena*
feeding on sucrose limited *Aerobacter*. At very high dilution rates the predator became extinct, consistent with the
predator isocline being too far to the right to intersect
the prey isocline. At somewhat lower dilution rates apparently damped oscillations occurred, and at yet lower dilution rates critically damped (non-oscillatory) steady states
occurred. This sequence is directly opposite to the predictions of the equations and graphical models above. The
authors suggested that predator lysis and a 'refractory'
carbohydrate were needed to explain the results, but these
incorporated into a new model produced only modest improve-
ment.

Jost *et al* (1973a) also studied *Tetrahymena* but on
glucose limited *Azotobacter* and/or *E. coli*. Sustained
oscillations were observed at higher dilution rates and
damped oscillations were observed at lower dilution rates
(where sustained oscillations were predicted). This sequence is directly opposite to the predictions and the
graphical models. The authors suggest (Jost, *et al*, 1973b)
that a 'multiple saturation' predation model may account for
the discrepancy, and a new model did seem to rectify the
predictions.

In apparently less complete data sets, Bazin (Bazin *et al*,
1974; Dent *et al*, 1976; Curds and Bazin, 1977; Bazin and

Saunders, 1978) has observed damped oscillatory behavior in
Dictyostelium feeding on glucose limited *E. coli*. Also observed were abrupt changes in behavior, suggested by the
authors as developmental changes in the slime mold amoebae
to their aggregation phase; this will be further discussed
below.

In an apparently single chemostat run, van den Ende (1973)
observed possibly damped oscillations in a *Tetrahymena-
Klebsiella* culture. Wall growth was suggested as a means
of ensuring predator-prey coexistence, but no evidence was
offered.

A metazoan predator

Most recently Boraas (1979, 1980) has completed an extensive study of predation in my laboratory, with what may
be the first single stage continuous culture of a metazoan
predator. The predator was the rotifer *Brachionus* feeding
on the nitrate limited green alga *Chlorella*. At a high
dilution rate the predator population became extinct, indicating again the non-intersection of predator and prey
isoclines. At a lower dilution rate the predator population
underwent $1\frac{1}{2}$ - 2 cycles of a large amplitude oscillation,
then abruptly entered a steady state. At a yet lower dilution rate there were more typical damped oscillations
(Fig. 6) into a stable steady state. It is difficult to
classify these results with respect to concord or discord
with theory: If the results at intermediate flow are fundamentally sustained oscillations aborted by some life history
phenomenon, then the results are qualitatively at variance
with predictions. If the results are fundamentally critically damped with some unknown perturbation exerting an
influence on the transient phase, then the results are in
qualitative accord with predictions. Or neither of the
above.

The steady states are stable, and resilient (Boraas,1979).
A pulse of nitrate added to the steady state (more than
doubling the steady state value), or a pulse of prey population (about five times its steady state value) was rapidly
without oscillation damped out to the original steady state.

Step function changes of input nutrient C_0 or of dilution
rate k_0 produced smooth adjustments of steady state values
without oscillations. While these results indicate the
extreme stability of the rotifer-alga interaction, they tell
us little about agreement with theory, except that conditions always remained in a region of stable isocline intersection.

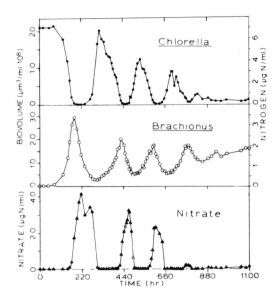

Fig. 6. Time course to steady state of a chemostat rotifer-alga predation interaction (from Boraas, 1979).

Simulations of the rotifer-algal interaction were more telling (Boraas, 1979). While saturation kinetics models similar to those presented here were adequate to account for two-stage continuous culture behavior, the saturation kin-etics model with independently estimated parameters was incapable of predicting the single stage outcomes. Again, predictions of sustained oscillations were realized as damped into a steady state. Simulations assuming (i) var-iable yield – roughly equivalent to a multiple saturation model, (ii) wall growth, and (iii) recycling of excreted rotifer nitrogen into algae all tended to stabilize the population simulations nearer to steady states.

Summary

Of the six sets of chemostat studies quoted here, only one came close to apparent confirmation of the model pre-dictions. The others showed qualitative contradictions to the model predictions. On the other hand the two non-chemostat results were in apparent agreement with the model predictions.

DISCUSSION

There may be an element of historical irony in all the
foregoing. Gause's and others' early experiments indicated
that simple laboratory predator-prey systems were far less
stable than predicted by the Lotka-Volterra equations.
Recent chemostat experiments indicate that the simple laboratory predator-prey systems are *more* stable than predicted by the saturation kinetics relationship explicitly
formulated for chemostat conditions: predicted sustained
oscillations are realized in most experiments as oscillations damped into a steady state or as non-oscillatory
approaches to a steady state. This is distressing in view
of the fact that the newer theory seemed to have resolved
the problems with the earlier experiments, as well as recent
ones of a similar nature (Maly, 1969; Luckinbill, 1973,
1974).

Where is the fault?

The fundamental question to be asked of the discrepancies is: *Are the simplifying environmental* (S.E.) *or the
simplifying biological* (S.B.) *assumptions at fault; or are
the anacalyptic assumptions* (A) *at fault?* That is, are the
species-or design-specific details, that are merely a
nuisance, clouding an observation of a dynamics that we
fundamentally understand? Or do we simply not understand
the basic dynamics? Both answers have been posed.

Canale *et al* (1972) suggest a heterogeneous prey nutrient
and a time dependent predator death: these call into
question S.E.2, S.B.2, S.B.6, and S.B.7, at the very least.
Jost *et al* (1973) suggest a different 'multiple saturation'
functional response for the predator: this calls into
question A.2. Subsequently Bonomi and Fredrickson (1976)
and Fredrickson (1977) repudiate this opinion, suggesting
wall growth as the major problem: this calls into question
S.E.1, and hence S.B.1 for the prey population. A model
including wall growth seems to work.

Curds and Bazin (1977) suggest that the abrupt changes
reported in Tsuchiya *et al* (1972), Dent *et al* (1976), and
Curds and Bazin (1977) are the result of life cycle changes
from the solitary to the aggregating phase of the slime mold
amoebae: this calls into question S.B.2. On the other hand
this interpretation may be subject to doubt because a very
similar situation occurred in our metazoan rotifer cultures

(Boraas, 1979) where no comparable life cycle event can be identified. Abrupt changes such as these cannot be predicted from any reasonably simple model with constant parameters. I suggest, however, that they are exactly the sorts of apparently abrupt changes predicted from consideration of evolutionary (selective) changes occurring in co-evolving predator-prey systems (Wilcox and MacCluer, 1979): this calls into question S.B.2 with a vengeance. However, the suggestion by van den Ende (1973) that prey wall growth and smaller predator size during starvation are *evolutionary* responses is totally without foundation in the evidence.

The anacalyptic assumption concerning predation rate (A.2) is also called into question for *Dictyostelium*. Bazin and Saunders (1978) analyze damped predator-prey isoclines by catastrophe theory to infer that the critical variable in predation rate is the prey: predator *ratio*, not simply prey density. This is in accord with observations by Curds and Cockburn (1968) and by Bazin *et al* (1974) who also suggest an overshoot in the functional response before it settles back into a saturating value (a maximum somewhere to the left in Fig. 1). This should be *destabilizing* (Williams, 1977 and unpublished) and cannot thus account for the observed discrepancies. It remains a puzzle, however. I suggest that it may be the result of predator anoxia at high prey densities; if so it would not be seen in predators on photosynthetic prey.

For our rotifer-algal dynamics, Boraas (1979) suggested that a variable yield function roughly equivalent to the multiple saturation model of Jost *et al* (1973b) might help explain the results: despite the form in which it was simulated, this calls into question S.B.8. Other suggestions are that excreted nitrogen from the predator may feed back as prey nutrient: this calls into question S.B.7, and possibly S.B.6.

Environmental heterogeneity?

Finally Boraas (1979) also suggests that wall growth of algae may account for the results: this calls into question S.E.1. and hence S.B.1. This situation was analyzed graphically by Rosenzweig and MacArthur (1963), who showed that a prey refuge stabilizes by bending the prey isocline upward at the lower end; an example is shown in Fig. 7. Since wall growth was not apparent in algal chemostats without predators, Boraas suggests that wall growth is a 'dynamic refuge' in response to predator presence. This is consistent with

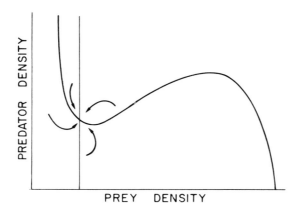

*Fig. 7. Isoclines for a system with prey refuge; stabili-
zation occurs via the vertical line on the left hand side
of the prey isocline (after Rosenzweig and MacArthur, 1963).*

earlier observations that predator presence caused prey
clumping: predators cause prey to become 'sticky'. Boraas
supports this idea by construction of the isoclines for his
data. In Fig. 8 is shown the isocline set for the rotifer-
alga population shown in Fig. 6. The predator isocline
falls on a refuge-like rise in the prey isocline, and stable
behavior is expected.

However, many other phenomena could cause a rise in the
lower end of the prey isocline: decreased predator efficiency
as in the multiple saturation model, or a change in prey
chemical composition at low densities, for example.

Finally, as Curds and Bazin (1977) point out, the assump-
tion that functional and numerical responses are identical
(S.B.5) is certainly incorrect (c.f. Williams, 1971).
Structured models may be required.

Conclusion

Thus we see that there is no dearth of candidates to explain the theoretical inadequacy of the models, and it is
unclear whether simplifying or anacalyptic assumptions are
at fault. But it is most difficult to countenance the fact
that *we don't know whether we understand or not.*

Overall it is interesting that the theory now seems to
work for non-chemostat situations - with changes in

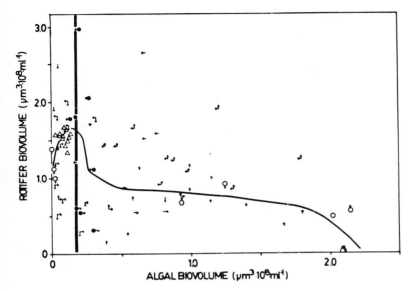

*Fig. 8.   The isocline set constructed by Boraas from the
data in Fig. 6.   The hump to left of the prey isocline
illustrates his notion of a 'dynamic refuge' (from Boraas,
1979).*

enrichment level and predatory efficiency. The theory does
not work for chemostat situations - where attention has
almost exclusively focused on dilution rate changes. Perhaps there is a totally different and yet unrecognized flaw
associated with chemostat dilution changes.

The ways to understanding predator-prey interactions
have been rocky ones, and we have a long way yet to go.

ACKNOWLEDGMENTS

I gratefully acknowledge the support of the National
Science Foundation (NSF BMS 75-18749 and DEB 77-24904). I
also owe a great deal to ongoing discussions with my many
fine graduate students, especially the one so often quoted
herein, Dr. Martin Boraas.

REFERENCES

Bazin, M.J. and Saunders, P.T. (1978). Determination of critical variables in a microbial predator-prey system by catastrophe theory, *Nature* 275, 52-54.

Bazin, M.J., Rapa, V. and Saunders, P.T. (1974). The integration of theory and experiment in the study of predator-prey dynamics. *In* "Ecological Stability" (Ed. M.B. Usher and M.H. Williamson), Chapman and Hall, London.

Bonomi, A. and Fredrickson, A.G. (1976). Protozoan feeding and bacterial wall growth, *Biotechnology and Bioengineering* 18, 239-252.

Boraas, M.E. (1979). Interactions of Nitrate, Algae and Rotifers in Continuous Culture: Experiments and Model Simulations. Ph.D. thesis, Pennsylvania State University, U.S.A.

Boraas, M.E. (1980). A chemostat system for the study of rotifer-algal-nitrate interactions. *In*: A.S.L.O. Special Symposium III: "The Evolution and Ecology of Zooplankton Communities" (Ed. W.C. Kerfoot), University Press of New England, Hanover.

Bungay, H.R. and Bungay, M.L. (1968). Microbial interactions in continuous culture, *Advances in Applied Microbiology* 10, 269-290.

Canale, R.P. (1969). Predator-prey relationships in a model for the activated process, *Biotechnology and Bioengineering* 11, 887-907.

Canale, R.P. (1970). An analysis of models describing predator-prey interaction, *Biotechnology and Bioengineering* 12, 353-378,

Canale, R.P., Lustig, T.D., Kehrberger, P.M. and Salo, J.E. (1973). Experimental and mathematical modeling studies of protozoan predation on bacteria, *Biotechnology and Bioengineering* 15, 707-728.

Curds, C.R. (1971). A computer simulation study of predator-prey relationships in a single stage continuous culture system, *Water Research* 5, 793-812.

Curds, C.R. and Bazin, M.J. (1977). Protozoan predation in batch and continuous culture. *Advances in Aquatic Microbiology* 1, 115-176.

Curds, C.R. and Cockburn, A. (1968). Studies on the growth and feeding of *Tetrahymena pyriformis* in axenic and monoxenic culture, *Joural of General Microbiology* 54, 343-358.

Dent, A.E., Bazin, M.J. and Saunders, P.T. (1976). Behaviour of *Dictyostelium discoideum* amoebae and *Escherichia coli* grown together in chemostat culture, *Archives for*

Microbiology 109, 187-194.
Fredrickson, A.G. (1977). Behaviour of mixed cultures of organisms, *Annual Review of Microbiology* 31, 63-87.
Fredrickson, A.G. and Tsuchiya, H.M. (1977). Microbial kinetics and dynamics. *In* "Chemical Reactor Theory" (Ed. L. Lapidus and N.R. Amundson), Prentice Hall, New York.
Gause, G.F. (1934). "The Struggle for existence". Williams and Wilkins, Baltimore.
Holling, C.S. (1965). The functional response of predators to prey density and its role in mimicry and population regulation, *Memorial Entomological Society of Canada* 45, 5-60.
Huffaker, C.B. (1958). Experimental studies on predation: dispersion factors and predator-prey oscillations, *Hilgardia* 27, 343-383.
Hunt, H.W., Cole, C.V., Klein, D.A., and Coleman, D.C. (1977). A simulation model for the effect of predation on bacteria in continuous culture, *Microbial Ecology* 3, 259-278.
Jost, J.L., Drake, J.F., Fredrickson, A.G. and Tsuchiya, H.M. (1973a). Interactions of *Tetrahymena pyriformis, Escherichia coli, Azotobacter vinelandii,* and glucose in a minimal medium, *Journal of Bacteriology* 113, 834-840.
Jost, J.L., Drake, J.F., Tsuchiya, H.M. and Fredrickson, A.G. (1973b). Microbial food chains and food webs, *Journal of Theoretical Biology* 41, 461-484.
Lotka, A.J. (1925). "Elements of physical biology" Williams and Wilkins, Baltimore.
Luckinbill, L.S. (1973). Coexistence in laboratory populations of *Paramecium aurelia* and its predator *Didinium nasutum, Ecology* 54, 1320-1327.
Luckinbill, L.S. (1974). The effects of space and enrichment on a predator-prey system, *Ecology* 55, 1142-1147.
Maly, E.J. (1969). A laboratory study of the interaction between the predatory rotifer *Asplanchna* and *Paramecium, Ecology* 50, 59-73.
May, R.M. (1977). "Theoretical Ecology, Principles and Applications". Saunders, New York.
Monod, J. (1942). "Recherches sur la croissance des cultures bacteriennes." Hermann et Cie, Paris.
Monod, J. (1950). La technique de culture continue; theorie et applications, *Annales de l'Institute Pasteur* 79, 390-410.
Novick, A. and Szilard, L. (1950). Description of the chemostat, *Science* 112, 715-716.
Odum, H.T. (1957). Trophic structure and productivity of Silver Springs, Florida, *Ecological Monographs* 27, 55-112.

Proper, G. and Garver, J. (1966). Mass culture of the protozoa *Colpoda steinii*, *Biotechnology and Bioengineering* 8, 287-296.

Ricklefs, R.E. (1973). "Ecology" Chiron, Portland.

Rosenweig, M.L. (1969). Why the prey curve as a hump, *American Nature* 103, 81-87.

Rosenweig, M.L. (1971). Paradox of enrichment: destabilisation of exploitation ecosystems in ecological time, *Science* 171, 385-387.

Rosenweig, M.L. (1977). Aspects of biological exploitation, *Quarterly REviews in Biology* 52, 371-380.

Rosenweig, M.L. and MacArthur, R.H. (1963). Graphical representation and stability conditions of predator-prey interactions, *American Nature* 97, 209-223.

Salt, G.W. (1974). Predator and prey densities as controls of the rate of capture by the predator, *Ecology* 55, 434-439.

Tsuchiya, H.M., Drake, J.F., Jost, J.L. and Fredrickson, A.G. (1972). Predator-prey interactions of *Dictyostelium discoideum* and *Escherichia coli* in continuous culture, *Journal of Bacteriology* 110, 1147-1153.

van den Ende, P. (1973). Predator-prey interactions in continuous culture, *Science* 181, 562-564.

Volterra, V. (1926). Variazioni e fluttuazioni del numero d'individui in specie animali conviventi, *Memorial Academy Lincei* 2, 31-113.

Williams, F.M. (1971). Dynamics of microbial populations. *In* "Systems analysis and simulation in ecology" (Ed. B.C. Patten), Academic Press, New York.

Williams, F.M. (1972). Mathematics of microbial populations with emphasis on open systems. *In* "Growth by Intussusception - Ecological Essays in Honor of C. Evelyn Hutchinson." (Ed. E.S. Deevey), Archon Books, Hamden Conn.

Williams, F.M. (1973). Mathematical modelling in microbial populations. *In* "Modern Methods in the Study of Microbial Ecology". (Ed. T. Rosswall), Stockholm.

Williams, F.M. (1977). Destabilisation of predator-prey interactions by changes in prey body size, *Bulletin of the Ecological Society of America* 58, 11.

APPENDIX. Three predator functional responses.

Time-limited predation

Total time T_t is divided into search time T_s plus a handling time T_h for each predator caught. The number caught N_a is proportional to the search time and the prey density M_1 (after Holling, 1965):

$$\frac{N_a}{T_t} = \frac{(1/T_h)\ M_1}{(1/aT_h) + M_1} = \frac{F_2 M_1}{K_2 + M_1}$$

Satiation-limited predation

Gut volume V is divided into food volume V_f and empty gut volume V_e. Intake is proportional to the 'motivation' V_e and the prey density M_1. Defecation is proportional to food volume:

$$dV_f/dt = aV_e M_1 - bV_f$$

Using a quasi-steady state assumption on V_f and rearrangement, the predation rate is:

$$aV_e M_1 = \frac{bVM_1}{(b/a) + M_1} = \frac{F_2 M_1}{K_2 + M_1}$$

Filter feeding predation

A 'filter' has N pores of which n may be plugged by n food particles. Water passes over filter with velocity proportional to empty pores $v = a(N-n)$ such that the volume filtered ('filtration rate') is $V = vA$ where A is filter area. At some characteristic time τ a 'wiper' (e.g., abreptor) removes the food for ingestion. Hence $dn/dt = aA(N-n)M_1 - n/\tau$. With a quasi steady state assumption and rearrangement, the feeding rate is:

$$\frac{n}{\tau} = \frac{(N/\tau)M_1}{(1/aA\tau)+M_1} = \frac{F_2 M_1}{K_2+M_1}$$

All three can be represented as rectangular hyperbolae, identical in final form to the Monod function, but with different mechanisms represented in the parameters F_2 and K_2.

THE ECOLOGY OF ALGAL-INVERTEBRATE SYMBIOSES

Graham W. Gooday and Shelagh A. Doonan

*Department of Microbiology, University of Aberdeen
Aberdeen AB9 1AS, Scotland*

THE PLANT-ANIMALS

Microbes have found niches in some of the world's most hostile environments, such as polar sea-ice, deserts, hot springs and the deep sea. What concerns us here are those microbes that make their homes in a potentially very inhospitable environment, that is inside animal cells. Microbes inhabiting other organisms are 'symbiotic', in the original sense as used by Anton de Bary just over one hundred years ago. He defined symbiosis as 'the living together of dissimilarly named organisms'. Algal-invertebrate symbiosis is an example of symbiosis involving transfer of nutrients between living cells of the two partners (other examples include the algal-fungus symbiosis of lichens, nitrogen-fixing bacteria and plants, mycorrhizal fungi and plants, and digestive tract symbionts in ruminants and wood-eating insects). The range of algal-invertebrate symbioses and their metabolic interrelationships are reviewed by Smith (1978). The hosts represent nearly ninety genera, chiefly found in the warmer seas, the most important being the reef-building corals. In contrast, the algal symbionts are of very few types. The most common are the "zooxanthellae" (dinoflagellates), with just the one species *Gymnodinium microadriaticum* being found in all reef-building corals and most other tropical coelenterates. *Chlorella* species are found in freshwater hosts; cyanobacteria, "cyanellae", in protozoa and some marine sponges; chloroplasts from siphonaceous seaweeds in many saccoglossan sea slugs; and a small number of other algae, some unidentified, in a range of hosts. The organism to be discussed here, *Convoluta roscoffensis* GRAFF (Class Turbellaria, Order Acoela), is symbiotic with the green

alga, *Platymonas convolutae* PARKE ET MANTON (Class Prasino-
phyceae).

Convoluta roscoffensis is of particular interest, as it
appears to be alone among the examples of algal metazoan
symbiosis in that the host no longer ingests food once the
symbiosis is established. This plant-animal is thus an
autotroph, with the host completely dependent on its symbionts
for nourishment (Keeble, 1910).

The algal-invertebrate symbioses have presumably arisen
from feeding relationships. Unicellular algae form an
important part of the food of many marine invertebrates.
Many of the algal-invertebrate symbioses have to be re-established each generation by the feeding of symbiont-free
juveniles on the "correct" algae. The step from feeding to
symbiosis will involve the avoidance of digestion by the
algae. With the algae surviving within the animal, other
intimate relationships must come into play, to allow the
orderly multiplication of the algae (except in the case of
the chloroplast symbioses), so that a regulated algal:animal
ratio is established and maintained. As the major observable
metabolic activity involved in these symbioses is a flow of
nutrients from alga to animal, we must also invoke mechanisms
that regulate this transfer of chemicals. Free-living algae
excrete metabolites, (Fogg, 1975a; Smith and Wiebe, 1976),
and this excretion might have been a factor originally
favouring evolution of some of these symbioses.

There has been a considerable amount of fruitful research
into algal-invertebrate symbioses, especially in the last 10-
20 years. Most of this work has investigated the structural
and metabolic interactions involved in the establishment and
maintenance of the symbiosis (Trench, 1979). This work and
the contemporary studies of other symbioses show us that
symbiosis is not a Curiosity of Nature - it is part of a
continuum of the relationships between living organisms.

As most of the work on algal-invertebrate symbiosis has
been laboratory-based, this gives us a good groundwork with
which to go back to the field, to try to assess the importance
of the symbiosis in the ecology of the organisms.

OCCURRENCE AND BIOLOGY OF *CONVOLUTA ROSCOFFENSIS*

The majority of algal invertebrate symbioses are found in
the tropical seas where, from the animal's point of view, the
relatively low concentrations of nutrients and oxygen and the
increased requirement for removal of waste products and for
oxygen might all be factors favouring the development of a
symbiosis with an alga (cf. Droop, 1963; Lewis, 1974).
Convoluta roscoffensis, however, which is confined chiefly to the
Brittany coast and to the Channel Islands. Isolated colonies

have been reported on the northern Spanish coast and recently the first record appeared from mainland Britain, in Aberthaw, South Wales (Mettam, 1979). There are reports for other green *Convoluta* species elsewhere, such as *Convoluta macnaei* from South Africa, and *Convoluta psammophila* from the Gulf of Naples and the Black Sea, but their relationship to *Convoluta roscoffensis* and its symbiosis remains unclear (Doonan, 1979).

Convoluta roscoffensis is an acoelous turbellarian, 2–4mm long, living gregariously in conspicuous green patches in the intertidal zone of sandy beaches. When the tide recedes, the animals emerge from the sand and expose themselves to the sunshine. They quickly form a translucent mucilage, which helps to prevent individuals from being lost downstream in the small pools or the run-off rivulets in the sand in which the colonies form (Holligan and Gooday, 1975). Although appearing very exposed to pollution on the intertidal open beach, colonies of *Convoluta roscoffensis* at Dunes de Ste. Marguerite, Brittany, quickly re-established, apparently almost unscathed, following the full onslaught of the oil from the Amoco Cadiz a few kilometres away (Gooday, 1980).

The symbionts of *Convoluta roscoffensis* are not inherited, and have to be ingested by the newly hatched juveniles. The natural symbiont in all samples from the Channel Islands and from Brittany that have been tested is the green prasinophycean alga, *Platymonas convolutae* (Parke and Manton, 1967). (The symbionts from samples from Aberthaw, however, appear to be a mixture of *Platymonas convolutae* and a closely related *Prasinocladus* species (A.E. Dorey, personal communication). Newly hatched juvenile animals feed on cells of *Platymonas convolutae* and the algae rapidly divide in the animal tissues (A.E. Douglas, personal communication). The view of Oschman (1966) is that the algae come to lie intercellularly, between muscle cells. This interpretation has been favoured by most workers (Muscatine, Pool and Trench, 1975; Smith, 1978), but the exact nature of the structural relationship between the animal and algal cells is controversial. Other evidence from electron microscopical studies including techniques involving penetration of membrane stains suggests that the algae may instead be intracellular, lying in ramifications of the original digestive syncitium which penetrate between the muscle cells (Dorey, 1965; A.E. Dorey, personal communication).

Once "infected" the animals cease to feed and grow apace with the division of the algae. The symbiotic animals have been grown in the laboratory through many generations in seawater containing nitrate, phosphate, minerals and vitamins (Dorey, 1965; Provasoli, Yamasu and Manton, 1968), but it has

not yet been possible to keep aposymbiotic juveniles alive
to determine if they can be grown heterotrophically to give
aposymbiotic adults

FIELD STUDIES OF *CONVOLUTA ROSCOFFENSIS* – SEASONAL VARIATIONS

In order to assess the importance of symbiosis in the life
of *Convoluta roscoffensis* in the field, particular colonies
have been studied over a long period on Shell Beach, Herm,
Channel Islands.

The size of the convoluta population in the study area
(a square grid, 20 x 20 m) fluctuated widely over the period
of investigation (Table 1). Numbers were lowest in early
summer and highest in early autumn and these fluctuations were
reflected in the size of the colonies. However, while the
density of worms, that is, "worm-spacing", remained relatively
constant (Table 1), the biomass as measured by chloropyll
a content showed a seasonal pattern of fluctuation which was
similar to that of the changes in population size, that is,
the lowest in early summer, highest in early autumn.

TABLE 1

Seasonal Variations in Convoluta population on Herm

	Sept 1976	Feb. 1977	May 1977	Sept 1977	Mar. 1978	May 1978
No. on beach $(10^{-5}$ x total in 20 x 20 m grid)	N.D.	731.2	18.4	1024.4	484.4	6.8
Density in colony $(10^{-5}$ x No. $m^{-2})$	11.1	8.5	N.D.	10.8	8.5	7.5
Chlorophyll *a* per worm (μg)	N.D.	0.30	0.10	0.32	0.20	0.12
Light, mean daily PAR $(E\ m^{-2})$	N.D.	11.2	55.9	15.5	18.7	44.2

Notes: All estimations performed according to Doonan (1979).
Mean daily Photosynthetically Active Radiation calculated as
Einsteins m^{-2} from meterological data from Jersey Airport.

A range of environmental factors was measured: temperature
and nutrient levels of beach run-off water and incident light
– to determine if any correlated with the changes in the
Convoluta population. Only light levels showed any coinci-
dence – highest light coincided with lowest Convoluta numbers

and lowest chlorophyll levels (Table 1). The drop in
chlorophyll content with high light could be in part the
result of a regulation of chlorophyll content within the
algae (as the chlorophyll content of many algae is inversely
proportional to the amount of light received during growth
(Meeks, 1974), and in part the result of photo-destruction
of chlorophylls by high light intensities (Kok, 1956; Pardy,
1976).

PRIMARY PRODUCTIVITY OF *CONVOLUTA ROSCOFFENSIS*

   Primary productivity has been measured in Convoluta
collected at different times and at different ambient temperatures. The amount of carbon fixed showed considerable
consistency in relation to the amount of light energy received;
ranging between 1·1 to 1·7 μg C fixed per worm at 1 E m^{-2}
(Table 2). Both Convoluta and cultured cells of *Platymonas
convolutae* became light-saturated at light intensities
between 1 and 2 E m^{-2}. The cultured Platymonas however
became photo-inhibited above 2 E m^{-2}, whereas higher values
were needed before Convoluta showed photo-inhibition. These
values appeared to vary with season of collection:

TABLE 2

Photosynthesis of Convoluta roscoffensis

Date	Temperature ($^{\circ}$C)	μg C fixed per worm at 1 E m^{-2}	Mean max. photosynthetic rate μg C worm^{-1} h^{-1})
22 Sept 1976	22	1.1	0.66
8 Feb 1977	11	1.5	0.45
5 Oct 1977	14	1.3	0.33
9 Mar 1978	13	1.7	0.49
10 Mar 1978	10	1.7	0.53
1 Jun 1978	18	1.1	0.40

Notes: Photosynthesis measured in natural light under a
series of neutral density filters by counting incorporation
of ^{14}C from NaH^{14}CO$_3$ (Doonan, 1979).

for example, 10 E m^{-2} being required to photo-inhibit worms on
8 June 1977 compared to 3 E m^{-2} on 29 September 1977. This
phenomenon of photo-inhibition is typical for most natural

phytoplankton populations in surface waters (Stengel and
Soeder, 1975). Strickland and Parsons (1972) suggest that
the level of illumination required for photo-inhibition may
be determined by previous light conditions, or the "history
of illumination", and so a direct comparison between values
for cultured cells and field-collected material may be
misleading. A comparison, however, between the light levels
measured in the field and those giving photo-inhibition in
the laboratory experiments suggests that, in the field,
saturating light levels of 1 - 2 E m^{-2} are likely to be
received by Convoluta for all of the year, and photo-inhibition
is likely to occur in the eight lighter months of the year.
It is possible that the phenomenon of photo-inhibition might
play a role in the symbiosis by resulting in an increased
flow of photosynthate from alga to animal, since Fogg (1975b)
suggests that there is increased excretion of metabolites
from natural populations of phytoplankton under conditions
of photo-inhibition.

The mean maximum photosynthetic rate measured during these
experiments with Convoluta ranged from 0.28 to 0.66 μg C
worm^{-1} h^{-1} (Table 2). The average value, from 14 estimations
from animals collected at different times of year, was 11.1
pg C (algal cell)$^{-1}$ h^{-1}, which may be compared with a value
of 11.0 pg C (algal cell)$^{-1}$ h^{-1} for cultured cells of
Platymonas convolutae grown to a density of 5.0 x 10^4 algae
ml^{-1} in a medium of low nutrient status.

Assimilation numbers in these experiments ranged from 1.1
to 3.0 mg C (mg chlorophyll a)$^{-1}$ h^{-1} for Convoluta, compared
with values of 1.6 and 2.3 measured for *Platymonas convolutae*
cultured in media of high and low nutrient levels respectively.
These estimates of assimilation numbers are close to those
for a range of natural populations of algae (Table 3).

In general, the results from these experiments indicate
that the intact algal-invertebrate unit has very similar
characteristics of primary production to those of the free-
living algae.

To enable comparisons to be made with other systems,
estimates have been made of the annual production of *Convoluta
roscoffensis* on Herm in 1977 (Table 4). To obtain these
values, measurements of carbon fixation per worm at various
times in the year were multiplied by the appropriate times
of exposure to daylight and by the appropriate figures for
worm density to give estimates of daily carbon fixation.
These were plotted and the area under the curve integrated
to give an annual estimate. These estimates of primary
production are based on the ^{14}C technique, which Steeman
Nielsen (1964) suggests will give a value somewhere between
true gross and net production, but nearer to the net production

TABLE 3

*Assimilation numbers for photosynthesis by Convoluta,
Platymonas, and natural ecosystems*

System	Temperature (^{o}C)/ Site/Season	Assimilation No. (mg C (mg chlorophyll a)$^{-1}$ h^{-1})	Source
Convoluta roscoffensis	11 – 22	1.1 – 3.0	Doonan (1979)
Platymonas convolutae	19	1.6 – 2.3	Doonan (1979)
Sandy beach, Loch Ewe	Scotland, June–Dec	0.77 – 1.78	Steele and Baird (1968)
Temperate surface water	Summer	4.0 – 4.2	Yentsch and Lee (1966)
Lake Kinneret	Israel	2.8	Dubinsky and Berman (1976)
Pacific seawater	Tropical	1.15 – 6.19	Thomas (1970)

(that is, much of the carbon dioxide respired during the
experiment is likely to be radioactive and so is not counted).
In Convoluta, 37–58% of photosynthate passes from alga to
animal (Smith, 1978), but the rate of carbon turnover in the
symbiosis is not known. In many phytoplankton populations,
turnover is extremely rapid, and so estimates for annual
production will obviously not correspond to annual amounts of
accumulated carbon (c.f. Talling, 1975). Nevertheless, the
value estimated for Convoluta per unit area of beach is
similar to those obtained in the richest areas of the ocean,
while the figure for the colony is approaching that for
coral reefs.

The habits of *Convoluta roscoffensis* are such that it can
photosynthesise only when the tide is out. At the study site

TABLE 4

Estimates for annual primary production

Systems	Production (g C m^{-2} yr^{-1})	Source
Convoluta roscoffensis in colony	872.9	Doonan (1979)
Convoluta roscoffensis on beach	208.8	Doonan (1979)
Algae in sand, Loch Ewe	4.9 – 9.2	Steele and Baird (1968)
Phytoplankton, open sea	25 – 75	Rythen (1969)
Phytoplankton, upwelling areas	300	Rythen (1969)
Phytoplankton, enriched inlet, Nova Scotia	220	Platt (1975)
Reef corals	1500 – 3500	Taylor (1973)

Notes: Estimates for Convoluta are for the 20 x 20 m experimental grid on Herm, and for a typical colony within this.

on Herm, its exposure to daylight varied from 6.8 h (neap tides) to 8.7 (spring tides). Thus it spends anything up to 17.5 hours in darkness, when presumably it slowly respires food reserves, perhaps mannitol in the alga and glycogen in the animal.

The high levels of primary production per unit area of beach carrying Convoluta colonies are a consequence of the high chlorophyll biomass. This is achieved by the dense packing of algae inside the animals (of the order of 6×10^7 algae ml^{-1} of worm) which themselves occur in dense patches (mean density 9.3×10^5 worms m^{-2}). The measured chlorophyll biomass of Convoluta colonies on Herm ranged from 90 to 320 mg chlorophyll a m^{-2}. These values are comparable to the highest theoretical values for free-living phytoplankton in the euphotic zone, of 200 to 300 mg chlorophyll a m^{-2} (Talling, Wood, Prosser and Baxter, 1973). This maximum theoretical value represents the biomass of algae distributed in the water column below 1 m^{-2} of sea surface and involves the extinction of light associated with the algal population density. That is, phytoplankton biomass would not be expected to exceed 200 - 300 mg chlorophyll a m^{-2} because of self-shading. Observation of Convoluta colonies shows that the worms are in continual movement, gliding above and below each other, but dividing animal density by animal "area" shows that the spacing in the colonies is such that the animals would not shade each other if they were spread out evenly.

RELATIONSHIPS IN CONVOLUTA BETWEEN PHOTOSYNTHESIS, PRODUCTION AND BIOMASS

Photosynthesis is the manufacture of organic material, giving primary production. The rate of this process is the primary productivity. Biomass or standing crop is the mass of organisms at a particular moment per unit space. In the simplest grazing food-chain, the primary producers are eaten by the primary consumers, the herbivores. The same measurements, production, productivity and biomass, can be made for the consumers. Such considerations have given rise to the construction of biomass pyramids and energy pyramids, with broad bases of primary producers supporting successively smaller tiers of primary, secondary and tertiary consumers. Where does *Convoluta roscoffensis* stand with such consideration? Can the animal, with its endosymbiotic algae, be thought of as a herbivore with its primary producer inside it, so that the symbiosis represents a compressed food chain? Or does the fact that the animal partner receives photosynthate without physical disruption of the algae mean that the association does not correspond to a herbivore consuming plant food; rather that the association itself functionally corresponds to a unit of primary producers?

Convoluta roscoffensis as a compressed food-chain

To examine the first contention, viewing the symbiosis as a compressed food-chain, requires estimates of biomass or calorific value for both the primary producer and the consumer so that the appropriate pyramid can be constructed. One approach has been to estimate the protein biomass ratio of algal to animal tissue in *Convoluta roscoffensis*. This involves measurement of :

   i)   protein content per whole Convoluta,
   ii)   algal number per Convoluta,
   iii) mean protein contents of cultured algae (the
assumption being that protein content per symbiotic alga is
equal to that per cultured alga).

       Values of algal protein for worms kept in differing
conditions in the laboratory varied from 4.2 to 7.8% of total
protein, with a mean value of 6.4%. This gives a ratio of
1:14.6 for algal:animal protein.

   This gives us the picture of one unit of algal biomass
"supporting" fifteen units of animal biomass. Clearly the
concept of the traditional biomass pyramid has little
relevance to this symbiosis. The relative biomasses themselves
can tell us nothing of the relative activities of alga and
animal. In addition, the biomass ratios used for investigating
producer-herbivore relationships do not consider transfer of
soluble materials – the situation that we have here in
Convoluta. As yet, methods have not been developed successfully to allow an energy pyramid to be constructed for
Convoluta. They would have to measure relative calorific
values and energy transformations between algae and animal
tissues. If this could be done, the resultant values should
be of more value than those for biomass in comparing
symbioses with conventional plant-animal relationships.

Convoluta roscoffensis as a primary producer

   All of the measurements reported here indicate that
Convoluta roscoffensis is a highly efficient primary producer.
Thus for 1977, when Convoluta on Herm were exposed by the
tide to 62.4% of the available sunlight, the estimations that
they fixed 872.9 g carbon per square metre of colony, or
208.8 g carbon per square metre of beach where they occurred
most abundantly (Table 4), place them among the most
productive marine ecosystems.

   This then raises the problem: what is the fate of the
results of this productivity? Convoluta appears to enjoy a
remarkable lack of predation. Keeble (1910) records worms
being eaten by another turbellarian, *Plagiostomum* sp., on
the beaches in Northern Brittany, and one of our collections
from Landeda, Brittany, contained some small swimming crabs,
Bathynectes longipes which ate the worms and their mucus
but these seem to be the only such reports. The animals
have an unpleasant smell when crushed, attributed to trimethyl-
amine by Keeble, who suggests that this may make them
unpalatable to potential predators. (For example, we have
never observed shore birds feeding on Convoluta). Thus they
form no obvious part in any food chain. Individuals are lost

from colonies by being swept downstream but observations of
several colonies indicate that this loss is insignificant,
being less than 0.1% of the colony per hour. However, the
mucilage that is formed rapidly as the colony emerges in the
wake of the ebbing tide is lost from the colony as they swim
down into the sand just before being covered by the incoming
tide. It thus contributes to the organic detritus on the
beach to supply substrates for heterotrophic bacteria, as
does mucus from coral reefs (Coles and Strathmann , 1973).
Holligan and Gooday (1975), however, consider the Convoluta
mucilage to be of minimal mass and so this is probably tiny
input into the beach as a whole. Perhaps a considerable part
of the photosynthetically fixed energy is expended in
activities such as surviving the long dark periods each day
and restlessly swimming whilst being illuminated. Perhaps
maintenance of the *status quo* of the symbiosis requires very
rapid respiration of the carbohydrate substrates, as is the
case for lichens (Farrar, 1973).

CONCLUSIONS

   The ecology of algal-invertebrate symbioses can be
considered both at the population level and at the individual
level. As illustrated by our study of *Convoluta roscoffensis*
investigations at the population level are concerned with
factors influencing the dispersion and changes in abundance
of the whole organism *en masse*. Although it can tolerate
a wide range of daily and annual fluctuations in external
conditions, Convoluta is confined to a very few sites. The
reasons for this remain unknown, in spite of extensive surveys
of the abiotic features of these sites by Guérin (1960) and
Doonan (1979). In tropical seas, however, algal-invertebrate
symbioses are widespread and, in reef ecosystems, are the
rule rather than the exception. It seems likely that, under
the conditions of low oxygen concentration and low nutrient
status characteristic of many tropical waters (for example,
Thomas, 1970), natural selection favours symbiotic associations.
Also, as stated earlier, at tropical temperatures, host organisms
will have higher metabolic rates than their temperature
relatives and thus have a greater requirement for removal of
waste by algal partners (Droop, 1963).

   At the individual level, we are concerned with the
"internal ecology" of the symbiotic association - the structural
and metabolic interrelationships and the factors responsible
for the integrated growth of host and symbionts. In Convoluta,
the consistency of the ratio of algal protein to host protein
(range 0.04 - 0.08) could be regarded as a sign of this
integrated growth, and it would be very instructive to compare

these values for Convoluta with values for other symbiotic
systems when they become available.
    We have seen that the Convoluta symbiosis can in one
sense be viewed as a herbivore which already contains its
food, and so, does not have to expend energy searching it
out. In this "food-chain" in Convoluta there is thought to
be tight recycling of nutrients (Smith, 1978), unlike
conventional food-chains where there is a loss of organic
nitrogen at each link in the chain (Bougis, 1976).
    This way of looking at the internal ecology of algal-
invertebrate symbioses emphasises our belief stated earlier
that these associations are not Curiosities of Nature, but
represent a variation on the theme of microbe: higher
organism relationships.

REFERENCES

Bougis, P. (1976). "Marine Phytoplankton Ecology", North
    Holland, American Elsevier, New York.
Coles S.L. and Strathmann, R. (1973). Observations on coral
    mucus "flocs" and their potential trophic significance,
    Limnology and Oceanography 18, 673-678.
Doonan, S.A. (1979). "Ecological Studies of Symbiosis in
    Convoluta roscoffensis" Ph.D. Thesis, University of
    Aberdeen.
Dorey, A.E. (1965). The organisation and replacement of the
    epidermis in acoelous turbellarians, *Quarterly Journal
    of Microscopic Science* 106, 147-172.
Droop, M.R. (1963). Algae and invertebrates in symbiosis,
    Symposium of the Society for General Microbiology 13,
    171-199.
Dubinsky, F. and Berman, T. (1976). Light utilisation
    efficiencies of phytoplankton in Lake Kinneret (Sea of
    Galilee), *Limnology and Oceanography* 21, 226-230.
Farrar, J.F. (1973). Lichen physiology: progress and pitfalls.
    In "Air Pollution and Lichens" (Ed. B.W. Ferry, M.S.
    Baddeley and D.C. Hawksworth), pp. 238-282. Athlone Press,
    London.
Fogg, G.E. (1975a). "Algal Cultures and Phytoplankton
    Ecology", University of Wisconsin Press.
Fogg. G.E. (1975b). Biochemical pathways in unicellular
    plants. *In* "Photosynthesis and Productivity in Different
    Environments" (Ed. J.P. Cooper), pp. 437-457. Cambridge
    University Press.
Gooday, G.W. (1980). *Convoluta roscoffensis* and the Amoco
    Cadiz oil spill, *Marine Pollution Bulletin*, in press.
Guérin, M. (1960). Observations ecologiques sur le *Convoluta
    roscoffensis* Graf, *Cahiers de Biologie Marine* 1, 205-220.

Holligan, P. and Gooday, G.W. (1975). Symbiosis in
   *Convoluta roscoffensis, Symposium of the Society for
   Experimental Biology* 29, 205-227.
Keeble, F. (1910). "Plant-Animals: A Study in Symbiosis"
   Cambridge University Press.
Kok, B. (1956). On the inhibition of photosynthesis by
   intense light, *Biochimica et Biophysica Acta* 21, 234-244.
Lewis, D.H. (1974). Microorganisms and plants: the evolution
   of parasitism and mutualism, *Symposium of the Society for
   General Microbiology* 24, 367-392.
Muscatine, L., Pool, R.R. and Trench, R.K. (1975). Symbiosis
   of algae and invertebrates: aspects of the symbiont
   surface and the host-symbiont interface, *Transactions of
   the American Microscopical Society* 94, 450-469.
Meeks, J.C. (1974). Chlorophylls. *In* "Algal Physiology and
   Biochemistry" (Ed. W.D.P. Stewart), pp. 161-175. Blackwell,
   Oxford.
Mettam, C. (1979). A northern outpost of *Convoluta
   roscoffensis* in South Wales, *Journal of the Marine
   Biological Association, U.K.* 59, 251-252.
Oschman, J.L. (1966). Development of the symbiosis of
   Convoluta roscoffensis Graff and *Platymonas* species
   Journal of Phycology 2, 105-111.
Pardy, R.L. (1976). The production of aposymbiotic hydra
   by the photodestruction of green hydra zoochlorellae,
   Biological Bulletin 151, 225-235.
Parke, M. and Manton, I. (1967). The specific identity of
   the algal symbiont of *Convoluta roscoffensis, Journal of
   the Marine Biological Association, U.K.* 47, 445-464.;
Platt, T. (1975). Analysis of the importance of spatial and
   temporal heterogeneity in the estimation of animal
   production by phytoplankton in a small, enriched, marine
   basin, *Journal of Experimental Marine Biology and
   Ecology* 18, 99-109.
Provasoli, L., Yamasu, T. and Manton, I. (1968). Experiments
   on the resynthesis of symbiosis in *Convoluta roscoffensis*
   with different flagellate cultures, *Journal of the Marine
   Biological Association, U.K.* 48, 465-479.
Ryther, T.H. (1969). Photosynthesis and fish production in
   the sea, *Science* 166, 72-76.
Smith, D.C. (1978). Photosynthetic endosymbionts of
   invertebrates. *In* "Companion to Microbiology" (Ed. A.T.
   Bull and P.M. Meadow), pp. 387-414. Longman, London.
Smith, D.F. and Wiebe, W.J. (1976). Constant release of
   photosynthate from marine phytoplankton, *Applied and
   Environmental Microbiology* 32, 75-79.
Steele, J.H. and Baird, I.E. (1968). Production ecology of

a sandy beach, *Limnology and Oceanography* 13, 14-25.

Steeman Nielsen, E. (1964). Recent advances in measuring and understanding marine primary productivity, *Journal of Animal Ecology (Supplement)* 33, 119-130.

Stengel, E. and Soeder, C.J. (1975). Control of photosynthetic production in aquatic ecosystems. *In* "Photosynthesis and Productivity in Different Environments" (Ed. J.P. Cooper), pp. 645-672. Cambridge University Press.

Strickland, J.D.H. and Parsons, T.R. (1972). "A Practical Handbook of Seawater Analysis". Second Edition. Fisheries Research Board of Canada, Bulletin No. 167, Ottawa.

Talling, J.F. (1975). Primary production of freshwater microphytes. *In* "Photosynthesis and Productivity in Different Environments" (Ed. J.P. Cooper), pp. 225-247. Cambridge University Press.

Talling, J.F., Wood, R.B., Prosser, M.V. and Baxter, R.M. (1973). The upper limit of photosynthetic productivity by phytoplankton: evidence from Ethiopian soda lakes, *Freshwater Biology* 3, 53-76.

Taylor, D.L. (1973). Symbiotic pathways of carbon in coral reef ecosystems. Present status and future prospects, *Helgoländer wissenschaft liche Meeresunters uchungen* 24, 276-283.

Thomas, W.H. (1970). On nitrogen deficiency in tropical Pacific oceanic phytoplankton: photosynthetic parameters in poor and rich water, *Limnology and Oceanography* 15, 380-385.

Trench, R.K. (1979). The cell biology of plant-animal symbiosis, *Annual Review of Plant Physiology* 30, 485-531.

Yentsch, C.S. and Lee, R.W. (1966). A study of photosynthetic light reactions, and a new interpretation of sun and shade phytoplankton, *Journal of Marine Research* 24, 319-337.

PROKARYOTIC FORM AND FUNCTION

Crawford S Dow and Roger Whittenbury

*University of Warwick,
Coventry, CV4 7AL, England*

INTRODUCTION

The shape and size of a bacterium, its appendages and internal structures, and the alternative cell forms (for example, spores and cysts) to which it may give rise (Figure 1), must be the evolutionary response of that microbe to the particular environment of which it is a natural inhabitant. Or to put it another way round, it seems reasonable to suppose that the selection and stabilization of expression of cell forms reflects pressures imposed during the evolution of the natural environment. Whatever the process it seems safe to assume that the form(s) of bacteria - as they now are - reflect a function related to their existence and persistence in their natural habitat. The problem is, of course, defining that functional relationship to the environment in biochemical and physiological terms.

Over the years, microbiologists have become conditioned (until very recently) to defining and describing microbes and their roles as viewed through laboratory conditions of monocultures grown in batch culture in artificial media, none of which usually bear any resemblance to the environmental circumstances in which those microbes naturally exist. Even though it has been accepted that the real nature of a microbe should be ascertained by a study of its functioning in the natural environment - a revealed truth as it were - defining the natural environment and then simulating it in the laboratory is quite another matter. However, the realisation of this state of affairs has helped to direct more sensibly the experimental approaches to such questions and has underlined the caution necessary in the interpretation of functional relationships and the roles of individual microbial species.

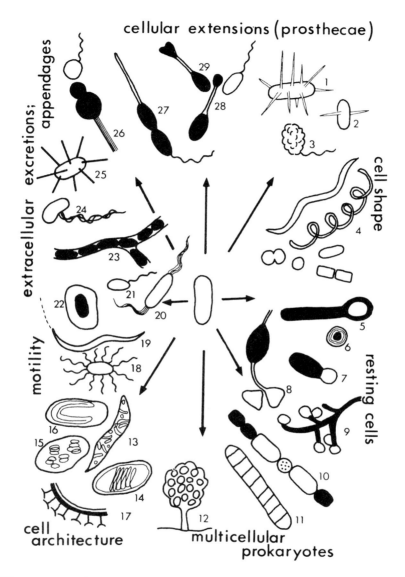

Figure 1. Diagrammatic representation of the diversity of the prokaryotic form.

1. Ancalomicrobium and Prosthecomicrobium (Staley, 1968; Whittenbury and Dow, 1977).

2. *Asticcacaulis biprosthicum (Pate, Porter and Jordan (1973)*
3. *Phenotypic variation in Ancalomicrobium sp. (Whittenbury and Dow, 1977).*
4. *Cell shape variation. Correlation with surface area/volume ratio? for example, Arthrobacter sp.*
5. *Endospore formation, for example, Clostridium sp.*
6. *Cyst formation, for example, Azotobacter sp.*
7. *Exospore formation by methylotrophs.*
8. *Exospore formation by Rhodomicrobium sp. (Whittenbury and Dow, 1977).*
9. *Actinomycete spores, for example, Micromonospora sp.*
10. *Heterocyst and akinete formation, for example, Anabaena cylindrica.*
11. *Caryophanon sp.*
12. *Fruiting body formation, for example, Myxobacteria sp.*
13. *Membrane system of Rhodospirillum rubrum.*
14. *Methylococcus internal membranes (Davies and Whittenbury, 1970).*
15. *Methylomonas, disc shaped vesicles (Davies and Whittenbury, 1970).*
16. *Lamellar membranes, for example, Rhodopseudomonas palustris.*
17. *Variation in cell wall architecture.*
18. *Peritrichous flagellation.*
19. *'Gliding' bacteria, for example Cytophaga sp.*
20. *Flagellar tufts (Strength and King, 1971).*
21. *Polar or sub-polar flagellation.*
22. *Capsule formation/excretion of polysaccharides.*
23. *Hydroxide deposit, for example in Sphaerotilus sp. filament.*
24. *Hydroxide ribbons, for example, in a Gallionella sp.*
25. *Bacterial spines (Easterbrook, McGregor-Shaw and McBride, 1973).*
26. *Planctomyces sp. stalk.*
27. *Caulobacter sp.*
28. *Rhodomicrobium sp. (Dow and France, 1980) and Hyphomicrobium sp. (Harder and Attwood, 1978).*

Another problem that needs to be recognised is that of singling out particular properties of cells when attempting to relate form to function. Such properties, whilst being a matter of amazement to us, might not be of overriding importance to that microbe's survival. For instance, the extreme heat resistance of some endospores may reflect more the physical properties of their packaging for survival in unsuitable environments, which may not be one where high temperature is the hazard. Obviously it is possible to identify a bit of extreme environment where the property of heat resistance ensures survival, but such a circumstance may be wholly irrelevant to the real circumstances appertaining to the normal life cycle and continuity of existence of that particular endospore-forming bacterium.

Despite this rather pessimistic appraisal of the problems confronting those who are attempting to relate form to function, some advances are being made. A particularly elegant example is that of Veldkamp and his colleagues (Kuenen, Boonstra, Schroder and Veldkamp, 1977) who, by studying populations of microbes developing in chemostats running at fast and slow dilution rates, have focussed attention upon the competitive advantage that spirilla have in very dilute media. This advantage appears to be that of an enhanced surface area to cell volume ratio compared to that of rod forms able to grow in the same medium, enabling the spirilla to compete more successfully for available nutrients.

In this contribution, we direct attention - in a somewhat speculative manner - to some particular cell forms, their evolution in some cases- and their possible function;

   i)   cell forms of microbes which adopt a polar rather than intercalary form of envelope growth;

   ii)   one cell form in particular arising from polarly growing cells - the swarm or I cell;

   iii) Polymorphism (in the sense of certain bacteria being able to adopt alternative cell forms which persist through cell cycles in defined circumstances);

   iv)   the role of prosthecae and membraneous lamellae in survival in the environment; and

   v)   colonial systems in aquatic environments.

Cell forms of microbes which adopt a polar rather than an intercalary form of envelope growth

The cell cycle of Escherichia coli

Before discussing envelope growth and its consequences, we need to review briefly the control and regulation of the cell cycle, and the cell forms which may arise. At present this is best understood at the molecular level in *Escherichia coli* (Fig. 2). Several, apparently invariable, cell cycle constants have been deduced from experimental data.

One of the most interesting invariants is that of cell proportions (ratio of length to breadth) which, although cell size varies with growth rate, remains constant at a given stage in the cell cycle under different growth conditions (Zalitsky, 1975). This implies that the ratio of surface area/volume will decrease as the cell grows faster, and that, consequently, the relatively increased surface area to cell volume ratio resulting under poor nutrient conditions may

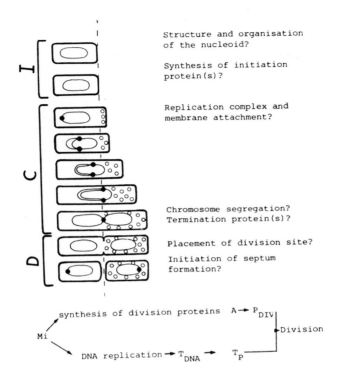

Structure and organisation
of the nucleoid?

Synthesis of initiation
protein(s)?

Replication complex and
membrane attachment?

Chromosome segregation?
Termination protein(s)?

Placement of division site?

Initiation of septum
formation?

synthesis of division proteins $A \rightarrow P_{DIV}$

Mi $\longrightarrow$ Division

DNA replication $\rightarrow T_{DNA} \rightarrow T_P$

*Figure 2. Diagrammatic representation of the cell cycle of
Escherichia coli growing with a generation time greater than
60 min. Growth is asymmetric and polar being from the
one growth point (Donachie and Begg, 1970; Begg and Donachie,
1977). The I period, the time preceding initiation of
chromosome replication, is variable and is thought to be the
time taken to synthesise initiation protein(s) or, alternatively,
the time taken to reach the critical cell mass (Mi) at which
initiation occurs (Donachie, 1968; Donachie, Begg and
Vincente, 1976). There is little experimental data on the
functional state of the cell during this phase. The C period
is the time taken for chromosome replication and has been
shown to be constant for a wide range of growth rates, that
is, the rate of DNA synthesis is constant. Obviously this
situation becomes more complex under conditions of nucleotide
limitation (Begg and Donachie, 1978). The D period is the
time required after termination of chromosome replication
for physical separation of the cells. In E. coli D = 20 min
and in Bacillus subtilis D = 150 min. Again this time is*

invariant and is not dependent on RNA or protein synthesis.
   Rather than being considered as a linear sequence of events,
the cell cycle is more correctly portrayed as two parallel
sequences, that of protein synthesis (presumably division
proteins) and DNA synthesis. The period of division protein
synthesis corresponds to the time taken for DNA synthesis,
that is, 40 min, however the events subsequent to this
protein synthesis do not require RNA, protein or DNA synthesis
($A - P_{DIV}$). This is thought to represent the time taken for
the formation of some septum "primordium". Before division
can occur, termination protein(s) must be synthesised –
these only appear on termination of chromosome replication.
Presumably interaction of these with the septum complex
precipitates cell division.

well be an adaptive response favouring increased powers of
nutrient absorption. Such an interpretation would seem to
be supported by the findings of Kuenen, Boonstra, Schroder
and Veldkamp (1977).
   Two other constants of importance (see Fig. 2) are those
concerning time taken to replicate the chromosome – C which
equals 40 min in *E. coli* – and the period post chromosome
replication which elapses prior to cell division – D which
equals 20 min in *E. coli*. But the constant of apparent
ecological/environmental importance (we will argue in respect
of swarm cells later on) is the I phase, which is the period
in the cell cycle, after cell division, before the chromosome
replicates. Many unanswered but very relevant questions are
raised by this phenomenon of the I phase: for instance, what
determines the length of time of the I phase? What is the
physiological state of the cells in that period? what is
the configuration of the non-replicating genome? How is
DNA synthesis suppressed? How is this DNA replication
suppression regulated and does the regulatory mechanism
relate to, or is it influenced by, external environmental
parameters?
   As the majority of ecological niches in which prokaryotes
find themselves will be those of limiting nutrients, that is,
doubling times will exceed the C + D time for those particular
microbes, a high percentage of the cells at any given moment
will be in the I phase – a stage of the cell cycle in which
swarm cells exist of the sort described in the section
following this one.

Cell envelope growth

   Studies of *E. coli* (Donachie and Begg, 1970; Begg and
Donachie, 1977) have shown that when the doubling time is
less than 60 min the growth points on the cell envelope are

multiple and that at division the resultant cells are
genuine siblings - being equivalent as far as can be determined
as regards their contents of old and new material. But at
doubling times in excess of 60 min envelope growth is polar,
that is, only one growth point/region of the cell is
activated and the cells resulting from the division process
are not equivalent siblings. As one cell is therefore older
than the other, and will increase in age at each subsequent
round of polar growth and division with respect to the second
cell, a parent-offspring (mother - daughter is the phrase
commonly used) relationship is a more accurate description
of the phenomenon. And one could argue that under the normal
environmental conditions in which *E. coli* is found (a 7 hour
doubling time seems characteristic in the human gut (Koch,
1971)), the polar mode of envelope growth is the natural one.
Whether or not the two cells arising from division of a
polarly growing *E. coli* culture follow an identical cell
cycle thereafter is unknown - but is certainly of interest
in relation to the discussion below which reveals that cells
arising from the division of obligately polarly growing
bacteria do not. Some of the consequences arising from the
adoption of an obligate polar mode of envelope growth can
be deduced from a study of the budding bacteria (Whittenbury
and Dow, 1977). It should be noted at this point that
"budding" and "binary fission" are not two separate forms
of cell multiplication; all budding bacteria undergo
binary fission as the terminal process in cell multiplication.
These are:

   i)   The mother cell ages beyond the division time - in
contrast to the intercalory-growing type of bacterium
producing identical siblings at division.

   ii)   The cells are oriented. This is clearly evident,
for example, in *Rhodopseudomas palustris* (Fig. 3).

   iii) Complex internal membrane systems (lamellae) of
asymmetric form (for example, *R. palustris*, Fig. 3) can
evolve as polar growth carries the new daughter cell and
the division point beyond the membraneous structure, so
avoiding the problem of dividing through it.

   iv)   Cell division can be asymmetric (for example,
R. palustris, Fig. 3).

   v)   There appears to have been a release on constraints
of cell morphology. Multiple growth point organisms which
divide to give equivalent siblings are morphologically simple
and yield cells of similar shape. The ability to grow
polarly, and therefore asymmetrically, can lead to a process
of morphogenetic evolution whereby the mother cell and
daughter cell are quite different in shape. A range of
polarly growing bacteria are depicted in Figure 4 in an

Rhodopseudomonas palustris

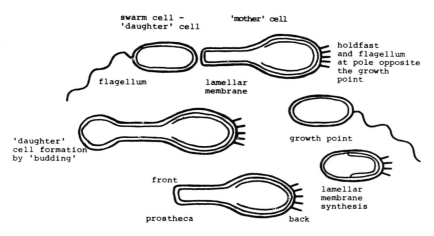

Figure 3. Diagrammatic representation of the cell types, cell orientation and membrane system in Rhodopseudomonas palustris (Whittenbury and McLee, 1967; Dow, Westmacott and Whittenbury, 1976). The lack of an extensive lamellar membrane system in the swarm cell ensures the membrane content of the resulting reproductive cell will be that required for the environmental situation the cell finds itself in, that is, the membrane system of the swarm cell is formed de novo and is not correlated with that of the "mother" cell.

order of increasing morphological complexity, starting with *E. coli*. The most complex of the budding organisms shown give expression to a range of cell types and cell cycles, only possible, it would seem, because these organisms have adopted a polar rather than intercalory form of envelope growth.

Swarm or I phase cells

For many years the term "swarm cell" has been used to describe the motile cell(s) formed at division by a variety of bacteria. A definition of "swarm cell" (its properties and role) has for the most part been left very incomplete; cells called swarm cells at one time or another would probably prove to be a heterogeneous collection of cell types. However, over the past few years more has become known about the biology and molecular biology of the swarm cells of some prosthecate bacteria (for example of *Caulobacter, Rhodo-*

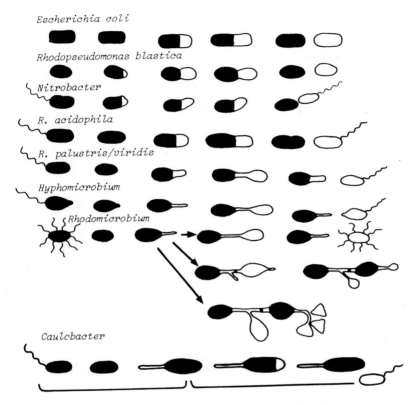

Escherichia coli

Rhodopseudomonas blastica

Nitrobacter

R. acidophila

R. palustris/viridis

Hyphomicrobium

Rhodomicrobium

Caulobacter

'maturation'   reproduction

*Figure 4. Budding bacteria ordered in degrees of morphological
and cell cycle complexity. E. coli is represented as a cell
having a generation time in excess of 60 min. Rhodopseudomonas
blastica (Eckersley and Dow, 1980) shows asymmetric polar
growth and symmetrical division. The two cells derived at
division are, however, physiologically distinct in that the
"older" cell initiates a new found of replication immediately,
whereas the "new" cell does so only after a lag period.
R. acidophila, R. palustris, Hyphomicrobium sp., Rhodomicrobium
sp., and Caulobacter sp., have cell cycles which are fundamentally identical, the major distinction being the increase
in morphological complexity. In addition, however, Rhodomicrobium sp., expresses one of three cell types, swarm cell,
chain cell or exospore depending on the applied environmental
stimulus. Caulobacter sp., is portrayed, it is important to
note, as the budding bacterium it is in low nutrient environ-*

ments (Swoboda, 1979).

pseudomonas, Rhodomicrobium and *Hyphomicrobium*), certainly
sufficient to place them in a   special functional category.
    Probably the most striking feature of these particular
"swarm cells" is their inability to proceed immediately into
the reproductive cycle, as the cells which give rise to
them are able to do (Fig. 5).   Instead, they undergo a
sequence of morphogenetic and molecular events not followed
by the mother cell which gave rise to them; these events
(Fig. 6) are only ever expressed once in the life of the
organism, being permanently switched off in the successive
rounds of multiplication of that bacterium.
    Such cells have been referred to as "immature" cells – for
the logical reason that they have to undergo development prior
to embarking upon a reproductive round of growth. However
it now seems that such cells have a special role, that of
being dispersal cells, and that the word "immature" is not a
wholly adequate adjective by which to describe them – as
will become clear in the ensuing paragraphs.
    The swarm cells of *Caulobacter, Hyphomicrobium* and
Rhodomicrobium strains so far examined differ from their
respective mother cells in a number of properties – in
particular that they are motile, lack prosthecae, do not
synthesise DNA, and synthesise RNA and protein at a much
reduced level. Subsequent development of these swarm cells
(Fig. 7 and 12) is influenced by the growth medium. Even
under the most favourable circumstances, development is
delayed by a period of time equivalent to about one third
of the mother cell cycle; but in exhausted media, dilute
media, or, as in the case of photosynthetic bacteria, in low
light, the swarm cell stage of this cycle can persist for many
hours or days even.   It is this property, the failure to
develop with all that that entails, which supports the real
role of swarm cells – to survive in inadequate environments
whilst actively seeking out a suitable niche – in other words,
they are specialised dispersal cells which remain, during
this time, in the I phase of the cell cycle.
    As yet, much remains to be done concerning the detailed
characterisation of these swarm of I phase cells. The ability
not to grow is worth exploration, raising in itself many
questions concerning the regulation (inhibition) of biosynthesis, gene expression and maintenance of the integrity
and function of a non-growing cell. Also, it would seem that
in the dispersal stage, chemoreceptors must play an important
role both in a negative and positive manner, guiding cells
away from unfavourable environments to favourable surfaces
where the cells can adhere and switch into the development

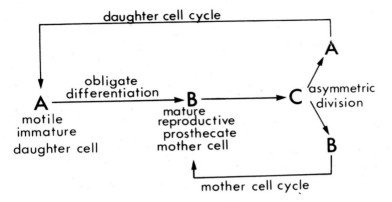

Figure 5. Generalised cell cycle characteristic of the dimorphic prosthecate budding bacteria.

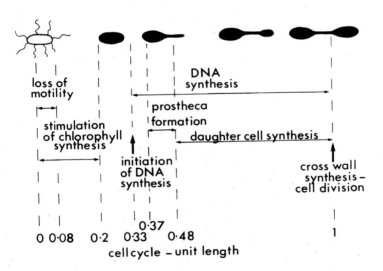

Figure 6. The cell cycle of Rhodomicrobium sp., showing the "landmark" events so far identified. These are fundamentally the same for Rhodopseudomonas palustris, Hyphomicrobium sp., and Caulobacter sp., that is, all of the dimorphic budding bacteria. As with Caulobacter sp., (Shapiro, 1976) this tempral sequence is constant and

independent of growth rate. From studies on isolated nucleoids of Caulobacter sp. (Evinger and Agabian, 1977; Swoboda, Dow and Vitkovic, 1980) and Geimsa stains and ultrastructural studies of Rhodomicrobium vannielii (Whittenbury and Dow, 1977), it has become apparent that the swarm cell nucleoid is in a highly condensed state (sedimentation coefficient of 7,000 S as opposed to 5,500 S for the stalked cell). The transcriptional consequences of this format are unclear but it is tempting to correlate this with the fact that DNA replication is suppressed in swarm cells and that replication is only initiated partway into the developmental sequence, that is, when the nucleoid configuration has altered significantly.

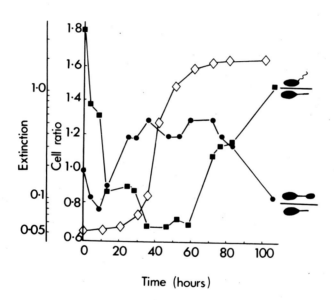

Figure 7. Ratios of the relative numbers of Hyphomicrobium swarm cells ( ), stalked cells ( ) and reproducing cells ( ) plotted against time.

■ , _____   • , _____   ◇ , *growth curve*

The relative numbers of each cell type were obtained from electron microscopic preparations and from cell volume distribution profiles (Coulter Electronic's electronic particle counter (ZB1) and cell volume distribution analyser

(C1000)). Similar data have been obtained for Caulobacter
sp., and for Rhodomicrobium sp., growing as the "simplified"
cell cycle (Dow and France, 1980). The data indicates that
during the exponential phase of batch growth the cell types
making up the population remains constant. With the onset
of the stationary phase the number of motile swarm cells
increases dramatically. These observations may be interpreted
as follows. When the system becomes limited for an essential
nutrient swarm cell production continues as long as possible.
However, the swarm cells produced are inhibited from going
into the differentiation sequence since, like many cell
differentiation systems, once the sequence is initiated it
must run to conclusion. Consequently, it would seem to be
detrimental to species survival not to have a development
switch linked to the nutrient status of the ecosystem, that
is, the swarm cell only initiates differentiation and growth
in a favourable environmental situation, an essential require-
ment for a dispersal phase. Further evidence for such a
developmental control is given in Figure 12.

phase leading to cell multiplication. Chemoreceptors and
their functioning in the development of the swarm cell is
also a topic worth exploring – particularly as an early event
in all the examples described so far is the jettisoning of
the flagella. Is the loss of flagella an event put in train
by a signal from the chemoreceptors? Do chemoreceptors
function *after* flagella are shed (so far, it has been
assumed that chemoreceptors express their function only via
cell movement)? Are chemoreceptors important as generators
of signals affecting subsequent development? These, and
other similar questions, need to be explored in defining the
nature of a swarm cell.

Finally, how widespread is the phenomenon of the swarm
cell? Clearly, the budding bacteria (for example, *Nitro-*
bacter sp., *Rhodopseudomonas acidophila, R. palustris*
R. viridis) form swarm cells of the type just described, as
also might *Bdellovibio* sp., (Fig. 8) and *Sphaerotilus* (Fig. 9)
species. It may also turn out that even familiar organisms,
such as *E. coli* and the pseudomonads, can give rise to
progeny with a prolonged I phase (and with some other
characteristics of swarm cells) when grown under conditions
more similar to their natural environment. *E. coli*, for
instance, in the human gut has a generation time of the order
of 7 hours – not the 1 hour or less achieved in artificial
culture in the laboratory. And as described earlier, *E. coli*
grown under conditions whereby the doubling time is in excess
of 1 hour, resorts to polar growth rather than multipoint

Bdellovibrio

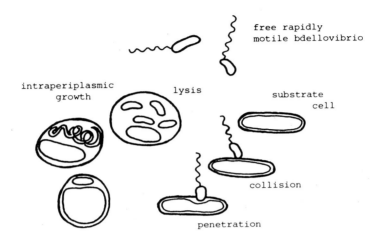

Figure 8.   Diagrammatic representation of the bdellovibrio cell cycle (Tomashow and Rittenberg, 1979).

envelope growth. However, the biology of such polarly-formed progeny of *E. coli*, and other bacteria, such as pseudomonads, has yet to be explored.

Sphaerotilus natans

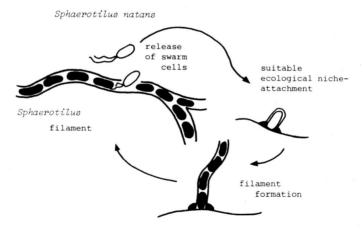

Figure 9.   Sphaerotilus natans colonial development and swarm cell release.

*The role of prosthecae, polymorphism and membraneous
lamellae in survival in the environment.*

Some bacteria clearly respond morphologically to changes
in the nutrient status of the environment or appear to have
adopted a particular form or internal ultrastructure as
a consequence of their chosen mode of existence.

The question to consider is whether these "responses"
indicate an evolutionary development to ameliorate difficulties
of growth in such circumstances. In particular, does the
development of prosthecae represent that bacterium's attempt
to increase its nutrient capturing net (by increasing its
surface area to volume ratio)? In the same context, is the
proliferation of membranes within the cell another type of
response to the same problem? There is very little evidence
to answer firmly either question; such as there is, is
considered below and it is conjectured that the two categories
of response actually serve to resolve different types of
nutrient problem — if the response is nutrient stimulated.

Polymorphic vegetative cell cycles

A phenomenon known now for a number of years is the curious
ability of certain bacterial species to adopt more than one
vegetative cell form. The cell forms are in themselves
stable and follow stable cell cycles. *Arthrobacter
crystallopoietes* is perhaps the most well known example
(Clark, 1973), one cell form is a rod, the other a sphere.
The shift from one form to the other follows a change in the
growth medium; the coccal form maintaining itself in
glucose-mineral salts by changing to the rod form when one
of a variety of amino and organic acids is added to the medium.
What purpose the two-phase morphology serves in the survival
of the organism is unknown; whether or not a change in
surface area to volume ratio, and hence the conferment of a
more competitive nutrient-acquiring ability to the coccal
form, is of significance remains untested.

Other examples of nutrient-induced polymorphism include
that of a *Geodermatophilus* strain (Fig. 10), a *Hyphomicrobium*
sp. (Fig. 11), *Rhodomicrobium vannielii* (Fig. 12), an
Ancalomicrobium sp. (Fig. 13) and *Chlorogloea fritschii*
(Fig. 14). Of these, only the *Ancalomicrobium* sp., example
appears easily explicable. Under very dilute conditions the
prosthecae are fully expressed whilst under nutrient rich
conditions they are fully suppressed — suggesting that the
role of the prosthecae, in this instance, is to greatly
increase the absorbtive area of the cell in relation to its
volume, so ensuring survival and growth (albeit at a very

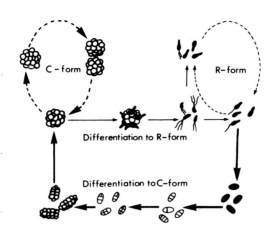

Figure 10. The polymorphic growth cycle of Geodermatophilus strain 22-68 (after Ishiguro and Wolfe, 1970). A variety of mono and divalent cations and organic amines induce R-form to C-form morphogenesis and differentiation. They also maintain the organism in the C-form (Ishiguro and Wolfe, 1974). The experimental data indicated that the uptake of cations was accompanied by the extrusion of intracellular protons and that the organic amines were taken up as the dissociated free base, in both cases resulting in an increase in intracellular pH - the net effect of which was to induce the morphogenetic transition.

slow rate) in near starvation conditions such as exist in oligotrophic waters (Fig. 15).

Morphological studies of microbial populations in oligotrophic and eutrophic waters would seem to support the above findings (Fig. 16), prosthecate bacteria of the *Ancalomicrobium* category, amongst others, being commonly detected (by morphology) in oligotrophic waters. However, even this apparently common-sense type of interpretation may prove to be an over simplified one. Particular nutrient shortages, or imbalances, may prove to be a trigger for morphogenetic responses, as is the case in some species of *Caulobacter* and budding bacteria which respond to nutrient shortages by dramatically increasing stalk length and budding tube lengths respectively. In these last two cases, there is no evidence to indicate that such responses enhance the

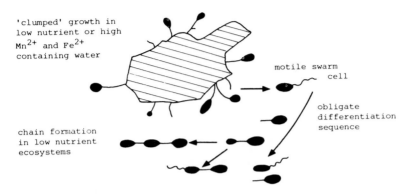

Hyphomicrobium

'clumped' growth in
low nutrient or high
Mn^{2+} and Fe^{2+}
containing water

motile swarm
cell

obligate
differentiation
sequence

chain formation
in low nutrient
ecosystems

'simple' (constitutive swarm cell
formation) cycle predominates in
rich medium

*Figure 11. Diagrammatic representation of cell expression
in Hyphomicrobium sp. In the presence of Mn^{2+} and Fe^{2+} this
organism becomes encrusted in a hydroxide deposit (Hirsch,
1967; Tyler, 1970) a characteristic which led to the
suggestion that the polar mode of growth was essential to
prevent total encrustation of the organism, that is, new
growth is always away from the deposit. In low nutrient
situations and in particular where Mn^{2+} and Fe^{2+} are present
Hyphomicrobium sp., grows as a multicellular system. However,
when nutrient concentrations are high, for example, as in
laboratory culture, the simple cycle (essentially constitutive
swarm cell formation) predominates.*

rate of nutrient uptake, as has been shown by Larson and
Pate (1976) for *Anticcacaulis bifresthecum*. The increased
length of prosthecae in these former cases therefore, may
merely reflect the increased doubling time, and that
whatever mechanism/substance regulates "switch-off" of
prostheca extension is delayed or does not reach an active
"threshhold" level as rapidly as occurs in nutrient-rich
circumstances, that is, different sorts of prosthecae serve
different functions.

Membrane invaginations

A number of bacterial species have the ability to extend
their cytoplasmic membrane into lamellae, tubes, patterns of

Rhodomicrobium vannielii

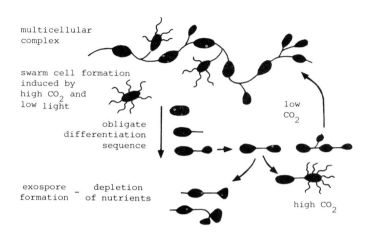

Figure 12. *Diagrammatic representation of cell cycle expression in Rhodomicrobium vannielii. The formation of swarm cells (formed sequentially from the filament tips) is induced by increasing the CO2 concentration and reducing the light intensity. If the available light intensity and nutrient levels are satisfactory the swarm cell initiates the developmental sequence, that is, colonisation of a new ecological situation. However, if the light intensity is very low, even when all other nutrient requirements are in excess, development is prevented, which results in a culture consisting of only two cell types – motile swarm cells and multicellular complexes – there are no intermediate stages.*

discs and so on; with the end result of increasing dramatically the membrane area of the organisms concerned. Examples include the phototrophs, the methane oxidizing bacteria, the ammonia oxidizers, the nitrite oxidizers, the lower-n alkane oxidizers and the methane producers. Speculative arguments about the principle role of such membranes in the growth and survival of these organisms will be limited to the first five categories of bacteria mentioned.

As a first point, it seems reasonable to exclude a general paucity of nutrients as being an influence on the formation of the networks of membrane. Only by increasing the size of the capturing net in relation to cell volume (as in the case of prosthecae production) would additional levels of

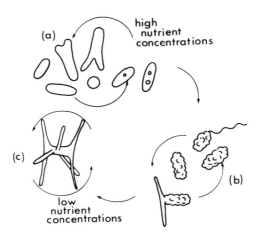

Figure 13. Phenotypic variations induced in a freshwater prosthecate bacterium by variation of the organic nutrient concentration. (i) High nutrient concentrations (< 100 µg ml^{-1}) are characterised by pleomorphic rods and ovoid cells. (ii) Transitional cell type which has a very characteristic knobbed appearance. Motility may be observed in this format. (iii) Low nutrient concentrations (< 20 µg ml^{-1}). Prosthecae synthesised, the number being dependent on the nutrient concentration.

nutrients be tapped.

What may be the key function dictating a demand for extra membrane? It could be the turnover rate of a nutrient involved directly or indirectly in energy production for biosynthesis.

In the instance of phototrophs, for example, the non-sulphur purple photosynthetic bacteria, light intensity is a factor influencing membrane development. In light saturation situations, membrane extension (and associated chlorophyll content) is minimal; whilst in light limiting conditions, membrane development (and associated chlorophyll content) is maximal. Therefore, the assumption made is that under light saturating conditions the photons striking the light harvesting chlorophyll (and/or carotenoid) molecules are doing so at a rate sufficient to provide the reaction centres with a more than adequate supply of energy needed for growth under the environmental conditions pertaining at the time. In light-limiting conditions, however, it is assumed that the light saturating levels of chlorophyll and

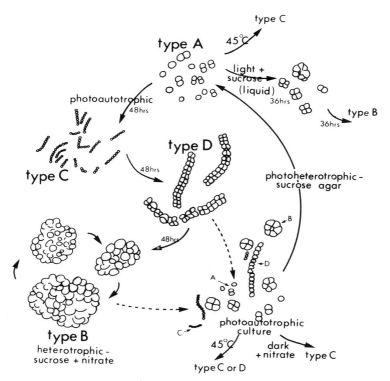

Figure 14. Summary of the variations in cell expression in Chlorogloea fritschii by environmental stimuli (Evans, Foulds and Carr, 1976). Cell type A - granulated cells (2 by 3 μm) existing either singly or as clumps containing two or more cells which arise from division in up to three planes. Cell type B - found in clumps which combine larger groups of cells surrounded by a mucilaginous sheath. Cell type C - small (1 μm) cells found in short filaments. Cell type D - larger than cell type C (1.5 μm) and found as filaments in the process of dividing.

reaction centres would be insufficient to ensure the continuity of energy supply needed. Not that this necessarily implies that extra sites of photon absorption ensures a more efficient harvesting rate of scarce photons, more probable is that in light-limiting situations bursts of photons arrive irregularly at the bacterium, shading being the real problem, and an ability to harvest these irregular but reasonably intense photon bursts, by increasing absorbtion sites, ensures

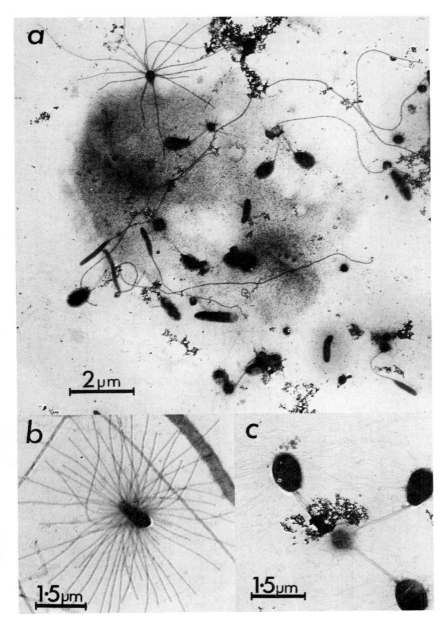

Figure 15. *(a) Microbial population characteristic of an*

a continuity of energy supply. If constant, but very low, photon intensity were the situation, merely increasing absorption sites within a confined area would not resolve the problem. Only by increasing the actual area of the light harvesting net would additional light be captured. Prosthecae formation with chlorophyll being located in the

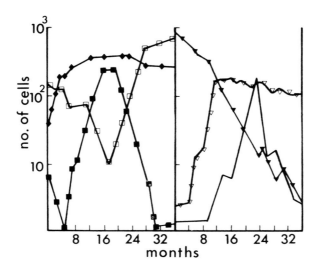

Figure 16. Population graph of a Draycote Water Reservoir sample over a three year period of static incubation at room temperature. In excess of one thousand cells were counted by electron microscopy and the number of each cell type was plotted against time. Hyphomicrobium sp., - □ -; Multi-appendaged cells - ▽ -; Caulobacter sp., - ◆ -; Gas vacuolated rods ———; Planctomyces sp., - ■ -; Others (rods, cocci, spirilla) - ▼ -. It is evident from this data that the prothecate bacteria have a considerable survival advantage as the nutrient content of the system is reduced. One interesting feature is that Planctomyces sp., competes very well initially but subsequently declines. This may reflect the fact that this organism has an extracellular fibrillar stalk and extensive pili (Fig. 15) as opposed to an integral cellular extension, for example, Caulobacter sp.

---

oligotrophic freshwater environment. (b) Fully expressed multi-appendaged cell. (c) Planctomyces sp. Note the fibrillar nature of the stalk, the holdfast and the extensive array of pili. Gold/palladium shadowed electron micrographs.

prosthecae would be a solution – and that could be an explanation for the existence of *Prothecochloris aestuarii* (Gorlenko, 1970) which is a highly prosthecate bacterium with chlorophyll located in its prosthecae.

In the case of nitrite-oxidzing bacteria, of which there are a number of species, the extensive membrane content may be related to the problem of energy production. Nitrite to nitrate oxidation is one of the lowest energy yielding redox couples exploited by bacteria and is reflected in the fact that nitrite turnover is quantitatively vast in comparison to cell yield. Besides the poor energy yield problem, there is an additional problem for these bacteria in that energy demands for biosynthesis (carbon dioxide fixation and reduced pyridine nucleotide synthesis at the expense of ATP and reduced cytochrome c) are very high. Consequently, it could be argued that even in the presence of ample nitrite, the provision of a large number of nitrite oxidizing sites – more than could be accommodated on a single cytoplasmic membrane – is an essential and logical way to overcome the imposition of a low energy-yielding mechanism, enhanced turnover rate compensating for the low energy yield. The one known nitrite-oxidizing bacterium not possessing an extensive complex of membranes, a *Nitrospira* species, appears to have compensated for this deficiency by its morphology; its cell surface (and therefore its cytoplasmic membrane area) to volume ratio being considerably greater than that of the membraneous nitrite oxidizers.

The last three examples to be considered, the methane oxidizers, the ammonia oxidizers and the lower n-alkane oxidizers, all share a common problem – a high demand for molecular oxygen. The energy yielding substrates in all cases are of low molecular weight and are fully reduced; all require oxygenation as an initial reaction prior to oxidation for energy yielding purposes. Therefore, on circumstantial grounds only, it seems reasonable to suppose that the oxygenase reaction, because of its turnover rate, possibly, dictates a multiplicity of sites to ensure the necessary flow of product for subsequent oxidation for energy and biosynthesis. Two pieces of evidence lend some credence to this idea. The first is that of Patt and Hanson (1978) who noted that the membrane content of the methane oxidizer *Methylobacterium organophilum* was directly related to oxygen tension. Under high oxygen tension the membrane content was approximately half that observed in organisms grown under low oxygen tention. No variation in membrane content appears to have been noted in organisms growing in conditions where methane is limiting or in considerable excess. The second

piece of evidence relates to the occurrence of membrane
complexes in organisms oxidizing C_2 + n-alkanes. Only those
organisms growing on the lower n-alkanes develop membrane
complexes, those growing on C_{10} or higher n-alkanes, for
instance, possessed only a cytoplasmic membrane. The point
of significance being that organisms growing on the higher
n-alkanes need to oxygenate considerably few molecules of
substrate per unit of energy and biosynthetic carbon
intermediate formed than do those organisms growing, say, on
ethane and propane indicating a demand, therefore, for fewer
oxygenation sites.

Colonial Systems in aquatic environments

A phenomenon which never seems to excite much comment in
relation to environmental adaptation is the ability of some
microbes to form colonies, by one means or another, and
presumably gain a nutritional advantage in unstable
circumstances. Obviously an aquatic environment is a good
example of one in which it would be of great advantage to a
microbe to retain its progeny within a confined area which
has proved to be nutritionally suitable.

A number of bacteria appear to have accomplished this feat
by ensuring that progeny remain linked, forming sheets (for
example, *Lampropedia* sp.), rings (for example, *Pelodictyon*
sp.), matrixes (for example *Rhodomicrobium* sp., and
Hyphomicrobium) or filaments (for example, *Leucothrix* sp.
Beggiatoa sp). Others have achieved a similar end by
confining themselves with a membrane/polysaccharide type of
sheath (for example, *Sphaerotilus* sp., *Thiocystis* sp., and
Thiocapsa sp.) Yet others have adopted the least organised
but still effective mode of colony formation of forming a
polysaccharide matrix in which they (and unrelated non-
polysaccharide forming species) are embedded.

Little is known of the biology of these colonies and the
life cycles of the organisms themselves. For instance:
How are new colonies started? What are the escape mechanisms
resulting in the release of cells to set up new colonies?
What changes in nutritional circumstances trigger off the
necessary differentiation processes to release dispersal
cells? In *Sphaerotilis* sp. for instance, swarm cells are
released but are these swarm cells specialist cells in the
sense described earlier? Is their release from the filament
fortuitous or deliberate? In the case of the trichome
forming species and the gonidia forming species, what triggers
off the event? Obviously, there are many facets of the
biology of colonial microbes which need revealing, but will
prove extremely difficult to determine because of the complex
nature of the parameters involved.

Perhaps the most interesting sort of colony to mention finally is that of certain filamentous cyanobacteria; that is, those which comprise differentiated cells. In particular, they contain cells which have differentiated into nitrogen-fixing heterocysts, so imparting on the colony the unique feature of being able to fix nitrogen whilst photosynthesis (with water as the electron donor) proceeds. Those and other features of the filament (for example, intracellular connections) point to the filament being in reality a *multicellular* prokaryote akin, in principle, to a multicellular eukaryote containing cells with different functions. Advantages of this sophisticated level of colonial organisation to those sorts of cyanobacteria is obvious from the evidence of their ability to successfully colonize barren sites where salts, water, light and gaseous nitrogen are the major nutrients.

REFERENCES

Begg, K.J. and Donachie, W.D. (1977). The growth of the *Escherichia coli* surface, *Journal of Bacteriology* 129, 1524-1535.

Begg, K.J. and Donachie, W.D. (1978). Changes in cell size and shape in thymine-requiring *Escherichia coli* associated with growth in low concentrations of thymine, *Journal of Bacteriology* 133, 452-458.

Clark, J.B. (1973). Morphogenesis in the genus *Arthrobacter*, *CRC Critical Reviews in Microbiology* 17, 521-544.

Davies, S.L. and Whittenbury, R. (1970). Fine structure of methane and other hydrocarbon-utilizing bacteria, *Journal of General Microbiology* 61, 227-232.

Donachie, W.D. (1968). Relationship between cell size and time of initiation of DNA replication, *Nature, London* 219, 1077-1079.

Donachie, W.D. and Begg, K.J. (1970). Growth of the bacterial cell, *Nature, London* 227, 1220-1224.

Donachie, W.D., Begg, K.J. and Vincente, M. (1976). Cell length, cell growth and cell division, *Nature, London* 264, 328-333.

Dow, C.S. and France, A.D. (1980). Simplified vegetative cell cycle of *Rhodomicrobium vannielii*, *Journal of General Microbiology* 117, 47-55.

Dow, C.S., Westmacott, D. and Whittenbury, R. (1976). Ultrastructure of budding and prosthecate bacteria. *In* "Microbial Ultrastructure" (Ed. R. Fuller and D.W. Lovelock) pp. 187-221. Academic Press, New York.

Easterbrook, K.B., McGregor-Shaw, J.B. and McBridge, R.P.

(1973). Ultrastructure of bacterial spines, *Canadian Journal of Microbiology* 19, 995-997.

Eckersley, K. and Dow, C.S. (1980). *Rhodopseudomonas blantica* sp. nov.: a member of the *Rhodospirillaceae*, *Journal of General Microbiology*, in press.

Evans, E.H., Foulds, I. and Carr, N.G. (1976). Environmental conditions and morphological variation in the blue-green alga *Chlorogloea fritschii*, *Journal of General Microbiology* 92, 147-155.

Evinger, M. and Agabian, N. (1977). Envelope associated nucleoid from *Caulobacter crescentus* stalked and swarmer cells, *Journal of Bacteriology* 132, 294-301.

Gorlenko, V.M. (1970). A new phototrophic green sulphur bacterium - *Prosthecochloris aestuarii* nov. gen. nov. spec. *Zeitschrift für Alleigname Mikrobiologie* 10, 147-149.

Harder, W. and Attwood, M.M. (1978). Biology, physiology and biochemistry of Hyphomicrobia. *Advances in Microbial Physiology* 16, 303-356.

Hirsch, P. (1968). Biology of Budding Bacteria IV. Epicellular deposition of iron by aquatic budding bacteria. *Archiv für Mikrobiologie* 60, 201-216.

Ishiguro, E.E. and Wolfe, R.S. (1970). Control of morphogenesis in *Geodermatophilus* ultrastructure studies, *Journal of Bacteriology* 104, 566-580.

Ishiguro, E.E. and Wolfe, R.S. (1974). Induction of morphogenesis in *Geodermatophilus* by inorganic cations and by organic nitrogeneous cations, *Journal of Bacteriology* 117, 189-195.

Koch, A.L. (1971). The adaptive response of *Escherichia coli* to a feast and famine existence. *Advances in Microbial Physiology* 6, 147-217.

Kuenen, J.G., Boonstra, J., Schroder, H.G.J. and Veldkamp, H. (1979). Competition for inorganic substrates among chemoorganotrophic and chemolithotrophic bacteria, *Microbial Ecology* 3, 119-130.

Larson, R.J. and Pate, J.L. (1976). Glucose transport in isolated prothecae of *Anticcacaulis biprosthecum*, *Journal of Bacteriology* 126, 282-293.

Pate, J.L., Porter, J.S. and Jordan, T.L. (1973). *Anticcacaulis biprosthecum* sp. nov. Life cycle, morphology and culture characteristics, *Antonie van Leeuwenhoek* 39, 569-583.

Patt, T.E. and Hanson, R.S. (1978). Intracytoplasmic membrane, phospholipid and sterol content of *Methylobacterium organophilum* cells grown under different conditions, *Journal of Bacteriology* 134, 636-644.

Shapiro, L. (1976). Differentiation in the *Caulobacter* cell cycle, *Annual Review Microbiology* <u>30</u>, 377-402.

Staley, J.T. (1968). *Prosthecomircobium* and *Amcalomicrobium:* new prosthecate freshwater bacteria, *Journal of Bacteriology* <u>95</u>, 1921-1942.

Strength, W.J. and King, N.R. (1971). Flagellar activity in an aquatic bacterium, *Canadian Journal of Microbiology* <u>17</u>, 1133-1137.

Swoboda, U. (1979). Studies on the growth and development of *Caulobacter*, Degree of Doctor of Philosophy, Open University.

Swoboda, U., Dow, C.S. and Vitkovic, L. (1980). Nucleoids of *Caulobacter crescentus* CB15 and their use in *in vitro* transcription studies, *Journal of General Microbiology* (in press).

Thomashow, M.F. and Rittenberg, S.C. (1979). The intraperiplasmic growth cycle - the life style of the *Bdellovibrios*, p. 115-138. *In* "Developmental Biology of Prokaryotes, Studies in Microbiology Volume 1" (Ed. J.H. Parish), Blackwell Scientific Publications.

Tyler, P.A. (1970). Hyphemicrobia and the oxidation of manganese in aquatic ecosystems, *Antonie van Leeuwenhoek* <u>36</u>, 567-578.

Whittenbury, R. and Dow, C.S. (1977). Morphogenesis and differentiation in *Rhodomicrobium vannielii* and other budding and prosthecate bacteria, *Bacteriological Reviews* <u>41</u>, 754-808.

Whittenbury, R. and McLee, A.G. (1967). *Rhodopseudomonas palustris* and *R. viridis* - photosynthetic budding bacteria, *Archiv für Mikrobiologie* <u>59</u>, 324-334.

Zaritzky, A. (1975). On dimensional determination of rod-shaped bacteria, *Journal of Theoretical Biology* <u>54</u>, 243-248.